Peptide Characterization and Application Protocols

METHODS IN MOLECULAR BIOLOGY™

John M. Walker, SERIES EDITOR

METHODS IN MOLECULAR BIOLOGY™

Peptide Characterization and Application Protocols

Edited by

Gregg B. Fields

Department of Chemistry and Biochemistry
Florida Atlantic University, Boca Raton, FL

HUMANA PRESS ✳ TOTOWA, NEW JERSEY

Preface

It has been over a century since the first peptide synthesis, as well as the creation of the term "peptide," was reported by Fischer and Fourneau. Prior volumes of Methods in Molecular Biology have included peptide synthesis (#35, #73, #87), analytical techniques that may be utilized to characterize synthetic peptides (#36, #60, #73, #146, #159, #211, #251, #276), combinatorial peptide libraries (#87), and production of antibodies using peptides (#151, chapter 26). However, this is the first volume of the series dedicated entirely to the characterization of peptides and their application for study of biochemical systems. The most recent advances in selected analytical techniques are presented. Among these is a comprehensive updating of high-performance liquid chromatography for purification and evaluation of peptides. Mass spectrometry is utilized for analysis of standard and modified peptides. Recent advances in the synthesis and characterization of membrane peptides closes out the first section of *Peptide Characterization and Application Protocols*.

A variety of specific applications of synthetic peptides compose the remainder of this volume. Conformationally constrained peptides are described as potential ligands, substrates, and enzyme inhibitors. Such peptides are also utilized for proteolytic profiling of melanoma. Drug and siRNA delivery is enhanced by peptides engineered for these purposes. Finally, peptides serve as imaging, anti-aggregatory, and antimicrobial agents. Overall, the collected works presented in this volume represent the systematic manner in which peptides can be constructed, analyzed, and applied to attack contemporary problems in biochemistry.

Peptide Characterization and Application Protocols is dedicated to the late Murray Goodman (1928–2004), a friend and mentor to so many in the peptide research community. His tireless research efforts covered many aspects of peptide synthesis, characterization, and application, and were inspirational.

Gregg B. Fields

Contents

Contributors

FERNANDO ALBERICIO • Barcelona Biomedical Research Institute and Department of Organic Chemistry, University of Barcelona, Barcelona, Spain

MOHAMMAD AL-GHOUL • Department of Chemistry & Biochemistry, Florida Atlantic University, Boca Raton, FL

CAROLYN J. ANDERSON • Mallinckrodt Institute of Radiology, Washington University School of Medicine, St. Louis, MO

JED P. AUCOIN • Department of Chemistry, Louisiana State University, Baton Rouge, LA

DIANE BARONAS-LOWELL • Department of Chemistry & Biochemistry, Florida Atlantic University, Boca Raton, FL

KEITH BREW • College of Biomedical Sciences and Center for Molecular Biology & Biotechnology, Florida Atlantic University, Boca Raton, FL

YUXIN CHEN • Department of Biochemistry and Molecular Genetics, University of Colorado at Denver and Health Sciences Center, Aurora, CO

MYRON CRAWFORD • WM Keck Foundation Biotechnology Resource Laboratory, Mass Spectrometry and Protomic Resources, Yale University, New Haven, CT

MARE CUDIC • Department of Chemistry & Biochemistry, Florida Atlantic University, Boca Raton, FL

PREDRAG CUDIC • Department of Chemistry & Biochemistry, Florida Atlantic University, Boca Raton, FL

SÉBASTIEN DESHAYES • Center de Recherches de Biochimie Macromoleculaire, Department of Molecular Biophysics & Thepeutics, Montpellier, France

GILLES DIVITA • Center de Recherches de Biochimie Macromoleculaire, Department of Molecular Biophysics & Thepeutics, Montpellier, France

NADIA J. EDWIN • Department of Chemistry, Louisiana State University, Baton Rouge, LA

MARCUS A. ETIENNE • Department of Chemistry, Louisiana State University, Baton Rouge, LA

JOSEP FARRERA-SINFREU • Barcelona Biomedical Research Institute, University of Barcelona, Barcelona, Spain

GREGG B. FIELDS • Department of Chemistry & Biochemistry and Center for Molecular Biology & Biotechnology, Florida Atlantic University, Boca Raton, FL

ERNEST GIRALT • Barcelona Biomedical Research Institute and Department of Organic Chemistry, University of Barcelona, Barcelona, Spain

ROBERT P. HAMMER • Department of Chemistry, Louisiana State University, Baton Rouge, LA

GARY M. HATHAWAY • The Beckman Institute, California Institute of Technology, Pasadena, CA

FRÉDÉRIC HEITZ • Center de Recherches de Biochimie Macromoleculaire, Department of Molecular Biophysics & Thepeutics, Montpellier, France

ROBERT S. HODGES • Department of Biochemistry and Molecular Genetics, University of Colorado at Denver and Health Sciences Center, Aurora, CO

DAVID R. KHAN • Department of Chemistry & Biochemistry, Florida Atlantic University, Boca Raton, FL

JAMES M. KOVACS • Department of Biochemistry and Molecular Genetics, University of Colorado at Denver and Health Sciences Center, Aurora, CO

JANELLE L. LAUER-FIELDS • College of Biomedical Sciences and Department of Chemistry & Biochemistry, Florida Atlantic University, Boca Raton, FL

JASON S. LEWIS • Mallinckrodt Institute of Radiology, Washington University School of Medicine, St. Louis, MO

COLIN T. MANT • Department of Biochemistry and Molecular Genetics, University of Colorado at Denver and Health Sciences Center, Aurora, CO

ROBIN L. MCCARLEY • Department of Chemistry, Louisiana State University, Baton Rouge, LA

WALTER MCMURRAY • WM Keck Foundation Biotechnology Resource Laboratory, Mass Spectrometry and Protomic Resources, Yale University, New Haven, CT

JANINE B. MILLS • Department of Biochemistry and Molecular Genetics, University of Colorado at Denver and Health Sciences Center, Aurora, CO

DMITRIY MINOND • Department of Chemistry & Biochemistry, Florida Atlantic University, Boca Raton, FL

MAY C. MORRIS • Center de Recherches de Biochimie Macromoleculaire, Department of Molecular Biophysics & Thepeutics, Montpellier, France

FRED NAIDER • Department of Chemistry, College of Staten Island, Staten Island, NY

LASZLO OTVOS • Sbarro Institute, Temple University, Philadelphia, PA

TRAIAN V. POPA • Department of Biochemistry and Molecular Genetics, University of Colorado at Denver and Health Sciences Center, Aurora, CO

EVONNE M. REZLER • Department of Chemistry & Biochemistry, Florida Atlantic University, Boca Raton, FL

MIRIAM ROYO • Combinatorial Chemistry Unit, University of Barcelona, Barcelona, Spain

PAUL S. RUSSO • Department of Chemistry, Louisiana State University, Baton Rouge, LA

FEDERICA SIMEONI • Center de Recherches de Biochimie Macromoleculaire, Department of Molecular Biophysics & Thepeutics, Montpellier, France

MACIEJ STAWIKOWSKI • Department of Chemistry & Biochemistry, Florida Atlantic University, Boca Raton, FL

KATHRYN L. STONE • WM Keck Foundation Biotechnology Resource Laboratory, Mass Spectrometry and Protomic Resources, Yale University, New Haven, CT

MATTHEW TIRRELL • College of Engineering, University of California, Santa Barbara, CA

BRIAN P. TRIPET • Department of Chemistry & Biochemistry and Biomedical Sciences and Center for Molecular Biology & Biotechnology, Florida Atlantic University, FL

RAYMOND TU • College of Engineering, University of California, Santa Barbara, CA

NANCY WILLIAMS • WM Keck Foundation Biotechnology Resource Laboratory, Mass Spectrometry and Protomic Resources, Yale University, New Haven, CT

KENNETH R. WILLIAMS • WM Keck Foundation Biotechnology Resource Laboratory, Mass Spectrometry and Protomic Resources, Yale University, New Haven, CT

ZHE YAN • Department of Biochemistry and Molecular Genetics, University of Colorado at Denver and Health Sciences Center, Aurora, CO

I

CHARACTERIZATION

1

HPLC Analysis and Purification of Peptides

Colin T. Mant, Yuxin Chen, Zhe Yan, Traian V. Popa, James M. Kovacs, Janine B. Mills, Brian P. Tripet, and Robert S. Hodges

Summary

High-performance liquid chromatography (HPLC) has proved extremely versatile over the past 25 yr for the isolation and purification of peptides varying widely in their sources, quantity and complexity. This article covers the major modes of HPLC utilized for peptides (size-exclusion, ion-exchange, and reversed-phase), as well as demonstrating the potential of a novel mixed-mode hydrophilic interaction/cation-exchange approach developed in this laboratory. In addition to the value of these HPLC modes for peptide separations, the value of various HPLC techniques for structural characterization of peptides and proteins will be addressed, e.g., assessment of oligomerization state of peptides/proteins by size-exclusion chromatography and monitoring the hydrophilicity/hydrophobicity of amphipathic α-helical peptides, a vital precursor for the development of novel antimicrobial peptides. The value of capillary electrophoresis for peptide separations is also demonstrated. Preparative reversed-phase chromatography purification protocols for sample loads of up to 200 mg on analytical columns and instrumentation are introduced for both peptides and recombinant proteins.

Key Words: Peptides; proteins; size-exclusion chromatography (SEC); anion-exchange chromatography (AEX); cation-exchange chromatography (CEX); mixed-mode hydrophilic interaction chromatography (HILIC)/cation-exchange chromatography (CEX); reversed-phase high-performance liquid chromatography (RP-HPLC); preparative RP-HPLC of peptides and proteins; amino acid side-chain hydrophilicity/hydrophobicity coefficients; amino acid α-helical propensity values; amino acid side-chain stability coefficients.

From: *Methods in Molecular Biology, vol. 386: Peptide Characterization and Application Protocols*
Edited by: G. Fields © Humana Press Inc., Totowa, NJ

1. Introduction

1.1. Scope of Chapter

The development of high-performance liquid chromatography (HPLC) packings and instrumentation over the past 25 yr has revolutionized the efficiency and speed of separation of molecules in general and peptides in particular. This development has also seen a tremendous output of published literature on the topic of HPLC of peptides, perhaps making the decision as to how best to approach a particular separation problem seem formidable to the novice, or even experienced HPLC user. Fortunately, regardless of whether high-performance approaches are utilized for routine peptide separations or for such state-of-the-art areas as proteomics, capillary methods, biospecific interactions, and so on, the fundamentals of chromatographic protocols remain the same. Thus, it is not the purpose of this chapter to present a comprehensive review of HPLC of peptides. Indeed, there is a wealth of relevant material accessible in the literature. For instance, several useful articles and reviews on HPLC of peptides can be found in **refs. *1–4***. In addition, **refs. *5–8*** represent excellent resource books in this area. Finally, **ref. *9*** offers an extensive source of information on the early development of HPLC of peptides.

This chapter is aimed at laboratory-based researchers, both experienced chromatographers and those with limited exposure to high-performance separation approaches, who wish to learn about peptide analysis by HPLC, based on representative examples from research carried out in our laboratory with general applicability. Standard analytical applications in HPLC of peptides will be stressed, together with novel approaches to separations and modest scale-up for preparative purification of peptides. In addition, the value of the complementary technique of capillary electrophoresis (CE) for peptide separations will be demonstrated. Finally, in order to maximize the "user friendliness" of this chapter, only nonspecialized columns, mobile phases, and instrumentation readily available and easily employed by the researcher are described.

1.2. Properties of Peptides and Proteins and Practical Implications

1.2.1. Properties of Amino Acids

The side-chains of amino acids are generally classified according to their polarity, i.e., nonpolar or hydrophobic vs polar or hydrophilic. Further, the polar side-chains are divided into three main groups: uncharged polar, positively charged, or basic side-chains, and negatively charged or acidic side-chains. Within any single group, there are considerable variations in the size, shape and properties of the side-chains. Peptides containing ionizable (acidic and basic) side-chains have a characteristic isoelectric point (pI) and the overall net charge and polarity of a peptide in aqueous solution will vary with pH. Thus, hydrophilicity/hydrophobicity, as well as the number of charged groups present, become important factors in the separation of peptides. Intrinsic amino acid side-chain hydrophilicity/hydrophobicity coefficients have been recently published *(10)*, and these reversed-phase (RP)-HPLC derived values at pH 2.0 and pH 7.0 are shown in **Table 1**. "Intrinsic" implies the maximum possible hydrophilicity/hydrophobicity of side-chains in the absence of nearest-neighbor or conformational effects that would decrease the full expression of the side-chain hydrophilicity/hydrophobicity when the side-chain is in a polypeptide chain. Such a scale is the fundamental starting point for determining the parameters that affect side-chain hydrophobicity for quanti-fying such effects in peptides and proteins (e.g., the quantitative evaluation of the contribution of a specific amino side-chain to ligand–protein and protein–protein interactions and to protein folding and stability) and in aiding the development of protocols for optimum separation of polypeptides. The RP-HPLC-based approach to determining these intrinsic coefficients is des-cribed under **Subheading 3.3.1.**

1.2.2. Peptide/Protein Conformation and Stability

Conformation can be an important factor in peptides as well as proteins and, thus, should always be a consideration when choosing the conditions for chromatography. Although secondary structure (e.g., α-helix) is generally absent even in benign (nondenaturing) aqueous conditions for small peptides (up to about 10 residues), the potential for a defined secondary or tertiary structure increases with increasing peptide length and, for peptides containing more than 20–30 residues, folding to internalize hydrophobic residues is likely to become a significant conformational feature. Zhou et al. *(11)* reported a complete and accurate scale of the intrinsic α-helical propensities of the 20 naturally occurring amino acids (**Table 2**) based on a synthetic model peptide approach whereby single amino acid substitutions were made in the center of the nonpolar face of an amphipathic α-helical peptide (**Fig. 1**) that is small, monomeric and noninteracting. As shown in **Table 2**, Ala has the highest side-chain α-helical propensity and Gly the lowest. A value for Pro could not be determined because the side-chain of this amino acid completely disrupts α-helical structure, even in the presence of helix inducing solvents such as trifluoroethanol (TFE). Such intrinsic propensity values provide a powerful tool for protein design as well as providing a guide to predicting the potential conformational status of a

Table 1
Hydrophilicity/Hydrophobicity Coefficients Determined at 25°C by Reversed-Phase High-Performance Liquid Chromatography of Model Peptides at pH 2.0 and 7.0[a]

Amino acid Substitution[b]	pH 2.0[c] $\Delta t_R(\text{Gly})$[d]	pH 7.0[c] $\Delta t_R(\text{Gly})$[d]
Trp	32.4	33.0
Phe	29.1	30.1
n-Leu	24.6	25.9
Leu	23.3	24.6
Ile	21.4	22.8
Met	15.7	17.3
n-Val	15.2	16.9
Tyr	14.7	16.0
Val	13.4	15.0
Pro	9.0	10.4
Cys	7.6	9.1
Ala	2.8	4.1
Glu[e]	**2.8**	**−0.4**
Lys	**2.8**	**−2.0**
Thr	2.3	4.1
Asp	**1.6**	**−0.8**
Gln	0.6	1.6
Arg	**0.6**	**4.1**
Ser	0.0	1.2
His	**0.0**	**4.7**
Gly	0.0	0.0
Asn	−0.6	1.0
Orn	**−0.6**	**−2.0**

[a] Taken from **ref. 10**.
[b] The amino acid substitutions at position X in the peptide sequence Ac-X-G-A-K-G-A-G-V-G-L-amide; n-Leu, n-Val and Orn denote norleucine, norvaline and ornithine, respectively.
[c] Conditions: *see* **Subheading 3.3.1.**
[d] $\Delta t_R(\text{Gly})$ denotes the change in retention time relative to the Gly-substituted peptide.
[e] Potentially charged residues (Asp, Glu, Arg, His, Lys, Orn) are in bold; the values between pH 2.0 and pH 7.0 that differ by more than 2 min are also in bold.

peptide during chromatography. The latter is particularly true of RP-HPLC because the nonpolar environment characteristic of this HPLC mode make it a strong inducer of α-helical structure in potentially α-helical amino acid sequences *(12–15)*.

Table 2
α-Helical Propensities of Amino Acid
Side-Chains in Peptides

Amino acid[a]	ΔΔG (kcal/mole)[b]
Ala	−0.96
Arg	−0.90
Leu	−0.81
Lys	−0.70
Met	−0.67
Gln	−0.61
Ile	−0.59
Trp	−0.49
Phe	−0.48
Tyr	−0.43
Cys	−0.43
Val	−0.42
Asn	−0.33
Ser	−0.33
His	−0.33
Glu	−0.32
Thr	−0.28
Asp	−0.21
Gly	0

[a] Single amino acid substitution were made in the center of the non-polar face of model amphipathic α-helical peptide shown in **Fig. 1**.

[b] ΔΔG, taken as a measure of amino acid α-helical propensity, is the difference in free energy of α-helix formation (ΔG) for each substituted amino acid residue relative to Gly (**ref. 11**).

Interactions between hydrophobic side-chains are the most important factor in polypeptide folding and the subsequent stability of the final polypeptide conformation. Thus, knowledge of the contribution of individual amino acid side-chains in the hydrophobic core of a folded protein to peptide/protein conformation and stability is also of importance when considering the effect of the peptide/protein solubility, conformation, and interactions with the HPLC matrix during chromatography. **Table 3** lists stability coefficients of amino acids generated from single amino acid substitutions in the "a" or "d" position of

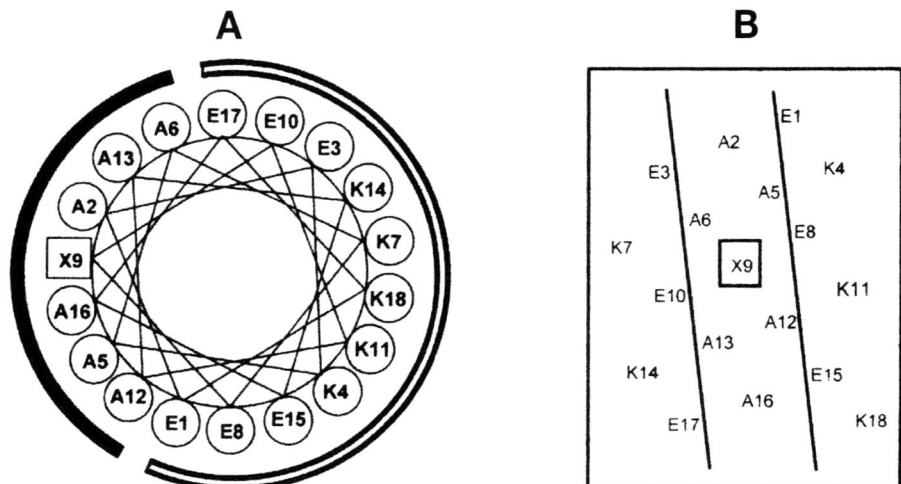

Fig. 1. Amino acid sequence of model amphipathic α-helical peptide used for evaluation of α-helical propensities of amino acids. **A**, Helical wheel representation; **B**, helical net representation. Amino acid substitutions were made at position X9, in the center of the nonpolar face of the α-helix. The α-helical propensities generated from this model peptide are shown in **Table 2**. (Reproduced from **ref. *11***, with permission from Bentham Science.)

the central heptad of model α-helical coiled coils (heptad positions are denoted "abcdefg," where positions "a" and "d" are the nonpolar positions responsible for the formation and stability of coiled-coils). Such coefficients allow an assessment of the effect of side-chain substitution in the hydrophobic core, as well as provide information on the effect of side-chain hydrophobicity and packing in the hydrophobic core. Although the values shown in **Table 3** were derived from a coiled-coil model *(16,17)*, we had previously shown that the relative contributions of nonpolar residues to protein stability in two-stranded α-helical coiled-coils showed an excellent linear correlation with the results obtained in globular proteins. In both protein types, the mutations were made in hydrophobic residues involved in the hydrophobic core, which is responsible for overall protein stability *(18)*. This demonstrates the general applicability of these results to proteins in general.

1.2.3. Peptide Detection

Peptide bonds absorb light strongly in the far ultraviolet (UV) (<220 nm), providing a convenient means of detection (generally at 210–220 nm). In

Table 3
Contribution of Amino Acid Side-Chains to Stability in the Hydrophobic Core of α-Helical Coiled-Coils

Amino acid[a]	$\Delta\Delta Gu$ (Ala)[b] (position a)	$\Delta\Delta Gu$ (Ala)[b] (position d)
Val	4.1	1.1
Ile	3.9	3.0
Leu	3.5	3.8
Met	3.4	3.2
Phe	3.0	1.2
Tyr	2.2	1.4
Asn	0.9	−0.6
Trp	0.8	−0.1
Thr	0.2	−1.2
Ala	0.0	0.0
Gln	−0.1	0.5
Lys	−0.4	−1.8
Arg	−0.8	−2.9
His	−1.2	−0.8
Ser	−1.3	−1.8
Orn	−1.9	−3.1
Glu	−2.0	−2.7
Gly	−2.5	−3.6
Asp	—[c]	−1.8

[a] Amino acid residue (denoted X) substituted at position 19a of the sequence Ac-CGGEVGALKAQVGALQAQXGALQKEVGALKKEVGA LKK-amide or at position 22d of the sequence Ac-CGGEVGALKAEVGAL KAQIGAXQKQIGALQKEVGALKK-amide; oxidation of the peptides formed a disulfide-bridged homo-two-stranded α-helical coiled-coil.

[b] $\Delta\Delta Gu$ (Ala) is the difference in the free energy of unfolding (ΔGu) relative to the Ala-substituted analog; a positive value indicates the substitution provides more stability relative to Ala; a negative value indicates the substitution is destabilizing relative to Ala.

[c] A value for the Asp side-chain could not be obtained as a result of its causing unfolding of the coiled-coil, i.e., it is more destabilizing than Gly.

addition, the aromatic side-chains of tyrosine, phenylalanine and tryptophan absorb light in the 250–290 nm UV range; it should be noted, however, that these aromatic residues are not present in all peptides.

1.3. HPLC Modes Used in Peptide Separations

The three major modes of HPLC traditionally employed in peptide separations utilize differences in peptide size (size-exclusion HPLC [SEC]), net charge (ion-exchange HPLC [IEX]), or hydrophobicity (RP-HPLC) *(1–8)*. Within these modes, mobile phase conditions may be manipulated to maximize the separation potential of a particular HPLC column. In addition, a novel mixed-mode approach to peptide separation termed hydrophilic interaction chromatography (HILIC)/cation-exchange chromatography (CEX) has also shown excellent potential as a complement to RP-HPLC in recent years *(19–30)*.

1.3.1. HPLC Supports

Silica-based packings remain the most widely used for the major modes of HPLC, the rigidity of microparticulate silica allowing the use of high flow rates of mobile phases. In addition, favorable mass transfer characteristics allow rapid analyses to be performed. Because most silica-based packings are limited to a pH range of 2.0–8.0 as a result of silica dissolution in basic eluents, high-performance column packings based on organic polymers with a broad pH tolerance (e.g., cross-linked polystyrene-divinylbenzene) are also being increasingly introduced.

1.3.2. Size-Exclusion HPLC

In the past, the range of required fractionation for peptides ($\sim$100–6,000 Da) has tended toward the low end of the fractionation ability of commercial columns, designed mainly for protein separations. However, such columns have still proven useful in the early stages of a peptide purification protocol (i.e., in the early stages of a multi-step purification protocol) or for peptide/protein separations *(2,17,31–34)*. The introduction of a size-exclusion column designed specifically for peptide separations (molecular weight range 100–7000) has raised the profile of this HPLC mode for peptide analysis in recent years *(35)*.

1.3.3. Ion-Exchange HPLC

IEX has proven extremely useful for peptide separations since HPLC packings capable of retaining both basic (net positive charge) or acidic (net negative charge) peptides have been introduced. Both anion-exchange (AEX) *(2,3,5,8,32,36–40)* and cation-exchange (CEX) *(2,3,5,8,31,32,34,36,38,41–45)* HPLC have been employed for peptide and protein separations, negatively charged and positively charged solutes, respectively, being retained by these ion-exchange modes. Common anion-exchange packings consist of primary, secondary, and tertiary (weak AEX) or quaternary amine (strong AEX) groups adsorbed or covalently bound to a support. These positively charged packings will interact with acidic (negatively charged) peptide residues (aspartic and glutamic acid above ~pH 4.0), as well as the negatively charged C-terminal α-carboxyl group. Common cation-exchange packings consist of carboxyl (weak CEX) or sulfonate (strong CEX) groups bound to a support matrix. These negatively charged packings will interact with the basic (positively charged) residues (histidine, pH < 6.0; arginine, pH < 12.0, and lysine, pH < 10.0), as well as the positively charged N-terminal α-amino group. Should a choice need to be made concerning the type of ion-exchange column for general peptide applications, a strong cation-exchange column is recommended. The utility of such a column lies in its ability to retain its negatively charged character in the acidic and neutral pH range, which is due to the strongly acidic sulfonate functionality characteristic of such a packing. Most peptides are soluble at low pH, where the side-chain carboxyl groups of acidic residues (glutamic acid, aspartic acid) and the free C-terminal α-carboxyl group are protonated (i.e., uncharged), thus emphasizing any basic, positively charged character of the peptides. Ion-exchange columns have proven particularly useful in a multistep protocol for peptide separations, particularly prior to a final RP-HPLC purification and desalting step.

1.3.4. Reversed-Phase High-Performance Liquid Chromatography

RP-HPLC remains the most widely used mode of HPLC for peptide separations *(1–8)*. It is generally superior to other HPLC modes in both speed and efficiency. In addition, the availability of volatile mobile phases makes it ideal for both analytical and preparative separations. The majority of researchers have tended to carry out HPLC below pH 3.0 to take advantage of acidic volatile mobile phases (particularly aqueous trifluoroacetic acid [TFA]/acetonitrile [CH_3CN] systems) *(1–8)*. Such mobile phase volatility is particularly useful when carrying out preparative purification of peptides or when employing RP-HPLC as a final desalting/purification mode in a multistep protocol. Finally, acidic pH values prevent undesirable ionic interactions between positively charged amino acid residues and any underivatized silanol groups (negatively charged above pH values of 3.0–4.0) on silica-based packings *(46)*, still the most widely used packing support for RP-HPLC of peptides. Favored RP-HPLC packings for the vast majority of peptide separations continue to be silica-based supports containing covalently bound octyl (C_8) or octadecyl (C_{18}) functionalities, with peptides being eluted from these hydrophobic stationary phases in order of increasing overall peptide hydrophobicity.

1.3.5. Hydrophilic Interaction/Cation-Exchange Chromatography

The term "hydrophilic interaction chromatography" was originally introduced to describe separations based on solute hydrophilicity *(47)*, with solutes being eluted from the HILIC column in order of increasing hydrophilicity, i.e., the opposite of RP-HPLC elution behavior. This concept was taken a step further by our laboratory by taking advantage of the inherent hydrophilic character of ion-exchange, specifically strong cation-exchange, columns. Thus, HILIC/CEX combines the most advantageous aspects of two widely different separation mechanisms, i.e., a separation based on hydrophilicity/hydrophobicity differences between peptides overlaid on a separation based on net charge *(19–30)*. Characteristic of HILIC/CEX separations is the presence of a high organic modifier concentration (generally, acetonitrile) to promote hydrophilic interactions between the solute and the hydrophilic/charged cation-exchange stationary phase. Peptides are then eluted from the column with a salt gradient. Generally, peptides are eluted in groups of peptides in order of increasing net positive charge. Within these groups, peptides are eluted in order of increasing hydrophilicity. Indeed, HILIC/CEX is basically CEX in the presence of high concentrations of acetonitrile (50–80%). HILIC/CEX is frequently an excellent complement to RP-HPLC. Indeed, it has rivaled or even exceeded RP-HPLC for specific peptide mixtures *(4,20,27,28)*.

1.4. Capillary Electrophoresis

Although, as noted above, RP-HPLC remains the favored separation approach for peptides, CE has also proven itself as a peptide separation tool in its own right *(48–56)*. CE, specifically capillary zone electrophoresis (CZE), exploits analyte charge or, more precisely, the mass-to-charge ratio. The value of these two peptide characteristics for peptide analysis in general, and peptide mapping in particular, has frequently been demonstrated *(57–59)*, with RP-HPLC and CZE also proving to complement each other for a multistep approach to peptide separations *(58,60–64)*.

2. Materials

2.1. Chemical and Solvents

1. Water is either obtained as HPLC-grade (BDH, Poole, UK; J. T. Baker, Phillipsburg, NJ; or EMD Chemical, Gibbstown, NJ) or purified by an E-pure water filtration device from Barnstead/Thermolyne (Dubuque, IA).
2. Reagent grade ortho-phosphoric acid (H_3PO_4) is obtained from Caledon Laboratories (Georgetown, Ontario, Canada) or Anachemia (Toronto, Ontario, Canada).

3. TFA is obtained from Hydrocarbon Products (River Edge, NJ) or Sigma-Aldrich (St. Louis, MO).
4. Pentafluoropropionic acid (PFPA) and heptafluorobutyric acid (HFBA) are obtained from Fluka (Buchs, Switzerland) or Sigma-Aldrich.
5. Triethylamine (TEA) is obtained from Anachemia (*see* **Note 1**).
6. Sodium chloride (NaCl) is obtained from Sigma-Aldrich or J.T. Baker.
7. Potassium chloride (KCl), potassium dihydrogen phosphate (KH$_2$PO$_4$), and reagent-grade urea are obtained from BDH. Extraction of UV-absorbing contaminants from analytical-grade phosphate-based buffer salts has occasionally been required. We have routinely prepared a stock solution (e.g., 1 L of 1 *M* to 2 *M* KH$_2$PO$_4$) and added a chelating resin (e.g., BioRad Chelex-100; BioRad Lab, Richmond, CA), stirring for 1 h. The phosphate solution is then aliquoted, diluted as desired, and filtered through a 0.22-μm filter. Reagent-grade urea (much cheaper than highly purified urea) may be purified to a level suitable for HPLC in a straightforward procedure: following preparation of a concentrated urea solution (often 6 *M* to 8 *M* range), the solution is stirred over a mixed-bed resin (e.g., BioRad AG 501-X8, 20 to 50 mesh) (10 g/L of solution) for 30–60 min; the resin is then removed by filtration through a sintered glass funnel and the supernatant subsequently filtered through a 0.22-μm filter.
8. Sodium perchlorate (NaClO$_4$) is obtained from Sigma-Aldrich or BDH (*see* **Note 2**).
9. HPLC-grade acetonitrile (CH$_3$CN) is obtained from Fisher Scientific (Pittsburgh, PA) or EM Science (Gibbstown, NJ).
10. Dithioerythritol (DTE) is obtained from Fisher Scientific.

2.2. Columns and Capillaries

Except where stated otherwise, all column packings are based on microparticulate silica supports.

2.2.1. Size-Exclusion HPLC

1. *Column 1:* Superdex Peptide HR 10/30 (300 × 10 mm inner diameter [ID]; non-silica-based support; separation range of 100–7000; Pharmacia Biotech, Baie d-Urfé, Quebec, Canada). This column (and Columns 2 and 3) are fast protein liquid chromatography (FPLC) columns from Pharmacia with stationary phases packed into glass rather than stainless steel columns. Such columns work excellently well on HPLC equipment (as opposed to FPLC equipment) if column pressure restrictions are noted.
2. *Column 2:* Superdex 75 HR 10/30 (300 × 10 mm ID; non-silica-based support; separation range of 3000–70,000 for globular proteins; Pharmacia Biotech, Baie d-Urfé, Quebec, Canada).

2.2.2. Ion-Exchange HPLC

1. *Column 3:* Mono S HR 5/5 strong CEX column (50×5 mm ID, non-silica-based support; $10 \mu m$; Pharmacia, Dorval, Canada).
2. *Column 4:* Polysulfoethyl A strong CEX column (200×2.1 mm ID; $5 \mu m$ particle size, $300 \text{\AA}$ pore size; PolyLC, Columbia, MD).

2.2.3. Reversed-Phase HPLC

1. *Column 5:* Kromasil C_{18} (150×2.1 mm ID; $5 \mu m$, $\text{\AA}$ 100; Hichrom, Berkshire, UK).
2. *Column 6:* Zorbax Eclipse XDB-C_8 (150×2.1 mm I.D.; $5 \mu m$, $80 \text{\AA}$; Agilent Technologies, Little Falls, DE); "XDB" denotes "extra dense bonding," these columns being designed to be particularly stable at neutral pH and above *(65,66)*, where dissolution of the silica matrix at neutral and (particularly) higher pH values has been previously problematic.
3. *Column 7:* Zorbax SB300-C_8 (150×2.1 mm ID; $5 \mu m$, $300 \text{\AA}$; Agilent Technologies); "SB" denotes "stable bond," these columns being designed to be particularly stable at highly acidic pH values (pH <3.0) by shielding the siloxane bonds between the alkyl chains (C_8, in this case) and silanol groups of the silica matrix from hydrolysis *(67,68)*.
4. *Column 8:* Zorbax SB300-C_8 (150×1 mm ID; $3.5 \mu m$, $300 \text{\AA}$; Agilent Technologies).
5. *Column 9:* Zorbax SB300-C_8 (250×9.4 mm ID; $6.5 \mu m$, $300 \text{\AA}$; Agilent Technologies).

2.2.4. Capillary Electrophoresis

Capillary 1: Uncoated capillaries of 60.2 cm $\times 50 \mu m$ ID (50 cm effective length, i.e., length from injection point to detection point) are provided by Beckman-Coulter (Fullerton, CA).

2.3. Instrumentation

2.3.1. HPLC

1. *Instrument 1:* The majority of analytical HPLC runs were carried out on an Agilent 1100 Series liquid chromatograph from Agilent Technologies.
2. *Instrument 2:* Older runs were carried out on a Varian Vista Series 5000 liquid chromatograph (Varian, Walnut Creek, CA) coupled to a Hewlett-Packard (Avondale, PA) HP1040A detection system, disc drive, HP2225A Thinkjet printer and HP7460A plotter.

3. *Instrument 3:* Preparative HPLC runs were carried out on a Beckman-Coulter instrument, comprised of a System Gold 126 Solvent Module and a System Gold 166 Detector.

2.3.2. Capillary Electrophoresis

Instrument 4: CE runs were carried out on a Beckman-Coulter Capillary Electrophoresis System controlled by 32 karat software (Version 5.0).

3. Methods
3.1. Size-Exclusion HPLC

Aqueous SEC is generally employed for peptide/protein separations and/or molecular weight determinations. Unique tertiary or quaternary structures can be demonstrated by molecular weight determinations in the presence and absence of denaturants in SEC *(17,35)*. Such applications require ideal SEC behavior, i.e., separations should be based solely on solute size. However, most modern high-performance SEC columns are anionic (i.e., carry a negative charge) to a greater or lesser extent *(69)*. Such a property may lead to inter-action with positively charged side-chains in peptides and proteins unless such undesirable electrostatic interactions are suppressed. Because electrostatic effects are minimized above an eluent ionic strength of about 0.05 M, aqueous phosphate buffers (pH 5.0–7.5) containing 0.1–0.4 M salts are commonly employed as the mobile phase for SEC of peptides and proteins *(2,3,5,8,69)*. Of course, peptide–protein or protein–protein interactions may be elimi-nated if the salt concentration is too high when electrostatic interactions are a dominating factor to the interaction. A mixture of five model synthetic peptide standards (10, 20, 30, 40, and 50 residues) with negligible secondary structure was developed both to detect nonideal retention behavior during SEC as well as to monitor suppression of such nonideal behavior with addition of salts *(69)*. Different column materials can exhibit dramatically different amounts of nonideal behavior. These SEC peptide standards can be obtained from the Alberta Peptide Institute, University of Alberta, Edmonton, Alberta, Canada.

3.1.1. SEC of Peptides

1. **Figure 2** (top panel) shows the elution profile of six peptides containing 4, 6, 8, 10, 14, and 20 residues on a Superdex Peptide column (Column 1) and Instrument 2.
2. The peptides are separated by isocratic elution with 50 mM aqueous KH_2PO_4, pH 7, containing 100 mM KCl at a flow rate of 0.5 mL/min and room temperature.

3. The excellent resolution of the six peptides illustrates the value of this column for small peptide separations. In addition, an initial fractionation of complex peptide mixtures by this column should simplify subsequent IEX and/or RP-HPLC steps.

4. **Figure 2** (bottom panel) shows the linear relationship between peptide size (number of residues) and distribution coefficient (K_d).

5. K_d values are calculated from the expression $K_d = V_e - V_o/V_t - V_o$, where V_e is the elution volume of the solute, V_o is the void volume of the packing (obtained from the elution volume of blue dextran), and V_t is the total accessible volume of the column (obtained from the elution volume of β-mercaptoethanol).

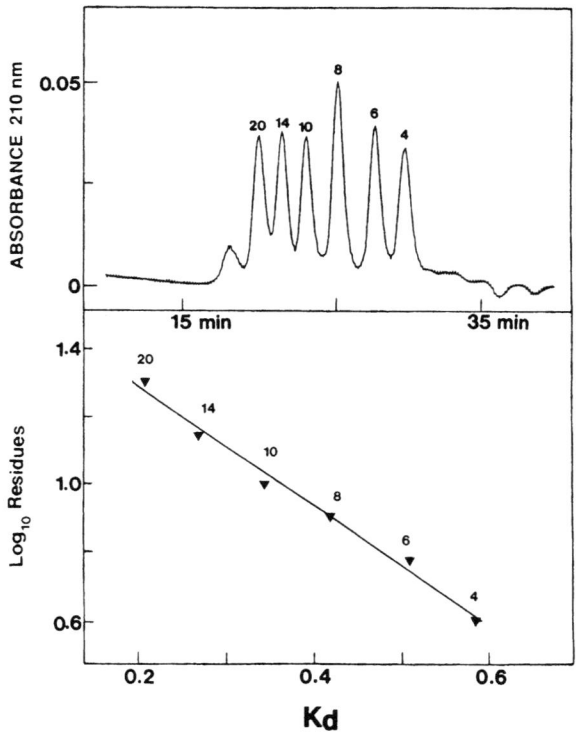

Fig. 2. Size-exclusion chromatography of small peptides. Column and conditions described under **Subheading 3.1.1. Top**, elution profile of peptides; **bottom**, relationship between Log_{10} of the number of residues and K_d (distribution coefficient; *see* text in **Subheading 3.1.1.** for calculation). The sequences of the peptides are FIPK (4 residues); Ac-GGTAGG-amide (6 residues); PQSPESVD-amide (8 residues); LKAETEALKA (10 residues); Ac-TDDPASPQSPESVD-amide (14 residues); and IEALKCEIEALKAEIEALKA-amide (20 residues). (Adapted from **ref. 35**, with permission from Elsevier Science.)

3.1.2. SEC Analysis of Peptide Oligomerization

1. **Table 4** presents sequences of native regions and analogs of regions within the SARS-Coronavirus Spike S glycoprotein that form coiled-coil structures (*70*).
2. HRN 916–950 and HRC 1150–1185 represent native sequences within the coiled-coils (where HRN and HRC denote heptad repeat regions) at the N- and C-termini of a domain within the coronavirus S fusion protein.
3. The linker is attached to the native HRN sequence via a Cys-Gly-Gly spacer extension at the N-terminal of the HRN 916–950 sequence. The linker and this HRN sequence are mixed at a molar ratio of 1:4 in phosphate-buffered saline at pH 7.0, followed by incubation at room temperature for 4 h during which a covalent bond is formed between the sulfhydryl group in the Cys side-chain of the peptide and the bromoacetyl groups in the linker, producing a three-stranded peptide. DTE (5 mM) is then added to the reaction mixture and incubated at room temperature for 10 min, resulting in reduction of any disulfide-bridged peptides. The three-stranded peptide and monomeric peptides are subsequently separated by RP-HPLC and the three-stranded peptide (denoted as "HRN with linker" in **Table 4**) characterized by mass spectrometry.
4. In **Table 4**, the peptide denoted HRN 916–950 (T923I, N937I) represents an analog of the native HRN sequence whereby Thr-923 and Asn-937 have been replaced by Ile in the hydrophobic core of the coiled-coil sequence.
5. In **Table 4**, HRN extended (902–950) represents the HRN 916–950 native sequence with the inclusion of two more native heptad sequences at the N-terminus.
6. The peptides in **Table 4** are subjected to SEC on a Superdex 75 column (Column 2 on Instrument 1) by isocratic elution with 50 mM aqueous phosphate buffer (NaH$_2$PO$_4$/Na$_2$HPO$_4$) containing 100 mM NaCl at a flow-rate of 0.25 mL/min and at room temperature; sample mixtures contained 0.2 mM of each peptide with 5 μL applied to the column.
7. **Figure 3A** illustrates the elution profiles of a mixture of HRN 916–950 (monomer), HRN linked 3-stranded molecule and the complex of HRC 1150–1185 with HRN 916–950 (hexamer). **Figure 3B** shows the elution profiles of a mixture of HRN 916–950 (monomer), HRN 916–950 (T923I, N937I) (dimer), and HRN extended (902–950) (trimer). **Figure 3C** represents the log MW vs retention time plot (*see* **Table 4** for values) resulting from observed peptide elution behavior.
8. As shown in **Fig. 3**, the native HRN 916–950 sequence is eluted as a random coil monomeric peptide during SEC. However, extending the native sequence by two heptads to form HRN extended (902–950) results in a stable, fully folded trimeric coiled-coil (**Fig. 3B**).
9. Attachment of the HRN 916–950 sequence to the linker (HRN with linker) also results in a stable, fully folded three-stranded coiled-coil (**Fig. 3A**).
10. Replacement of two polar residues, Thr and Asn, by the hydrophobic Ile residues in the native sequence (HRN 916–950) (T923I, N937I) results in a stable, fully folded

Table 4
Size Exclusion/High Performance Liquid Chromatography Analysis of Peptide Oligomerization State

Peptide name	Oligomerization state	Sequence[b]	MW	Retention time (min)
HRC 1150–1185	monomer	$\qquad$ a $\quad$ d $\quad$ a $\quad$ d $\qquad$ a $\quad$ d $\quad$ a $\quad$ d Ac-DISGINASVVNIQKEIDRLNEVAKNLNESLIDLQEL-amide	4050	
HRN 916–950	monomer	d $\quad$ a $\quad$ d $\quad$ a $\quad$ d $\quad$ a $\quad$ d $\quad$ a Ac-IQESLTTTSTALGKLQDVVNQNAQALNTLVKQLSS-amide	3756	53.6
HRN 916–950 (T923I, N937I)	dimer	Ac-IQESLTTISTALGKLQDVVNQIAQALNTLVKQLSS-amide	7534	48.2
HRN with linker[a]	3-stranded	Ac-CGGIQESLTTTSTALGKLQDVVNQAQALNTLVKQLSS-amide	12713	45.8
HRN 902–950	trimer	d $\quad$ a $\quad$ d $\quad$ a $\quad$ d $\quad$ a $\quad$ d $\quad$ a $\quad$ d $\quad$ a Ac-QKQIANQFNKAISQIQESLTTTSTALGKLQDVVNQNAQALNTLVKQLSS-amide	16068	43.2
HRC 1150–1185/ HRN 916–950	hexamer		23418	41.8

[a] The linker sequence is Ac-GK(X)GK(X)GK(X)G-amide, where X is –CO-CH$_2$-Br group attached to the ε-amino groups of lysine.
[b] In the heptad repeat, denoted abcdefg, positions a and d in the hydrophobic core of the coiled-coil are underlined.

dimeric coiled-coil (**Fig. 3B**). The Ile residues provide the additional stability in the hydrophobic core to stabilize the folded coiled-coil, but as a dimer.

11. Interestingly, although the native HRN 916–950 and HRC 1150–1185 sequences are eluted as unfolded, random coil peptides during SEC, a mixture of these two peptides produces a fully folded, α-helical six-helix bundle (i.e., a hexamer) (**Fig. 3A**).

12. Synthetic model amphipathic α-helical peptide standards designed to monitor the effect of SEC packings and mobile phases on peptide oligomerization have been previously described *(35)*.

3.2. Ion-Exchange HPLC

The retention time of a peptide in either AEX or CEX will depend on a number of factors including buffer pH and the nature and ionic strength of the anion or cation employed for displacement of acidic (negatively charged) or basic (positively charged) peptides, respectively. Most ion-exchange separations are carried out using sodium or potassium ion as the cationic counterion and chloride ion as the anionic counterion. High-performance IEX packings tend to exhibit hydrophobic characteristics to varying degrees, perhaps resulting in significant peak broadening or even nonelution of peptides due to undesirable interactions with nonpolar residues in the peptide, which can be suppressed through the addition of an organic modifier (generally, 10–20% CH_3CN) to the mobile phase *(42)*. Synthetic peptide standards are available to monitor such nonideal effects on CEX columns (sequences shown in **Fig. 4** legend) and to gauge the level of organic modifier required for ideal (i.e., predictable) elution behavior of peptides on a specific column *(42)*. These CEX peptide standards can be obtained from the Alberta Peptide Institute, University of Alberta, Edmonton, Alberta, Canada.

3.2.1. Monitoring of Peptide Chain Length Effects on CEX Using Synthetic Peptide Standards

1. Two series of synthetic, positively charged peptide standards were subjected to CEX on a Mono S strong cation-exchange column (Column 3 on Instrument 2): a series of four 11-residue cation-exchange standards with net charges of +1, +2, +3, and +4; and a series of five synthetic size-exclusion standards of 10, 20, 30, 40, and 50 residues in length (net charge of +1, +2, +3, +4, and +5, respectively). Note that the +2 and +4 CEX standards both contain Tyr, allowing detection at 260–280 nm as well as 210 nm.

2. The peptides were eluted with a linear AB gradient (20 m*M* salt/min following 10-min isocratic elution with eluent A) at a flow rate of 1 mL/min and a temperature

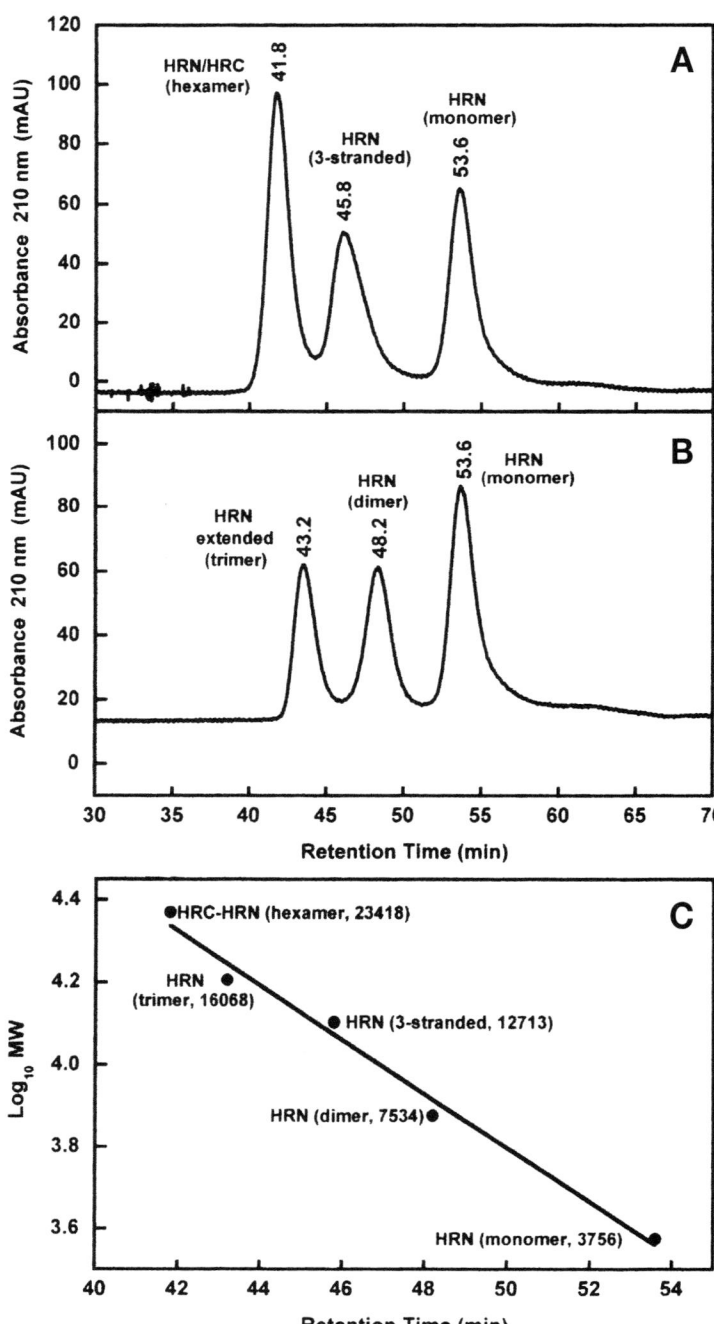

of 26°C, where eluent A is 5 mM KH$_2$PO$_4$, pH 6.5, and eluent B is eluent A containing 0.5 MNaCl, both eluents also containing 40% CH$_3$CN (*see* **Note 3**).

3. In **Fig. 4**, similarly charged species are not necessarily eluted at similar times: the 50-residue peptide (+5; **Fig. 4A**) is not retained as long as the 11-residue +3 and +4 peptides (**Fig. 4B**); the 40-residue peptide (+4; **Fig. 4A**) is eluted prior to the 11-residue +3 peptide (**Fig. 4B**).
4. The comparative retention behavior of the peptides of different chain length and charge density is linearized by an elution time vs net charge/1nN plot (N = number of residues) (**Fig. 4C**). This simple linearization approach is important for the prediction of peptide retention behavior where the net charge is known.

3.3. Reversed-Phase HPLC

In the authors' experience, the best approach to most analytical peptide separations is to employ aqueous TFA to TFA/CH$_3$CN linear gradients (pH 2.0) at room temperature. Peptide resolution can be optimized by varying the steepness of the acetonitrile gradient (generally, 0.5–2.0% CH$_3$CN/min) and the volatility of TFA eliminates the need for subsequent sample desalting. TFA is also effective in separating complex peptide mixtures because of its ion-pairing properties. Peptides are charged molecules at most pH values and the presence of different counterions will influence their chromatographic behavior. Thus, anionic counterions (e.g., TFA$^-$, phosphate) will interact with the proto-nated basic (i.e., positively charged) residues of a peptide. In comparison, a cationic counterion (e.g., trimethylammonium, triethylammonium, tetrabuty-lammonium) will show an affinity for ionized carboxyl (i.e., negatively charged) groups. Recently, we determined the optimum TFA concentration for peptide separations to be 0.2–0.25% TFA, significantly higher than the traditionally employed concentration range of 0.05–0.1% (*71*).

3.3.1. Determination of Intrinsic Hydrophilicity/Hydrophobicity of Amino Acid Side-Chains by RP-HPLC of Model Peptides

1. A model peptide sequence is designed for quantitation of amino acid side-chain hydrophilicity/hydrophobicity in the absence of nearest-neighbor or conformational effects: Ac-X-G-A-K-G-A-G-V-G-L-amide, where X is substituted by the 20 amino acids plus norvaline (*n*-Val), norleucine (*n*-Leu) and ornithine (Orn) (*10*).

Fig. 3. Size-exclusion chromatography analysis of oligomerization states of peptides. Column and conditions shown in **Subheading 3.1.2. A,B**, elution profiles of peptides; **C**, relationship between Log$_{10}$ molecular weight of eluted species and their retention times. The sequences of the peptides are shown in **Table 4**.

2. The peptides are eluted from a Kromasil C_{18} column (Column 5 on Instrument 1) at pH 2.0 by a linear AB gradient (0.25% CH_3CN/min) at a flow-rate of 0.3 mL/min and a temperature of 25°C, where eluent A is 20 mM aqueous TFA, pH 2.0 containing 2% acetonitrile, and eluent B is 20 mM TFA in CH_3CN (**Fig. 5A**). Some

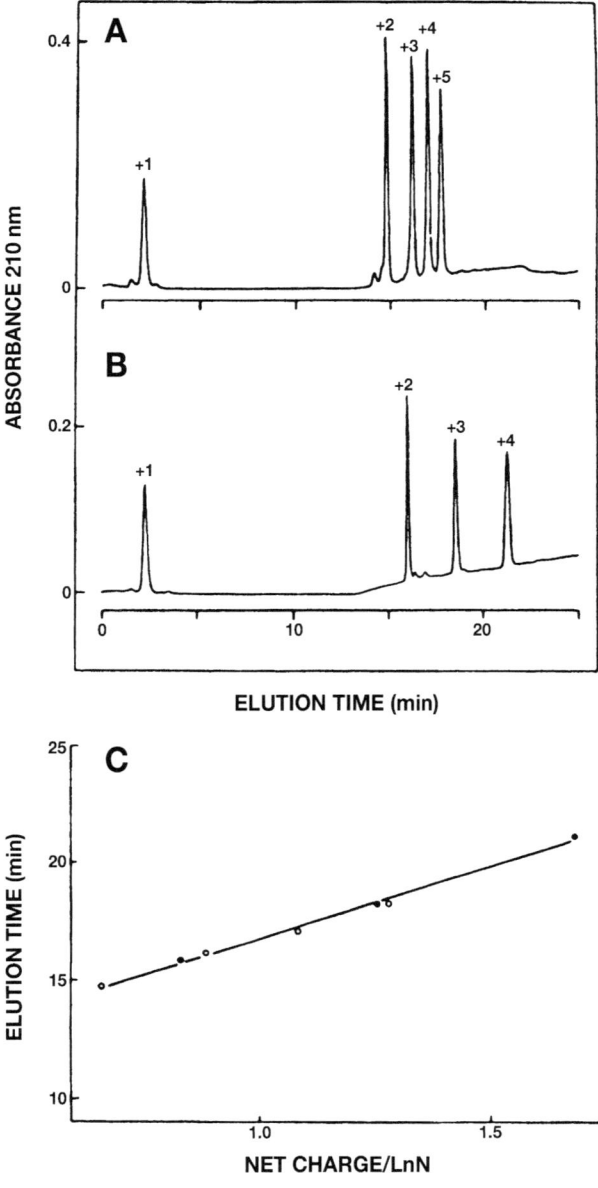

"wetting" of the stationary phase, particularly highly hydrophobic and high-ligand-density phases such as the Kromasil C_{18} packing, is occasionally required to ensure reproducible results.

3. These same peptides are eluted from a Zorbax XDB-C_8 column (Column 6) at pH 7.0 by a linear AB gradient (0.25% CH_3CN/min) at a flow-rate of 0.3 mL/min and a temperature of 25°C, where eluent A is 10 mM aqueous NaH_2PO_4, pH 7.0 and eluent B is eluent A containing 50% CH_3CN, both eluents also containing 50 mM NaCl **(Fig. 5B)**.

4. The resulting intrinsic hydrophilicity/hydrophobicity coefficients are reported in **Table 1** as Δt_R (Gly) values, where Δt_R (Gly) denotes the change in retention time relative to the Gly-substituted peptide. These coefficients are independent of pH, buffer conditions, ion-pairing reagents, or whether a C_8 or C_{18} column was used for 17 (uncharged) side-chains and are dependent on pH, buffer conditions, and ion-pairing reagents for potentially charged side-chains (Orn, Lys, His, Arg, Asp and Glu) **(Fig. 5C)**. For a detailed discussion of the importance of these side-chain hydrophilicity/hydrophobicity coefficients in the peptide and protein field, together with validation of these values, see our recent publication *(10)*.

5. The correlation of the hydrophilicity/hydrophobicity values for the 17 uncharged side-chains at pH 2.0 and pH 7.0 was excellent as expected (R = 0.998) *(10)*. The relative hydrophobicities of the acidic side-chains (Asp and Glu) decrease with an increase in pH due to deprotonation (ionization) of these side-chains at pH 7.0 (the pK_a of these side-chains is ~4.0), that is, they become negatively charged and more hydrophilic. In contrast, the His side-chain is essentially deprotonated at pH 7.0 (pK_a ~6.0) and thus no longer positively charged, i.e., its hydrophobicity increases dramatically. The dramatic increase in the hydrophobicity of the Arg side-chain with the change in pH from 2.0 to 7.0 is due to differences in effectiveness of the counterions TFA and $H_2PO_4^{-2}$ anions to neutralize the positive charge of the Arg side-chain at pH 2.0 and pH 7.0, respectively. The $H_2PO_4^{-2}$ anion more effectively neutralizes the positive charge at pH 7.0, thus increasing the relative hydrophobicity of the Arg side-chain. In the case of the alkyl amino groups on the side-chains of

---

Fig. 4. Strong cation-exchange chromatography of synthetic peptide standards. Column and conditions shown in **Subheading 3.2.1. Top**, elution profile of five synthetic size-exclusion standards (sequences: Ac-(GLGAKGAGVG)$_n$-amide, where n = 1, 2, 3, 4 or 5; +1, +2, +3, +4 and +5, respectively); **middle**, elution profile of four synthetic cation-exchange standards (sequences: Ac-GGGLGGAGGLK-amide (+1), Ac-KYGLGGAGGLK-amide (+2), Ac-GGALKALKGLK-amide (+3) and Ac-KYALKALKGLK-amide (+4); **bottom**, plot of observed peptide elution time versus peptide net charge, divided by the logarithm of the number of residues (LnN). (Adapted from **ref. 42**, with permission from Elsevier Science.)

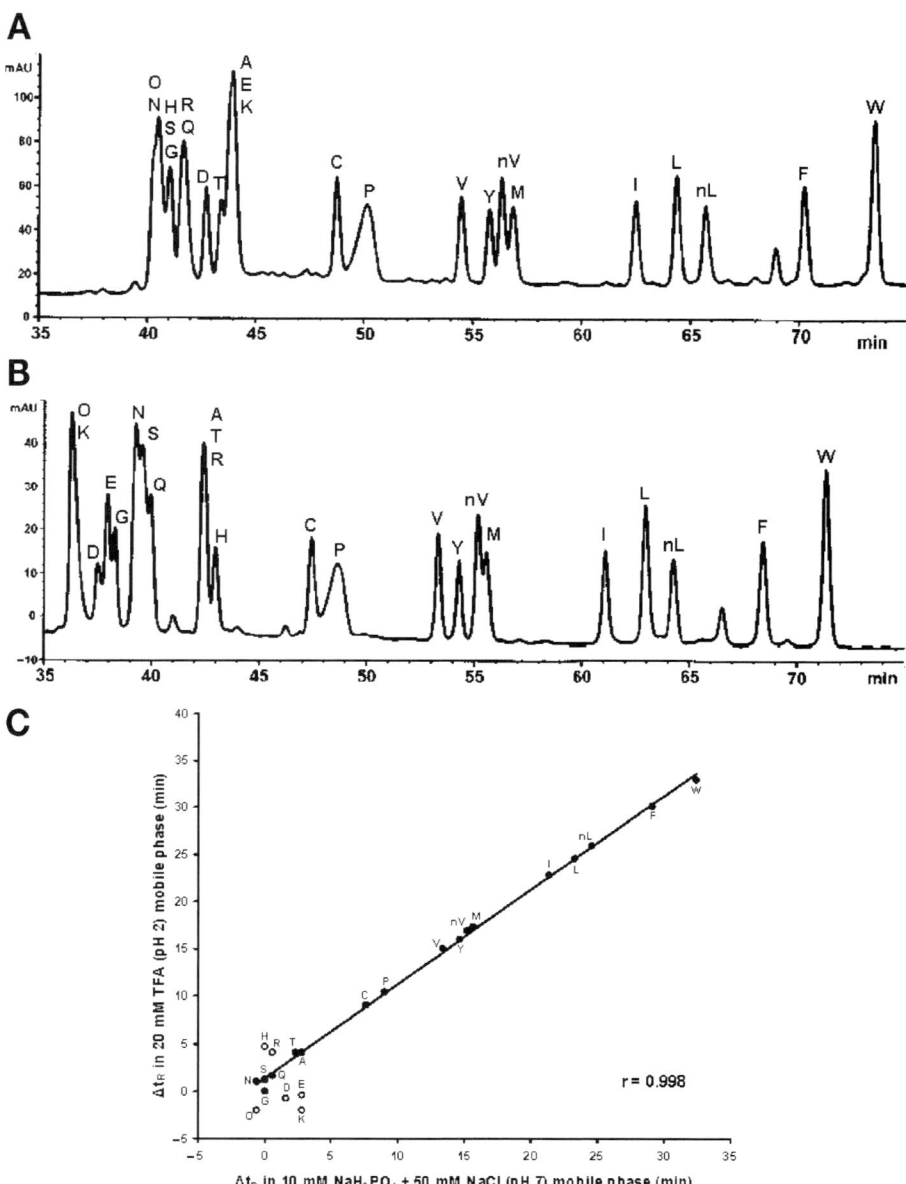

Fig. 5. Determination of intrinsic hydrophilicity/hydrophobicity of amino acid side-chains by reversed-phase high-performance liquid chromatography of model peptides. Column and conditions shown in **Subheading 3.3.1. A**, pH 2.0 elution profile; **B**, pH 7.0 elution profile; **C**, correlation of amino acid side-chain coefficients obtained at pH 2.0 vs pH 7.0. Peptide sequences shown under **Subheading 3.3.1.**

Orn and Lys, the increase in hydrophilicity (decrease in hydrophobicity) at pH 7.0 relative to pH 2.0 is again due to differences in the effectiveness of these two anions to ion-pair with these side-chains. The side-chains of Lys and Orn more effectively ion-pair with the TFA anion (hydrophobic ion-pairing reagent) than the $H_2PO_4^{-2}$; thus, these side-chains are more hydrophobic at pH 2.0 than pH 7.0. The physicochemical properties of the guanidinium group of Arg and the alkyl amino groups of Lys and Orn are quite different. These results clearly demonstrate that Arg and Lys side-chains, although both positively charged at pH 2.0 and pH 7.0, behave differently with buffer anions, which affects their hydrophilicities/hydrophobicities differently. Thus, Arg and Lys residues on the surface of proteins should not be considered equivalent at neutral pH.

3.3.2. Effect of Anionic Ion-Pairing Reagent Hydrophobicity on Selectivity of Peptide Separations

1. The perfluorinated homologous series of acids (TFA, PFPA, and HFBA) represents a useful series of anionic ion-pairing reagents used for peptide separations *(72,73)*.
2. The order of counterion hydrophobicity is TFA⁻ < PFPA⁻ < HFBA⁻.
3. RP-HPLC is applied to a mixture of three series of four synthetic peptides: a +1 series (i.e., each peptide in the series has a net charge of +1), a +3 series, and a +5 series (**Table 5**).

Table 5
Sequences and Denotions of Synthetic Peptide Standards (see Subheading 3.3.2)

Peptide Series[a]	Peptide Denotion	Peptide Sequence[b]
+1	1a	Ac- G G **G G** G L G L G K – amide
	1b	Ac- G G **A G** G L G L G K – amide
	1c	Ac- G G **A A** G L G L G K – amide
	1d	Ac- G G **V G** G L G L G K – amide
+3	3a	Ac- G R **G G** K L G L G K – amide
	3b	Ac- G R **A G** K L G L G K – amide
	3c	Ac- G R **A A** K L G L G K – amide
	3d	Ac- G R **V G** K L G L G K – amide
+5	5a	NH_3^+- R R **G G** K L G L G K – amide
	5e	NH_3^+- R R **V A** K L G L G K – amide
	5h	NH_3^+- R R **V V** K L G L G K – amide
	5i	NH_3^+- R R **I I** K L G L G K – amide

[a] The charge on the peptide is shown at pH 2.0.
[b] The different amino acid substitutions are shown in bold letters.

4. As shown in **Table 5**, within each peptide series there is only a subtle increase in hydrophobicity between adjacent peptides: 1a < 1b < 1c < 1d (+1 series); 3a < 3b < 3c < 3d (+3 series); 5a < 5e < 5h < 5j (+5 series); this is particularly true of the +1 and +3 series, where there is a difference of only one carbon atom between adjacent peptides.

5. The mixture of peptides is eluted from a Zorbax SB300-C_8 column (Column 7 on Instrument 1) by a linear AB gradient (0.5% CH_3CN/min) at a flow rate of 0.3 mL/min and a temperature of 25°C, where eluent A is 30 mM aqueous TFA, PFPA, or 10 mM aqueous HFBA, pH 2.0 and eluent B is the corresponding reagent concentration in CH_3CN.

6. As shown in **Fig. 6**, increasing counterion hydrophobicity (TFA$^-$ < PFPA$^-$ < HFBA$^-$) results in increasing peptide retention time; in addition, there is a

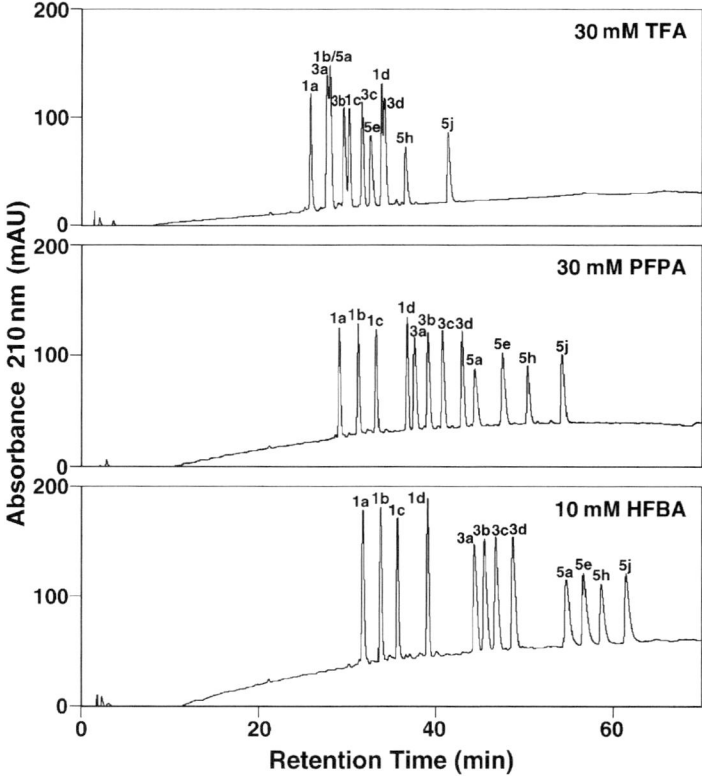

Fig. 6. Effect of anionic ion-pairing reagent hydrophobicity on reversed-phase high-performance liquid chromatography of positively charged peptides. Column and conditions described under **Subheading 3.3.2.** Peptide sequences shown in **Table 5**. (Adapted from **ref. 73**, with permission from Elsevier Science.)

general overall peak shape improvement with increasing counterion concentration (not shown).

7. As the hydrophobicity of the counterion increases, the relative hydrophobicity of the peptides is increasing in the order of $+1$ peptides $< +3$ peptides $< +5$ peptides, resulting in a change in peptide elution order with increasing counterion hydrophobicity and culminating in the excellent resolution of all 12 peptides in the presence of $10 \, \text{m}M$ HFBA whereby the peptides are separated by charged groups and hydrophobicity within these groups.

3.3.3. Effect of Temperature on RP-HPLC of Peptides

1. The effect of temperature on the elution behavior of a mixture of nine synthetic random coil peptides is shown in **Fig. 7.**

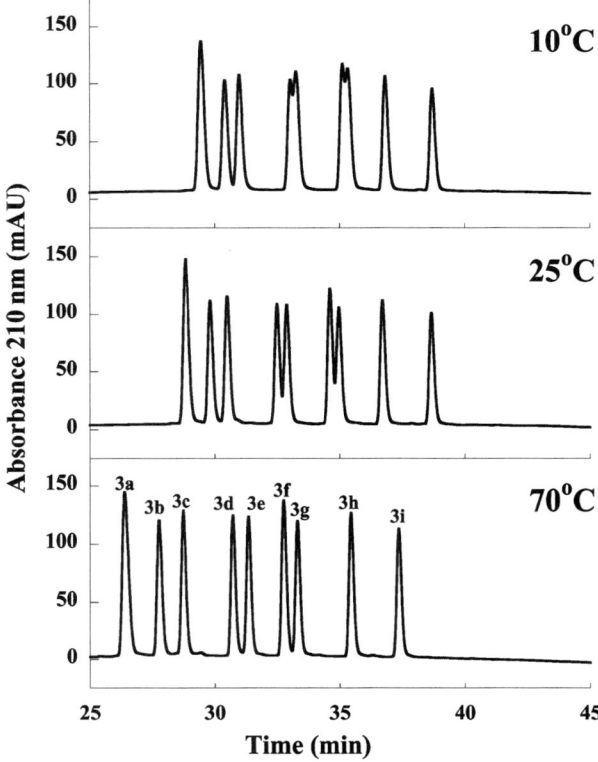

Fig. 7. Effect of temperature on reversed-phase high-performance liquid chromatography of peptides ($+3$ net charge). Column and conditions described under **Subheading 3.3.3**. Peptide sequences shown in **Table 6**.

2. The nine 10-residue peptides each have a net charge of +3 with just a subtle variation in hydrophobicity between adjacent peptides (**Table 6**). Indeed, there is generally a change of just one carbon atom between adjacent peptides of the +3 peptides; the only exceptions are peptides 3e and 3f, which have identical numbers of carbon atoms.
3. The peptides are eluted from a Zorbax SB300-C_8 microbore column (Column 8 on Instrument 1) by a linear AB gradient (1% CH_3CN/min) at a flow rate of 0.1 mL/min and a temperature of 10°C, 25°C, and 70°C, where eluent A is 0.2% aqueous TFA, pH 2.0, and eluent B is 0.2% TFA in CH_3CN.
4. The elution order of the peptides is based on their increasing hydrophobicity (3a < 3b < 3c < 3d < 3e < 3f < 3g < 3h < 3i).
5. As shown in **Fig. 7**, increasing temperature results in decreasing peptide retention time and a decrease in peak width. Such results reflect the general effects of increasing temperature, i.e., increased solubility of the solute in the mobile phase as the temperature rises as well as an increase in mass transfer between the mobile and stationary phases (*74,75*).
6. As shown in **Fig. 7**, an increase in temperature clearly improves overall peak resolution; indeed two pairs of peptides barely resolved as doublets at 10°C are baseline resolved at 70°C: 3d/3e and 3f/3g.

3.3.4. Effect of Temperature on RP-HPLC of Random Coil vs α-Helical Peptides

1. **Figure 8** compares the RP-HPLC separation of random coil and α-helical L- or D-peptides around the optimum temperatures for overall separation of the peptide mixtures (*76*).
2. The synthetic random coil peptides have the sequence Ac-X-L-G-A-K-G-A-G-V-G-amide, where X is substituted with the 19 L- and D-amino acids and Gly.
3. The synthetic amphipathic α-helical peptides have the sequence Ac-E-A-E-K-A-A-K-E-X-E-K-A-A-K-E-A-E-K-amide, where X is substituted with the 19 L- and D-amino acids and Gly.
4. Mixtures of the peptides are subjected to RP-HPLC on a Zorbax SB300-C_8 column (Column 7 on Instrument 1) using a linear AB gradient (0.5% CH_3CN/min) at a flow rate of 0.25 mL/min at various temperatures, where eluent A is 0.05% aqueous TFA and eluent B is 0.05% TFA in CH_3CN.
5. The hydrophobicity of the RP-HPLC column, together with the nonpolar organic modifier (acetonitrile) in the mobile phase, will induce α-helical structure in peptides with that inherent α-helical potential (*12–15*). Thus, unlike random coil peptides, amphipathic α-helical peptides will exhibit preferred binding for their nonpolar face with the hydrophobic stationary phase (*12*) (*see* **Note 4**).
6. Both random coil and helical peptide analogs exhibited the trend of a reduction in retention time with increasing temperature (10–80°C). However, temperature has a greater effect on the L-/D-helical peptide analogs compared to the

Table 6
Sequences of Two-dimensional Cation-Exchange/Reversed-Phase Chromatography Peptide Standards

Peptide Standard	Net Charge	Change in Carbon atom Content	Peptide Sequence	Mass	Change in Mass
1a	+1	0	Ac-Gly-Gly-Gly-Gly-Leu-Gly-Leu-Gly-Lys-amide	814	0
1b	+1	1	Ac-Gly-Gly-Ala-Gly-Gly-Leu-Gly-Leu-Gly-Lys-amide	828	14
1c	+1	2	Ac-Gly-Gly-Ala-Ala-Gly-Leu-Gly-Leu-Gly-Lys-amide	842	28
1d	+1	3	Ac-Gly-Gly-Val-Gly-Gly-Leu-Gly-Leu-Gly-Lys-amide	856	42
1e	+1	4	Ac-Gly-Gly-Val-Ala-Gly-Leu-Gly-Leu-Gly-Lys-amide	870	56
1f	+1	4	Ac-Gly-Gly-Ile-Gly-Leu-Gly-Leu-Gly-Lys-amide	870	56
1g	+1	5	Ac-Gly-Gly-Ile-Ala-Gly-Leu-Gly-Leu-Gly-Lys-amide	884	70
1h	+1	7	Ac-Gly-Gly-Ile-Val-Gly-Leu-Gly-Leu-Gly-Lys-amide	912	98
1i	+1	8	Ac-Gly-Gly-Ile-Ile-Gly-Leu-Gly-Leu-Gly-Lys-amide	926	112
2a	+2	0	Gly-Gly-Gly-Gly-Leu-Gly-Gly-Leu-Gly-Lys-amide	772	0
2b	+2	1	Gly-Gly-Ala-Gly-Gly-Leu-Gly-Gly-Leu-Gly-Lys-amide	786	14
2c	+2	2	Gly-Gly-Ala-Ala-Gly-Leu-Gly-Gly-Leu-Gly-Lys-amide	800	28
2d	+2	3	Gly-Gly-Val-Gly-Gly-Leu-Gly-Gly-Leu-Gly-Lys-amide	814	42
2e	+2	4	Gly-Gly-Val-Ala-Gly-Leu-Gly-Gly-Leu-Gly-Lys-amide	828	56
2f	+2	4	Gly-Gly-Ile-Gly-Leu-Gly-Gly-Leu-Gly-Lys-amide	828	56
2g	+2	5	Gly-Gly-Ile-Ala-Gly-Leu-Gly-Gly-Leu-Gly-Lys-amide	842	70
2h	+2	7	Gly-Gly-Ile-Val-Gly-Leu-Gly-Gly-Leu-Gly-Lys-amide	870	98
2i	+2	8	Gly-Gly-Ile-Ile-Gly-Leu-Gly-Leu-Gly-Lys-amide	884	112

Table 6
(Continued)

Peptide Standard	Net Charge	Change in Carbon atom Content	Peptide Sequence	Mass	Change in Mass
3a	+3	0	Gly-Gly-Gly-Lys-Leu-Gly-Leu-Gly-Lys-amide	843	0
3b	+3	1	Gly-Gly-**Ala**-Gly-Lys-Leu-Gly-Leu-Gly-Lys-amide	857	14
3c	+3	2	Gly-Gly-**Ala-Ala**-Lys-Leu-Gly-Leu-Gly-Lys-amide	871	28
3d	+3	3	Gly-Gly-**Val**-Gly-Lys-Leu-Gly-Leu-Gly-Lys-amide	885	42
3e	+3	4	Gly-Gly-**Val-Ala**-Lys-Leu-Gly-Leu-Gly-Lys-amide	899	56
3f	+3	4	Gly-Gly-**Ile**-Gly-Lys-Leu-Gly-Leu-Gly-Lys-amide	899	56
3g	+3	5	Gly-Gly-**Ile-Ala**-Lys-Leu-Gly-Leu-Gly-Lys-amide	913	70
3h	+3	7	Gly-Gly-**Ile-Val**-Lys-Leu-Gly-Leu-Gly-Lys-amide	941	98
3i	+3	8	Gly-Gly-**Ile-Ile**-Lys-Leu-Gly-Leu-Gly-Lys-amide	955	112

Ac denotes N^{α}-acetyl; amide denotes C^{α}-amide. Standards in the +2 and +3 groups have a free α-amino group. Variations in composition of the peptide analogs are indicated in bold where the amino acid residues differ from 1a, 2a or 3a.

L-/D-random peptide analogs, likely due to differences in peptide structural changes with temperature variations (unfolding of the α-helix with increasing temperature) *(78)*.
7. As shown in **Fig. 8**, excellent separations of mixtures of L- and D-peptides are obtained at 21°C and 62.5°C, respectively, in contrast to the presence of co-eluted peaks at higher and lower temperatures, indicated by the arrows.

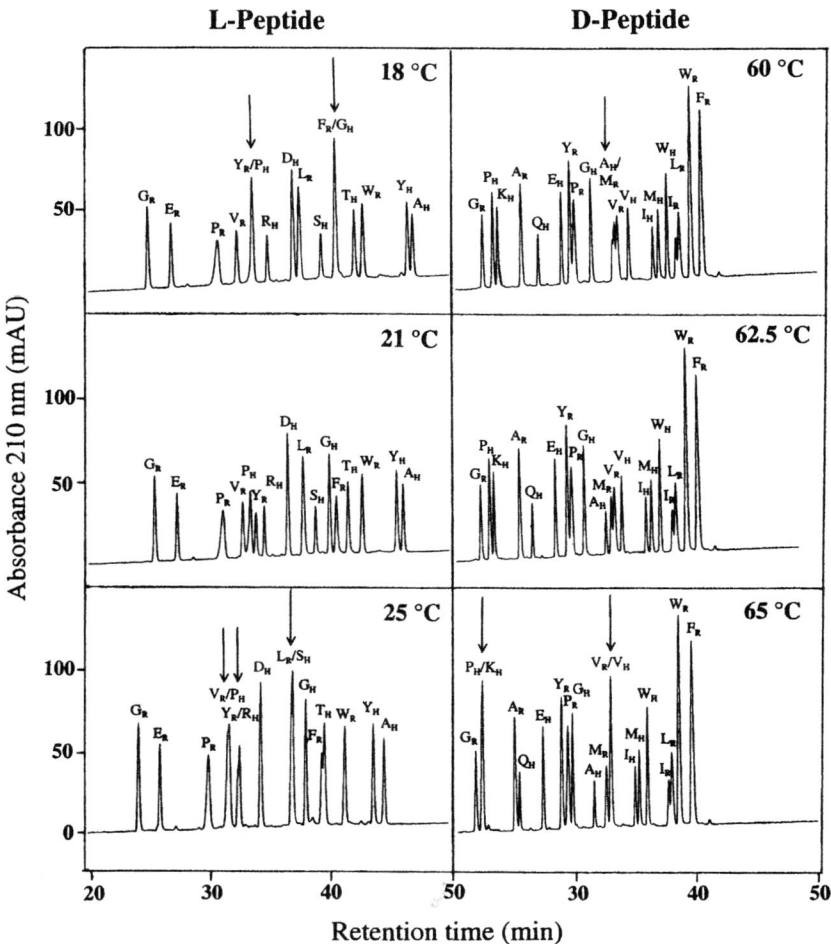

Fig. 8. Effect of temperature on reversed-phase high-performance liquid chromatography of random coil vs α-helical peptides. The subscript R and H denote random coil and helical peptides, respectively. Column, conditions, and peptide sequences described under **Subheading 3.3.4**. (Reproduced from **ref. 76**, with permission from Elsevier Science.)

8. Thus, in contrast to the optimum elution profiles in **Fig. 8**, the chromatograms at higher or lower temperature represent the sensitivity of a temperature variation approach to influence the selectivity of RP-HPLC for separation of peptides with conformational differences.

3.4. HILIC/CEX

It is important to note that different ion-exchange packings exhibit differing degrees of hydrophobic characteristics *(42)*. In order to gain the full benefit of peptide separations by the HILIC mode in mixed-mode HILIC/CEX, it is important to overcome unwanted hydrophobic properties of the matrix with as low a level of organic modifier (CH_3CN) as possible, i.e., the ion-exchange matrix should be as hydrophilic as possible. In this way, there is a greater organic modifier range open to the researcher to effect mixed-mode HILIC/CEX peptide separations. Concomitant with this hydrophilic character, it is desirable for the column to retain even weakly charged species ($+1$) to obtain full benefit of retention based on ion-exchange as well as hydrophilic characteristics.

3.4.1. HILIC/CEX of Proteomic Peptide Standards

1. HILIC/CEX is applied to the separation of a mixture of three groups of synthetic model peptides designed as two-dimensional CEX/RP-HPLC peptide standards, i.e., proteomics standards (**Table 6**): $+1$, $+2$ and $+3$ groups of peptides, each containing nine peptides, where peptide hydrophobicity increases in the order a < b < c < d < e < f < g < h < i.
2. From **Table 6**, peptides within the $+1$, $+2$ and (as noted above under **Subheading 3.3.3.**) $+3$ groups vary only subtly in hydrophobicity: in general, there is just a one-carbon difference between adjacent peptides except for peptide pairs 1e/1f, 2e/2f and 3e/3f which contain identical numbers of carbon atoms and for peptide pairs 1g/1h, 2g/2h and 3g/3h which differ by two carbon atoms.
3. The 27-peptide mixture was subjected to HILIC/CEX on a Polysulfoethyl A strong CEX column (Column 4 on Instrument 1) using a linear AB gradient ($1\,mM\,NaClO_4$/min) at a flow-rate of 0.3 mL/min and a temperature of 25°C, where eluent A is $5\,mM$ aqueous triethylammonium phosphate, pH 4.5, and eluent B is eluent A plus $200\,mM$ $NaClO_4$, both eluents also containing 60% (v/v) CH_3CN.
4. $NaClO_4$ is particularly useful for this mixed-mode approach as a result of its high solubility in organic modifier, which allows the use of high levels of organic modifier.
5. As shown in **Fig. 9**, an excellent separation of the 27-peptide mixture is obtained, with just the 1e/1f peptide pair not being completely resolved. Indeed, this separation is superior to that obtained by both RP-HPLC and CE (*see* **Subheading 3.5**).
6. The peptide separation shown in **Fig. 9** is achieved by a mixed-mode or bidimensional mechanism: the three groups of peptides are separated by an ion-exchange mechanism ($+1$ < $+2$ < $+3$); within these groups of peptides, the peptides are eluted

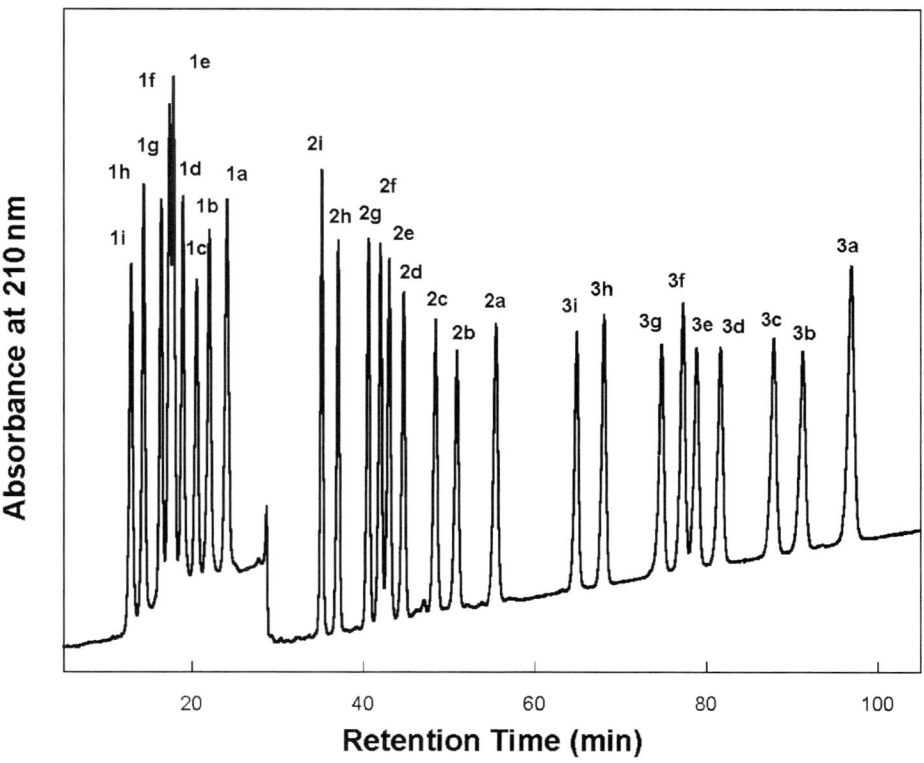

Fig. 9. Hydrophilic interaction/cation-exchange chromatography of proteomic peptide standards. Column and conditions described under **Subheading 3.4.1**. Peptide sequences shown in **Table 6**.

in order of increasing hydrophilicity (decreasing hydrophobicity) by the HILIC mechanism. The high acetonitrile concentration during CEX promotes hydrophilic interactions with the hydrophilic CEX matrix. Thus, within each charged group of peptides, the most hydrophobic peptide is eluted first and the most hydrophilic peptide is eluted last, i.e., the opposite elution order to RP-HPLC. Compare the separation of the +3 group by RP-HPLC (**Fig. 7**) vs HILIC/CEX (**Fig. 9**).

3.4.2. Comparison of HILIC/CEX and RP-HPLC for Separation of Amphipathic α-Helical Peptides

1. **Figure 10** compares RP-HPLC with HILIC/CEX for the separation of model amphipathic α-helical peptides with the sequence Ac-K-W-K-S-F-L-K-T-F-K-X-A-V-K-T-V-L-H-T-A-L-K-A-I-S-S-amide, where X (in the center of the nonpolar face of the amphipathic α-helix) is substituted by various L- and D-amino acids (*29*).

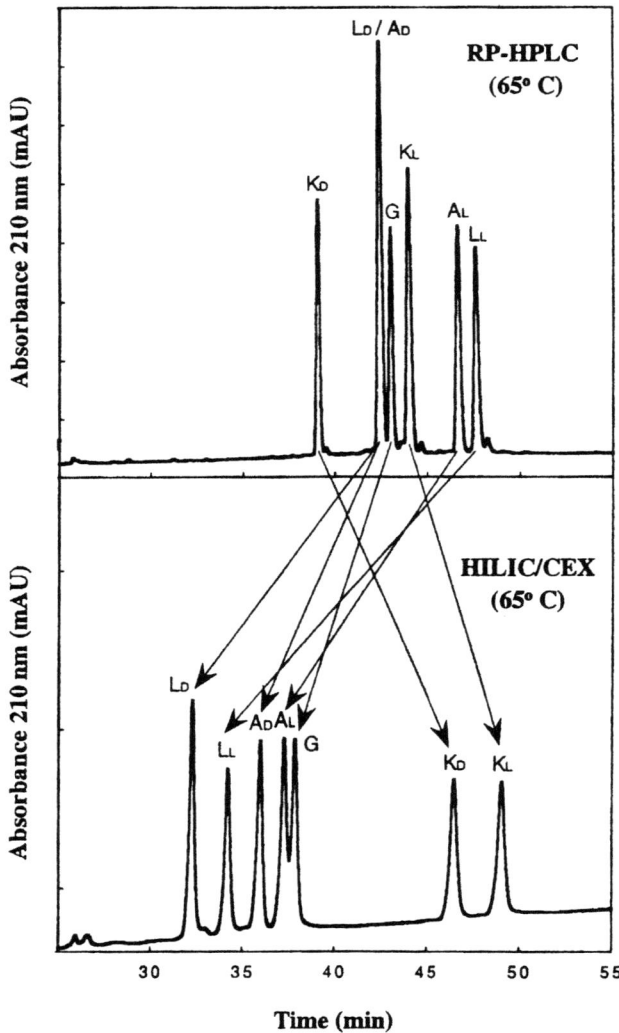

Fig. 10. hydrophilic interaction/cation-exchange chromatography vs reversed-phase high-performance liquid chromatography of diastereomeric amphipathic α-helical peptides. Columns, conditions, and peptide sequences described under **Subheading 3.4.2.** (Adapted from **ref. 29**, with permission from Elsevier Science.)

2. RP-HPLC (**Fig. 10,** top) is carried out on a Zorbax SB300-C_8 column (Column 7 on Instrument 1) by a linear AB gradient (1% CH_3CN/min) at a flow rate of 0.3 mL/min and a temperature of 65°C, where eluent A is 0.05% aqueous TFA, pH 2, and eluent B is 0.05% TFA in CH_3CN.

3. HILIC/CEX (**Fig. 10,** bottom) is carried out on a PolySulfoethyl A strong cation-exchange column (Column 4 on Instrument 1), by a linear AB salt gradient (5 mM NaClO$_4$ to 250 mM NaClO$_4$ in 50 min, i.e., 4.9 mM NaClO$_4$/min) at a flow-rate of 0.3 mL/min and temperature of 65°C, where eluent A is 5 mM aqueous triethylammonium phosphate (TEAP), pH 4.5, containing 5 mM NaClO$_4$ and eluent B is 5 mM aqueous TEAP, pH 4.5, containing 250 mM NaClO$_4$, both eluents also containing 70% (v/v) CH$_3$CN (*see* **Note 5**).

4. During RP-HPLC, the amphipathic α-helical peptides will interact with the hydrophobic stationary phase through preferential binding with their hydrophobic faces; during HILIC/CEX, the peptides would be expected to interact with the hydrophilic stationary phase through preferential binding with their hydrophilic faces.

5. From **Fig. 10**, useful selectivity changes between RP-HPLC and HILIC/CEX of the peptides are apparent, underlining the complementary nature of these two HPLC modes.

6. In both RP-HPLC and HILIC/CEX, the peptides substituted by D-amino acids were always eluted earlier than their L-counterparts. This observation is likely due to disruption of the hydrophobic and hydrophilic preferred binding domains, respectively, by the introduction of a D-amino acid into an α-helix otherwise comprised solely of L-amino acids *(79,80)*.

3.4.3. Monitoring the Hydrophilicity/Hydrophobicity of Amino Acid Side-Chains in the Nonpolar and Polar Faces of Amphipathic α-Helical Peptides

1. **Figure 11** compares RP-HPLC and HILIC/CEX separations of amphipathic α-helical peptides with amino acid substitutions made on the center of the hydrophobic face or hydrophilic face of the helix: Ac-K-W-K-S-F-L-K-T-F-K-X$_1$-A-X$_2$-K-T-V-L-H-T-A-L-K-A-I-S-S-amide, where position X$_1$ (in the center of the hydrophilic face; X$_1$ = Ser) and position X$_2$ (in the center of the hydrophobic face; X$_2$ = Val) are substituted by various L-amino acids *(30)*.

2. The native peptide sequence, with Ser and Val at positions X$_1$ and X$_2$, respectively, is a biologically active amphipathic α-helix (denoted V681) with potent antimicrobial and hemolytic properties *(81,82)*.

3. The ability to monitor the hydrophilicity/hydrophobicity effects of amino acid substitutions in both the nonpolar and polar faces of potentially useful antimicrobial amphipathic α-helical peptides is critical in the design process for such molecules.

4. The peptides are subjected to RP-HPLC by a linear AB gradient (1% CH$_3$CN/min) on a Zorbax SB300-C$_8$ column (Column 7 on Instrument 1) at a flow-rate of 0.3 ml/min and room temperature, where eluent A is 0.05% aqueous TFA, pH 2, and eluent B is 0.05% TFA in CH$_3$CN.

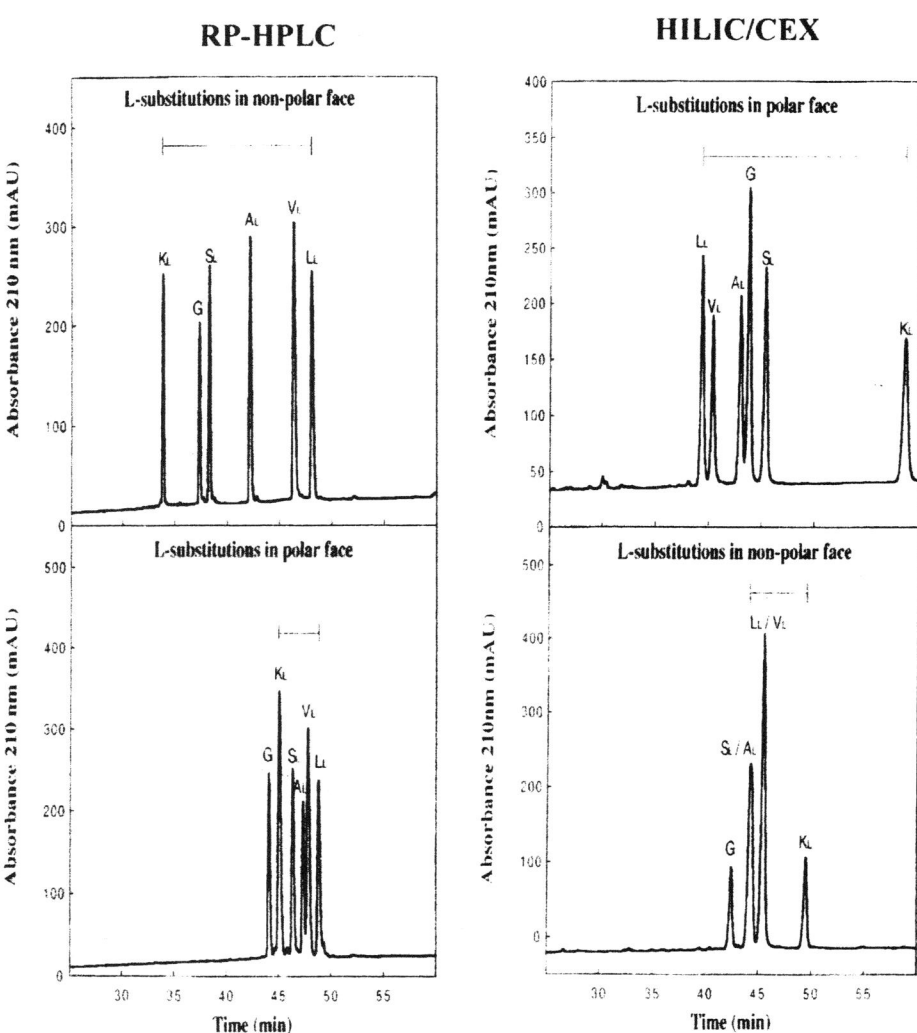

Fig. 11. Hydrophilic interaction/cation-exchange chromatography vs reversed-phase high-performance liquid chromatography for monitoring hydrophilicity/hydrophobicity of amino acid side-chains of amphipathic α-helical peptides. Column, conditions, and peptide sequences described under **Subheading 3.4.3.** (Adapted from **ref. *30***, with permission from Elsevier Science.)

5. The peptides are subjected to HILIC/CEX on a PolySulfoethyl A strong cation-exchange column (Column 4 on Instrument 1), by a linear AB gradient (5 m*M* NaClO$_4$ to 250 m*M* NaClO$_4$ in 60 min, i.e., 4 m*M* NaClO$_4$/min) at a flow-rate of 0.3 mL/min and 65°C, where eluent A is 5 m*M* aqueous TEAP, pH 4.5, containing

5 mM NaClO$_4$ and eluent B is 5 mM aqueous TEAP, pH 4.5, containing 250 mM NaClO$_4$, both buffers also containing 70% (v/v) CH$_3$CN.

6. As shown in **Fig. 11**, RP-HPLC and HILIC/CEX are best suited for resolving amphipathic peptides were substitutions are made in the nonpolar and polar faces, respectively. Concomitantly, RP-HPLC and HILIC/CEX are best suited as monitors of hydrophilicity/hydrophobicity variations where amino acid substitutions were made in these respective faces.

3.5. Capillary Electrophoresis

1. As noted above **(Subheading 3.4.1.)**, a mixture of 27 peptides is designed as synthetic proteomic peptide standards. These peptides are comprised of three groups of nine 10-residue peptides (with net charges of +1, +2, and +3 for all nine peptides within a group), the hydrophobicity of the nine peptides within a group varying only subtly between adjacent peptides. The sequences and characteristics of the peptide standards are shown and described in **Table 6** and under **Subheading 3.4.1**.

2. **Figure 12** shows the RP-HPLC separation of the peptide standards on a Zorbax SB300-C$_8$ column (Column 8 on Instrument 1) obtained by a linear AB gradient (0.5% CH$_3$CN/min) at a flow rate of 0.3 mL/min and a temperature of 70°C, where eluent A is 20 mM aqueous TFA, pH 2.0, and eluent B is 20 mM TFA in 80% (v/v) aqueous CH$_3$CN.

3. The conditions for RP-HPLC (an efficient ion-pairing reagent, relatively shallow gradient, high temperature, small packing particle size of 3.5 µm) are designed to optimize the separation. However, despite the reasonably satisfactory analytical elution profile, just 13 out of 27 peptides are well resolved. It is unlikely that this unidimensional approach to the separation of the peptides can be optimized further to achieve complete (or even near complete) resolution of all 27 peptides.

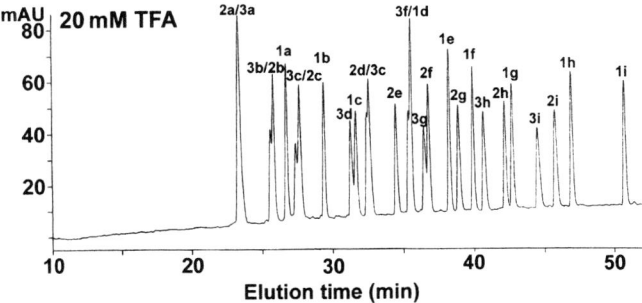

Fig. 12. Reversed-phase high-performance liquid chromatography of proteomic peptide standards. Column and conditions described under **Subheading 3.5**. Peptide sequences shown in **Table 6**. (Adapted from **ref. 56**, with permission from Elsevier Science.)

4. **Figure 13** shows the separation of the 27 peptides by CE in analogous conditions to that of RP-HPLC (i.e., with aqueous TFA as the background electrolyte [BGE] in the absence or presence of CH_3CN, albeit in the absence of a hydrophobic surface (uncoated capillary).

5. Conditions for CE in **Fig. 13**: uncoated capillary (Capillary 1 on Instrument 4); BGE of 10 mM aqueous TFA, adjusted to pH 2.0 with LiOH, with or without 25% (v/v) CH_3CN; applied voltage, 25 kV (direct polarity) with 5-min voltage ramp; temperature, 15°C; UV absorption at 195 nm.

6. As shown in **Fig. 13**, both in the absence (top panel) and presence (bottom panel) of 25% CH_3CN, the peptides are separated according to their charge-to-mass ratio (capillary zone electrophoresis (CZE) mechanism), each peak representing nine co-migrated peptides. The mobility of these three groups of peptides is according to net charge, i.e., as expected, the +3 group migrates faster than the +2 group, which migrates faster than the +1 group.

7. **Figure 13** represents a unidimensional CE separation, where peptide separation is achieved/optimized via a single peptide property (charge) and via a single charge-based mechanism (CZE). This laboratory believes that the introduction of a hydrophobicity-based mechanism, to produce a bidimensional separation, would enable an effective peptide separation.

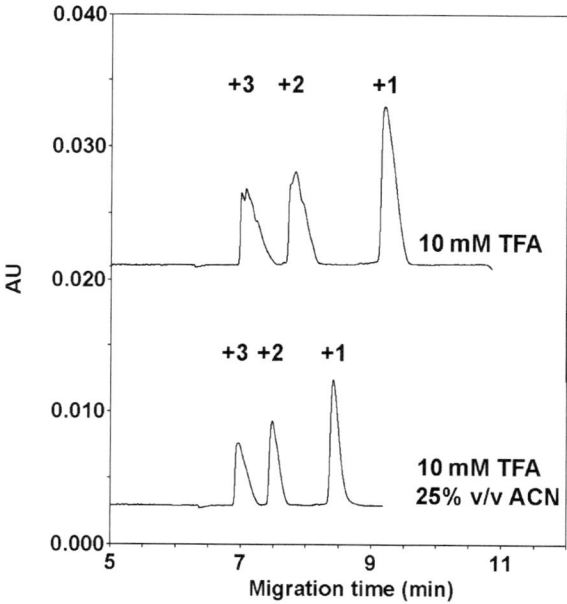

Fig. 13. Capillary zone electrophoresis of proteomic peptide standards. Capillary and conditions described under **Subheading 3.5.** Peptide sequences shown in **Table 6.** (Reproduced from **ref. 56**, with permission from Elsevier Science.)

8. The 27 peptides were again subjected to CE under the following conditions: BGE of 0.4 *M* aqueous TFA, PFPA, or HFBA, adjusted to pH 2.0 with LiOH; remaining conditions as above.

9. As shown in **Fig. 14**, an excellent separation of the 27 peptides is now achieved, with increasing ion-pairing reagent hydrophobicity (TFA < PFPA < HFBA) resulting in improved resolution.

10. These separations represent bidimensional separations: peptides of the same length but different nominal charge are separated in order of decreasing charge (i.e., +3 peptides migrate the earliest and +1 peptides migrate the last) according to a CZE mechanism; within each group of peptides, the peptides are separated in order of increasing hydrophobicity according to a hydrophobically mediated mechanism introduced by the anionic ion-pairing reagent, i.e., within each charged group of peptides, the most hydrophobic peptide is eluted last and the most hydrophilic peptide is eluted first., analogous to that seen in RP-HPLC (**Fig. 7**).

11. This novel CE approach has been termed ion-interaction (II)CZE and is clearly superior to RP-HPLC (**Fig. 12**) for separation of these peptide standards. This CE approach also exhibits a much larger peak capacity than RP-HPLC, as seen by the

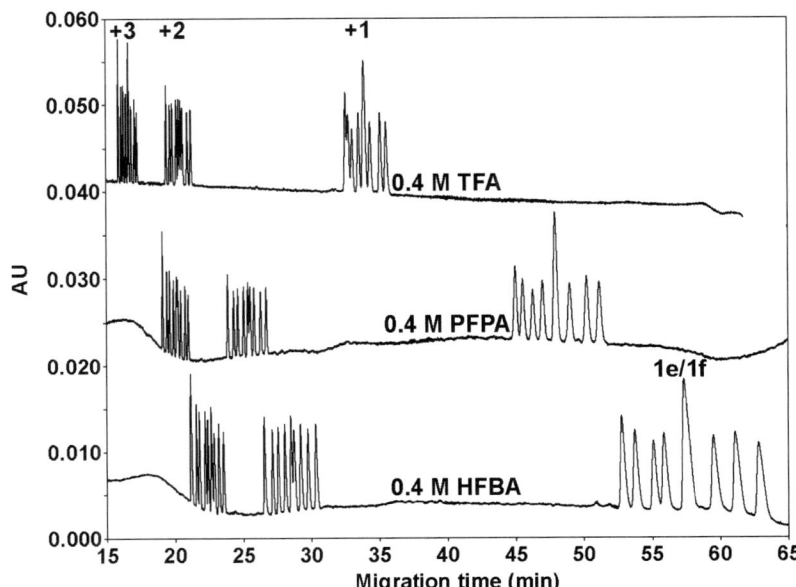

Fig. 14. Ion-interaction capillary zone electrophoresis of proteomic peptide standards. Capillary and conditions described under **Subheading 3.5.** Peptide sequences shown in **Table 6**. (Reproduced from **ref. 56**, with permission from Elsevier Science.)

large distance between the +2 and +1 peptide groups, where peptides of different charge-to-mass ratios could be located.

12. It should be noted that only one pair of peptides with identical charge-to-mass ratios (denoted 1e/1f) was not resolved in the presence of 0.4 *M* HFBA. However, two other peptide pairs with identical charge-to-mass ratios (one pair in each of the +2 and +3 peptides; 2e/2f and 3e/3f, respectively) are resolved under these conditions, a significant achievement when one considers that such separations were hitherto believed impossible.

13. Interestingly, as noted previously (**Subheading 3.4.1.**), the best overall resolution of the peptides within each group is achieved by HILIC/CEX. The high peak capacity of the II-CZE and HILIC/CEX approaches, coupled with their excellent resolution capabilities, again highlight their useful complementarity (and often superior separative effectiveness) to the ubiquitous RP-HPLC mode.

3.6. Preparative HPLC of Synthetic Peptides and Recombinant Proteins

The excellent resolving power and separation time of RP-HPLC, coupled with the availability of volatile mobile phases, has made this HPLC mode the favored method for preparative separations of peptides. Also, most researchers would likely wish to carry out both analytical and preparative peptide separations on analytical equipment and columns no longer than 250 mm and no larger diameters than 10 mm ID, avoiding the prohibitively expense scale-up costs in terms of equipment, larger columns and solvent consumption. It has long been a goal of this laboratory to develop novel one-step purification schemes to avoid loss of product yield frequently a feature of multistep protocols (e.g., SEC followed by IEX and RP-HPLC).

3.6.1. Preparative One-Step RP-HPLC of a Synthetic Amphipathic α-Helical Antimicrobial Peptide

1. **Figure 15** shows the analytical RP-HPLC profile of a crude synthetic 26-residue amphipathic α-helical antimicrobial peptide, with the sequence Ac-K-W-K-S-F-L-K-T-F-K-S-A-K-K-T-V-L-H-T-A-L-K-A-I-S-S-amide and all residues are D-amino acids *(83)*.

2. The analytical run was carried out on a Zorbax SB300-C_8 column (Column 7 on Instrument 1) by a linear AB gradient (1% CH_3CN/min) at a flow rate of 0.25 mL/min and room temperature, where eluent A is 0.2% aqueous TFA, pH 2.0, and eluent B is 0.2% TFA in CH_3CN.

3. As shown in **Fig. 15**, the synthesis of the peptide by the solid-phase approach is successful; however, purification of the desired product from both hydrophilic and hydrophobic impurities is clearly necessary.

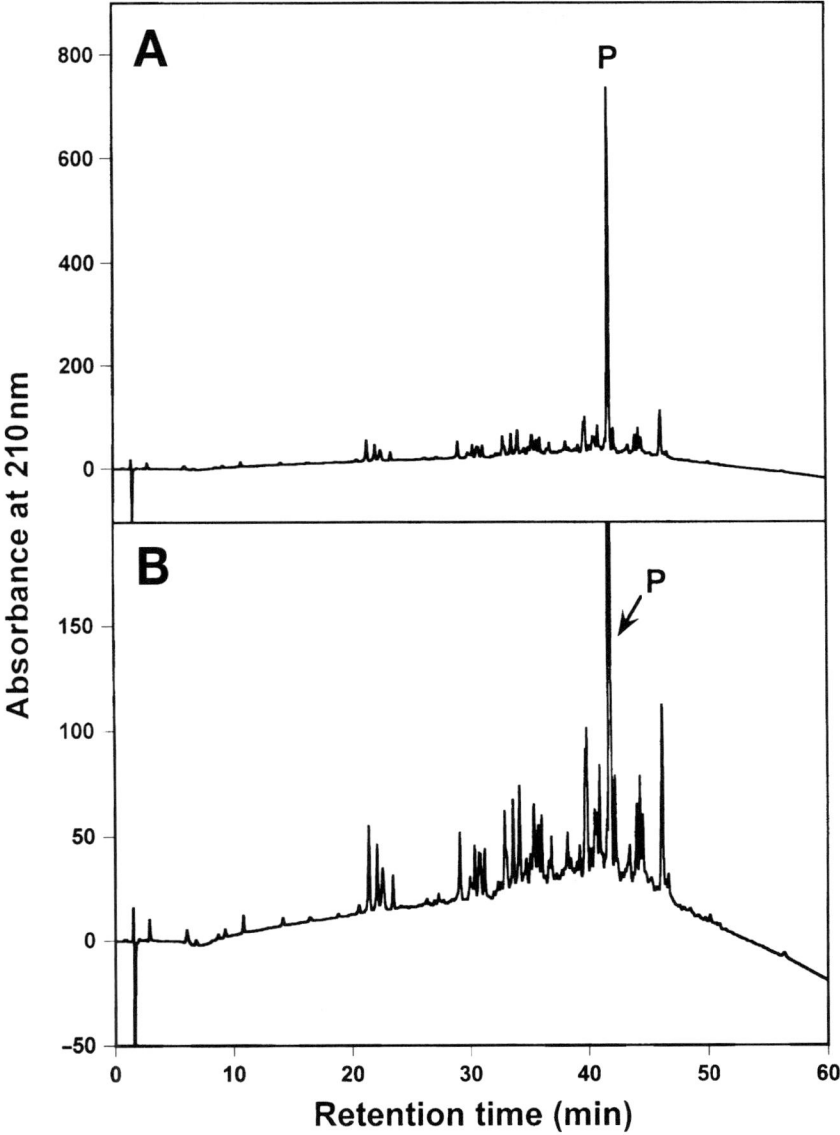

Fig. 15. Analytical reversed-phase high-performance liquid chromatography profile of crude synthetic 26-residue amphipathic α-helical antimicrobial peptide. Column, conditions, and peptide sequence shown in **Subheading 3.6.1. A**, On-scale elution profile; **B**, fourfold expansion of elution profile. P denotes desired product.

4. The synthetic crude peptide is applied (100 mg and 200 mg sample amounts) to a Zorbax SB300-C$_8$ semipreparative RP-HPLC column (Column 9 on Instrument 3). A linear AB gradient (1% CH$_3$CN/min for 30 min, followed by 0.1% CH$_3$CN/min for 150 min, followed by 1% CH$_3$CN/min for 30 min) at a flow rate of 2 mL/min and room temperature, where eluent A is 0.2% aqueous TFA, pH 2, and eluent B is 0.2% TFA in CH$_3$CN. To design the gradient method, a "rule of thumb" has been developed. The researcher determines the percentage acetonitrile required to elute the peptide of interest by running the crude peptide on an analytical column at 1% CH$_3$CN per min. In the present case, the peptide of interest is eluted at 42% CH$_3$CN. To establish the percentage CH$_3$CN at which the 0.1% CH$_3$CN gradient begins, the rule of thumb is to start 12% below that required to elute the peptide in the 1% gradient run. Thus, the gradient for the preparative run is 0 to 30% acetonitrile at 1% acetonitrile/min (30% is 12% below 42%), then begin the gradient of 0.1% CH$_3$CN per min for 150 min (increase of 15% CH$_3$CN) followed by 1% CH$_3$CN per min for 30 min to wash the remaining hydrophobic impurities off the column.

5. Fractions are collected every 2 min and fraction analysis carried out under the same conditions described for **Fig. 15**. Fractions are repooled as hydrophilic impurities, pure product and hydrophobic impurities (**Fig. 16A**, **B**, and **C**, respectively). These three pools are then lyophilized, redissolved in water (10 mL) and reanalyzed by RP-HPLC (5 µL sample volumes; same conditions as described for **Fig. 15**).

6. **Figure 16** shows the excellent product purity and yield obtained by this slow gradient approach to purification of the 200-mg sample load (**Fig. 16B**). Very little product is found in the hydrophilic fraction (**Fig. 16A**) and hydrophobic fraction (**Fig. 16C**).

7. **Figure 17** illustrates fraction analyses of adjacent fractions at the beginning and end of product appearance for both the 100 mg (top panels) and 200 mg (bottom panels) sample amounts. Note that only one fraction for each sample amount contained overlapping product and hydrophilic impurities (Fr. 63 and Fr. 54 for 100-mg and 200-mg runs, respectively) or overlapping product and hydrophobic impurities (Fr. 75 and Fr. 71 for 100-mg and 200-mg runs, respectively), underlying the effectiveness of this slow gradient approach for resolution of desired peptide product from even closely structurally related synthetic impurities.

8. It should be noted that the greater the sample load, the sooner the appearance of desired peptide product eluted from the column, i.e., Fr. 54 for a 200-mg load vs Fr. 63 for a 100-mg load. In addition, the larger the load, the greater the number of fractions needed for product elution (13 fractions for the 100 mg load and 18 fractions for the 200-mg load).

9. The RP-HPLC column has the capacity for an even larger sample load than 200 mg. However, there is not sufficient material available to determine the maximum load on this column.

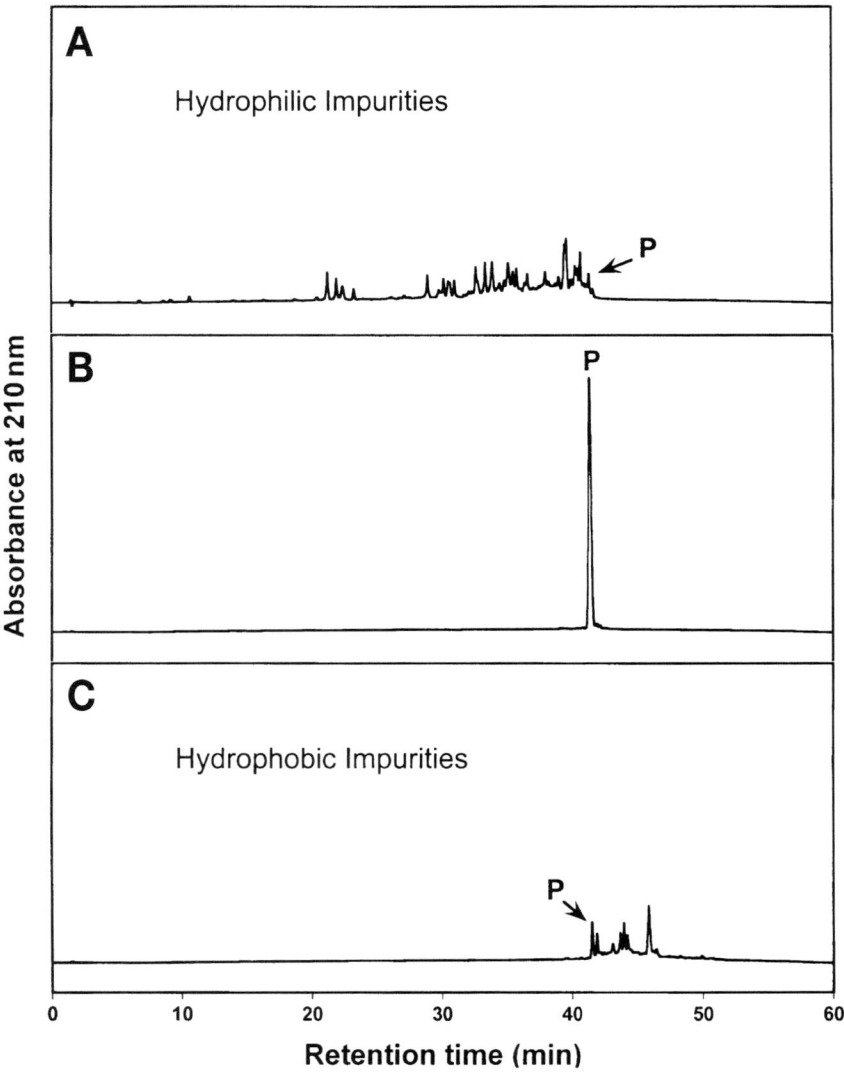

Fig. 16. Pooled fractions following reversed-phase (RP)-high-performance liquid chromatography (HPLC) purification of 100 mg of crude synthetic 26-residue amphipathic α-helical antimicrobial peptide. Column, conditions, and peptide sequence described under **Subheading 3.6.1.** Analytical RP-HPLC elution profiles of pooled fractions: **(A)** pool of all fractions containing hydrophilic impurities, **(B)** pool of all fractions containing pure product, denoted P, and **(C)** pool of all fractions containing hydrophobic impurities.

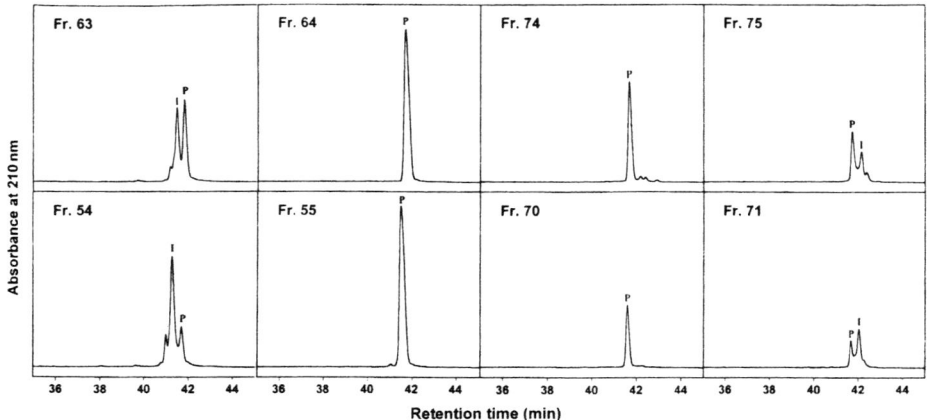

Fig. 17. Fraction analysis of adjacent fractions at beginning and end of product appearance following preparative reversed-phase (RP)-high-performance liquid chromatography (HPLC) of 100 mg (**top panels**) and 200 mg (**bottom panels**) of synthetic 26-residue amphipathic α-helical antimicrobial peptide. Column, conditions of preparative RP-HPLC and fraction analysis, and peptide sequence described under **Subheading 3.6.1.** P denotes desired product.

3.6.2. Preparative One-Step Purification of a Recombinant Protein From a Whole Cell Lysate

1. A truncated form (99 residues) of chicken skeletal α-tropomyosin (denoted Tm1-99) is prepared as fusion protein containing a T7 tag (14 amino acids in length) at the N-terminus *(84)*.
2. The construct is inserted into a pET3a vector and the fusion protein is subsequently overexpressed in *Escherichia coli.*
3. Following cell lysis, the cell contents are extracted with 0.1% aqueous TFA.
4. **Figure 18** (top) presents an analytical RP-HPLC elution profile of the cell lysate. This profile is obtained on a Zorbax SB300-C_8 column (Column 7 on Instrument 1) by a linear AB gradient (1% CH_3CN/min) at a flow rate of 0.3 mL/min and room temperature, where eluent A is 0.05% aqueous TFA, pH 2.0, and eluent B is 0.05% TFA in CH_3CN.
5. As shown in **Fig. 18** (top), the desired Tm1-99 product represented 3.2% of total contents of crude sample.
6. A sample volume of 4 mL is applied to a Zorbax SB300-C_8 narrow bore column (Column 7 on Instrument 3) and subjected to a linear AB gradient (2% CH_3CN/min up to 24% CH_3CN, followed by 0.1% CH_3CN/min up to 40% CH_3CN; a rapid rise of 4% CH_3CN/min up to 60% CH_3CN was then followed by an isocratic wash with

60% aqueous CH$_3$CN) at a flow rate of 0.3 mL/min and room temperature, where eluent A is 0.05% aqueous TFA, pH 2, and eluent B is 0.05% TFA in CH$_3$CN.

7. Fractions are collected every 2 min and analyzed by RP-HPLC on a Zorbax SB300-C$_8$ column (Column 7 on Instrument 1) under the same conditions as for the analytical profile of the crude cell lysates (**Fig. 18**, top), save for a linear AB gradient of 2% CH$_3$CN/min.

8. The fractions are pooled into hydrophilic impurities (**Fig. 18**, bottom left), purified product (**Fig. 18**, bottom middle), and hydrophobic impurities (**Fig. 18**, bottom right). These analytical profiles are obtained with the same column and conditions as the crude analytical profile (**Fig. 18**, top).

9. As shown in **Fig. 18**, the purified product had a purity of >95% and a yield of >90% recovery. This was in contrast with a 64% purity of Tm1-99 obtained by a one-step antibody affinity approach at one-tenth the recovery of the one-step RP-HPLC approach. Scale up of the affinity approach to produce enough semi-pure material (still requiring the final RP-HPLC step) of comparable purity and yield of the RP-HPLC-based protocol alone would certainly be prohibitively expensive.

10. Note that the same rule of thumb for the shallow gradient approach of 0.1% CH$_3$CN/min (as described for the 26-residue peptide (**Subheading 3.6.1.**) can be applied to the preparative purification of larger polypeptides and proteins.

3.6.3. Alternative Preparative RP-HPLC Purification Approaches on Analytical Columns and Instrumentation

The slow gradient approach to preparative RP-HPLC of peptides and proteins described under **Subheadings 3.6.1.** and **3.6.2.** produces excellent yields of purified samples. Indeed, we recommend this simple one-step RP-HPLC protocol for routine purification of relatively small sample quantities. However, for larger scale applications, where it is of particular importance to maximize, for cost purposes, yield of purified material per gram of RP-HPLC stationary phase, we wish to direct readers to our novel preparative approach termed sample displacement chromatography (SDC), characterized by the major separation process generally taking place in the absence of organic modifier concomitant with optimum utilization of RP-HPLC column capacity *(85–89)*.

3.6.3.1. ONE-STEP SDC OF PEPTIDES

1. Conventional reversed-phase SDC is conceived as a novel purification approach on high-performance analytical columns and instrumentation, where the only variable is sample load, i.e., without the addition of organic modifier or displacer (the latter being characteristic of traditional displacement chromatography *(90)*) to the mobile phase *(85–89)*.

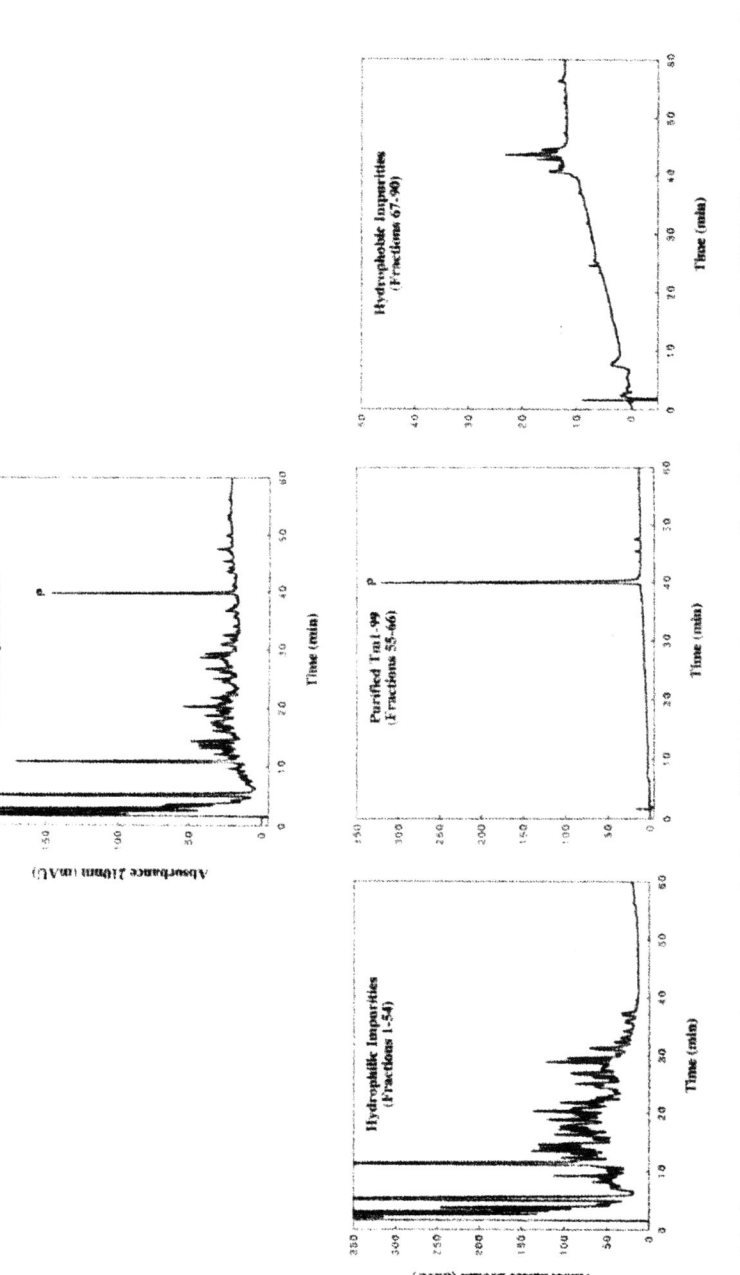

Fig. 18. One-step preparative reversed-phase (RP)-high-performance liquid chromatography (HPLC) purification of a recombinant protein from a whole cell lysate. **Top**, analytical RP-HPLC profile of crude sample mixture. Bottom: pooled fractions following preparative purification of recombinant protein Tm 1–99. Column, conditions of analytical RP-HPLC, preparative RP-HPLC, and fraction analysis described under **Subheading 3.6.2.** P denotes desired product. (Reproduced from **ref. 84**, with permission from Elsevier Science.)

2. Because peptides favor an adsorption-desorption mechanism of interaction with a hydrophobic stationary phase, under normal analytical load conditions an organic modifier is typically required for their elution from a reversed-phase column. However, when such a column is subjected to high loading of a peptide mixture dissolved in a 100% aqueous mobile phase, e.g., 0.1% aqueous TFA, there is competition by the sample components for the adsorption sites on the reversed-phase sorbent, resulting in solute–solute displacement during washing with 100% aqueous mobile phase. A more hydrophobic peptide component competes more successfully for these sites than a less hydrophobic component, which is thus displaced ahead of the more hydrophobic solute, i.e., the sample components acts as their own displacers.

3. The SDC approach is thus simply an application of the well established general principles of displacement chromatography without the need for a separate displacer. This mode of operation, a hybrid scheme of frontal chromatography followed by elution, is characterized by a marked reduction in solvent consumption, minimal elution volumes, and the collection of fewer fractions for product isolation than in conventional RP-HPLC, with consequence reductions in time and handling.

4. Several successful SDC separations have been reported *(85–89)*, including the purification of a therapeutically important synthetic peptide representing part of the sequence of luteinizing hormonereleasing hormone (LHRH) *(88)*.

5. SDC has also been adapted to modular solid-phase extraction (SPE) technology for development of a rapid, simple and cost-effective procedure for the efficient and parallel purification of multiple peptide mixtures *(91)*.

6. An interesting application of anion-exchange chromatography in sample displacement mode for protein purification has also been reported *(92)*.

3.6.3.2. Two-Step SDC of Peptides

1. SDC methodology has been developed further to carry out preparative separations on analytical equipment and columns (15 cm in length) for sample loads ≤200 mg *(93)*.

2. Following sample loading of 10- or 11-residue bradykinin antagonists in 100% aqueous solvent (e.g., 0.05% aqueous TFA) at a concentration of 7–10 mg/mL (sample loads varying from 67 mg to 200 mg) onto a small C_{18} column (150 × 4.6 mm ID, made up of 3 × 50-mm columns attached in series), isocratic elution with aqueous CH_3CN at two concentrations is applied.

3. The first (lower) CH_3CN concentration displaces hydrophilic impurities off the column; the second (higher) CH_3CN concentration displaces pure product from the column. Hydrophobic impurities remain trapped on the column.

4. This modified SDC approach promises to allow great flexibility in purifying peptides, at high yield of pure product (>99% purity), and encompassing a range of sample hydrophobicities and sample loads.

Acknowledgment

This work was supported by a National Institutes of Health grant to R.S. Hodges (R01GM61855).

4. Notes

1. Although highly pure TEA can be readily purchased, the authors' laboratory has regularly used lesser-grade reagent (following distillation over ninhydrin, where necessary) with no discernible problems.
2. Occasionally, it has been necessary to remove UV-absorbing contaminants from reagent-grade $NaClO_4$ by passing a stock solution of the salt through a preparative (e.g., 1 cm ID) RP-HPLC column (following filtration through a 0.22-μm filter).
3. The unusually high concentration of acetonitrile (40%) is required to eliminate nonideal behavior because of the considerable hydrophobic characteristics of the column packing. Other CEX packings are available that are hydrophilic and require considerably less acetonitrile or no acetonitrile in the mobile phase. In general, it is preferable to have each HPLC mode utilize just a single mechanism, i.e., ion-exchange mechanism only, in the absence of any nonspecific hydrophobic interactions which could complicate interpretation of results.
4. Fifty percent of all α-helices in proteins are amphipathic *(77)*.
5. In a similar manner to the effect of the hydrophobic environment characteristic of RP-HPLC in inducing helical structure in potential α-helical peptides, the presence of a high concentration (70% in this case) of organic modifier in HILIC/CEX will also induce such structure.

References

1. Dorsey, J. G., Foley, J. P., Cooper, W. T., Barford, R. A., and Barth, H. G. (1992) Liquid chromatography: theory and methodology. *Anal. Chem.* **64**, 353–389.
2. Mant, C. T., Zhou, N. E., and Hodges, R. S. (1992) Amino acids and peptides, in *Chromatography,* 5th ed. (Heftmann, E., ed.). Elsevier, Amsterdam, The Netherlands: pp. B75–B150.
3. Mant, C. T. and Hodges, R. S. (1996) Analysis of peptides by HPLC. *Methods Enzymol.* **271**, 3–50.
4. Mant. C. T., Kondejewski, L. H., Cachia, P. J., Monera, O. D., and Hodges, R. S. (1997) Analysis of synthetic peptides by high-performance liquid chromatography. *Methods Enzymol.* **289**, 426–469.
5. Mant, C. T. and Hodges, R. S. (eds.) (1991) *HPLC of Peptides and Proteins: Separation, Analysis and Conformation.* CRC, Boca Raton, FL.
6. Hearn, M. T. W. (ed.) (1991) *HPLC of Proteins, Peptides and Polynucleotides: Contemporary Topics and Application.* VCH, New York, NY.
7. Cunico, R. L., Gooding, K. M., and Wehr, T. (eds.) (1998) *Basic HPLC and CE of Biomolecules.* Bay Bioanalytical Laboratory, Richmond, CA.

8. Gooding, K. M. and Regnier, F. E. (eds.) (2002) *HPLC of Biological Macro-molecules,* 2nd ed. Marcel Dekker, New York, NY.
9. Hancock, W. S. (ed.) (1984) *Handbook of HPLC for the Separation of Amino Acids, Peptides and Proteins,* Vols. I and II. CRC, Boca Raton, FL.
10. Kovacs, J. M., Mant, C. T. and Hodges, R. S. (2006) Determination of intrinsic hydrophilicity/hydrophobicity of amino acid side-chains in peptides in the absence of nearest-neighbor or conformational effects. *Biopolymers (Peptide Science),* **84,** 283–297.
11. Zhou, N. E., Monera, O.D., Kay, C. M. and Hodges, R. S. (1994) α-Helical propensities of amino acids in the hydrophobic face of an amphipathic α-helix. *Protein Pep. Lett.* **1,** 114–119.
12. Zhou, N. E., Mant, C. T., and Hodges, R. S. (1990) Effect of preferred binding domains on peptide retention behavior in reversed-phase chromatography: amphipathic α-helices *Peptide Res.* **3,** 8–20.
13. Blondelle, S. E., Ostresh, J. M., Houghten, R. A., and Pérez-Payá, E. (1995) Induced conformational states of amphipathic peptides in aqueous/lipid environments. *Biophys. J.* **68,** 351–359.
14. Purcell, A. W., Aguilar, M. I., Wettenhall, R. E. W., and Hearn, M. T. W. (1995) Induction of amphipathic helical peptide structures in RP-HPLC. *Peptide Res.* **8,** 160–170.
15. Steer, D. L., Thompson, P. E., Blondelle, S. E., Houghten, R. A. and Aguilar, M. I. (1998) Comparison of the binding of α-helical and β-sheet peptides to a hydrophobic surface. *J. Peptide Res.* **51,** 401–412.
16. Wagschal, K., Tripet, B., Lavigne, P., Mant, C., and Hodges, R. S. (1999) The role of position a in determining the stability and oligomerization state of α-helical coiled coils: 20 amino acid stability coefficients in the hydrophobic core of proteins. *Protein Sci.* **8,** 2312–2329.
17. Tripet, B., Wagschal, K., Lavigne, P., Mant, C. T., and Hodges, R. S. (2000) Effects of side-chain characteristics on stability and oligomerization state of a *de novo*-designed model coiled-coil: 20 amino acid substitutions in position "d." *J. Mol. Biol.* **300,** 377–402.
18. Hodges, R. S., Zhou, N. E., Kay, C. M. and Semchuk, P. D. (1990) Synthetic model proteins: contribution of hydrophobic residues and disulfide bonds to protein stability. *Peptide Res.* **3,** 123–137.
19. Zhu, B.-Y., Mant, C. T., and Hodges, R. S. (1991) Hydrophilic-interaction chromatography of peptides on hydrophilic and strong-cation-exchange columns. *J. Chromatogr.* **548,** 13–24.
20. Zhu, B.-Y., Mant, C. T., and Hodges, R. S. (1992) Mixed-mode hydrophilic and ionic interaction chromatography rivals reversed-phase chromatography for the separation of peptides. *J. Chromatogr.* **594,** 75–86.
21. Lindner, H., Sarg, B., Meranes, C., and Helliger, W. (1996) Separation of acetylated core histones by hydrophilic-interaction liquid chromatography. *J. Chromatogr. A* **743,** 137–144.

22. Lindner, H., Sarg, B., Meraner, C., and Helliger, W. (1997) Application of hydrophilic-interaction liquid chromatography to the separation of phosphorylated H1 histones. *J. Chromatogr. A* **782**, 55–62.
23. Mant, C. T., Litowski, J. R., and Hodges, R. S. (1998) Hydrophilic interaction/cation-exchange chromatography for separation of amphipathic α-helical peptides. *J. Chromatogr. A* **816**, 65–78.
24. Mant, C. T., Kondejewski, L. H., and Hodges, R. S. (1998) Hydrophilic interaction/cation-exchange chromatography for separation of cyclic peptides. *J. Chromatogr. A* **816**, 79–88.
25. Lindner, H., Sarg, B., Hoertnagl, B., and Helliger, W. (1998) The microheterogeneity of the mammalian H1° histone. Evidence for an age-dependent deamidation. *J. Biol. Chem.* **273**, 13,324–13,330.
26. Lindner, H., Sarg, B., Grunicke, H., and Helliger, W. (1999) Age-dependent deamidation of H1° histones in chromatin of mammalian tissues. *J. Cancer Res. Clin. Oncol.* **125**, 182–186.
27. Litowski, J. R., Semchuk, P. D., Mant, C. T., and Hodges, R. S. (1999) Hydrophilic interaction/cation-exchange chromatography for the purification of synthetic peptides from closely related impurities: serine side-chain acetylated peptides. *J. Peptide Res.* **54**, 1–11.
28. Mant, C. T. and Hodges, R. S. (2000) Liquid chromatography, in *Encyclopedia of Separation Science* (Wilson, I. D., Adland, T. R., Poole, C. F. and Cook, M., eds.), Academic: New York, pp. 3615–3626.
29. Hartmann, E., Chen, Y., Mant, C. T., Jungbauer, A. and Hodges, R. S. (2003) Comparison of reversed-phase liquid chromatography and hydrophilic interaction/cation-exchange chromatography for the separation of amphipathic α-helical peptides with L- and D-amino acid substitutions in the hydrophilic face. *J. Chromatogr. A* **1009**, 61–71.
30. Hodges, R. S., Chen, Y., Kopecky, E., and Hodges, R. S. (2004) Monitoring the hydrophilicity/hydrophobicity of amino acid side-chains in the non-polar and polar faces of amphipathic α-helices by reversed-phase and hydrophilic interaction/cation-exchange chromatography. *J. Chromatogr. A* **1053**, 161–172.
31. Mant, C. T. and Hodges, R. S. (1985) A general method for the separation of cyanagen bromide digests of proteins by high-performance liquid chromatography: rabbit skeletal troponin I. *J. Chromatogr.* **326**, 349–356.
32. Mant, C. T. and Hodges, R. S. (1989) Optimization of peptide separations in high-performance liquid chromatography. *J. Liq. Chromatogr.* **12**, 139–172.
33. Opiteck, G. J., Jorgenson, J. W., and Anderegg, R. J. (1997) Two-dimensional SEC-RPLC coupled to mass spectrometry for the analysis of peptides. *Anal. Chem.* **69**, 2283–2291.
34. Liu, H., Lin, D., and Yates, J. R. (2002) Multidimensional separations for protein/peptide analysis in the post-genomic era. *Biotechniques* **32**, 898–911.

35. Mant, C. T., Chao, H., and Hodges, R. S. (1997) Effect of mobile phase on the oligomerization state of α-helical coiled-coil peptides during high-performance size-exclusion chromatography. *J. Chromatogr. A* **791**, 85–98.
36. Andrews, P. C. (1988) Ion-exchange HPLC for peptide purification. *Peptide Res.* **1**, 93–99.
37. Adachi, T., Takayanagi, H., and Sharpe, A. D. (1997) Ion-exchange high-performance liquid chromatographic separation of protein variants and isoforms on MCI GEL ProtEx stationary phases. *J. Chromatogr. A.* **763**, 57–63.
38. Wagner, K., Miliotis, T., Marko-Varga, G., Bischoff, R., and Unger, K.K. (2002) An automated online multidimensional HPLC system for protein and peptide mapping with integrated sample preparation. *Anal. Chem.* **74**, 809–820.
39. Kato, Y., Nakamura, K., Kitasuma, T., Tsuda, T., Hasegawa, M., and Sasaki, H. (2004) Effect of chromatographic conditions on resolution in high-performance ion-exchange chromatography on macroporous anion-exchange resin. *J. Chromatogr. A* **1031**, 101–105.
40. Andersen, T., Pepaj, M., Trones, R., Lundanes, E., and Greibrokk, T. (2004) Isoelectric point separation of proteins by capillary pH-gradient ion-exchange chromatography. *J. Chromatogr. A* **1025**, 217–226.
41. Mant, C. T. and Hodges, R. S. (1985) Separation of peptides by strong cation-exchange high-performance liquid chromatography. *J. Chromatogr.* **327**, 147–155.
42. Burke, T. W. L., Mant, C. T., Black, J. A. and Hodges, R. S. (1989) Strong cation-exchange high-performance liquid chromatography of peptides. Effect of non-specific hydrophobic interactions and linearization of peptide retention behaviour. *J. Chromatogr.* **476**, 377–389.
43. Link, A. J., Eng, J., Schieltz, D. M., et al. (1999) Direct analysis of protein complexes using mass spectrometry. *Nat. Biotech.* **17**, 676–682.
44. Peng, J., Elias, J. E., Thoreen, C. C., Licklider, L. J., and Gygi, S. P. (2003) Evaluation of multidimensional chromatography coupled with tandem mass spectrometry (LC/LC-MS/MS) for large-scale protein analysis: the yeast proteome. *J. Proteome Res.* **2**, 43–50.
45. Kang, X. and Frey, D. D. (2003) High-performance cation-exchange chromatofocusing of proteins. *J. Chromatogr. A* **991**, 117–128.
46. Mant, C. T. and Hodges, R. S. (1987) Monitoring free silanols on reversed-phase supports with peptide standards. *Chromatographia* **24**, 805–814.
47. Alpert, A. J. (1990) Hydrophilic-interaction chromatography for the separation of peptides, nucleic acids and other polar compounds. *J. Chromatogr.* **499**, 177–196.
48. Thorsteinsdóttir, M., Beijersten, I., and Westerlund, D. (1995) Capillary electroseparations of enkephalin-related peptides and protein kinase A peptide substrates. *Electrophoresis* **16**, 564–573.
49. Kašička, V. (1999) Capillary electrophoresis of peptides. *Electrophoresis* **20**, 3084–3105.

50. Kašička, V. (2001) Recent advances in capillary electrophoresis of peptides. *Electrophoresis* **22**, 4139–4162.
51. Hu, S. and Dovichi, N. J. (2002) Capillary electrophoresis for the analysis of biopolymers. *Anal. Chem.* **74**, 2833–2850.
52. Popa, T. V., Mant, C. T., and Hodges, R. S. (2003). Capillary electrophoresis of synthetic peptide standards varying in charge and hydrophobicity. *Electrophoresis* **24**, 4197–4208.
53. Popa, T. V., Mant, C. T., and Hodges, R. S. (2004) Capillary electrophoresis of amphipathic α-helical peptide diastereomers, *Electrophoresis* **25**, 94–107.
54. Popa, T. V., Mant, C. T., and Hodges, R. S. (2004). Capillary electrophoresis of cationic random coil peptide standards: effect of anionic ion-pairing reagents and comparison with reversed-phase chromatography. *Electrophoresis* **25,** 1219–1229.
55. Popa, T. V., Mant, C. T., Chen, Y., and Hodges, R. S. (2004) Capillary zone electrophoresis (CZE) of α-helical diastereomeric peptide pairs using anionic ion-pairing reagents. *J. Chromatogr. A* **1043**, 113–122.
56. Popa, T. V., Mant, C. T., and Hodges, R. S. (2006) Ion-interaction-capillary zone electrophoresis of cationic proteomic peptide standards. *J. Chromatogr. A* **1111**, 192–199.
57. Kornfelt, T., Vinther, A., Okafo, G. N., and Camilleri, P. (1996) Improved peptide mapping using phytic acid as ion-pairing buffer additive in capillary electrophoresis. *J. Chromatogr. A* **726**, 223–228.
58. Issaq, H. J., Conrads, T. P., Janini, G. M., and Veenstra, T. D. (2002) Methods for fractionation, separation and profiling of proteins and peptides. *Electrophoresis* **23**, 3048–3061.
59. Simó, C. and Cifuentes, A. (2003) Capillary electrophoresis-mass spectrometry of peptides from enzymatic protein hydrolysis: simulation and optimization. *Electrophoresis* **24**, 834–842.
60. Moore, A. W. and Jorgenson, J. W. (1995) Rapid comprehensive two-dimensional separations of peptides via RPLC-optically gated capillary zone electrophoresis. *Anal. Chem.* **67**, 3448–3455.
61. Moore, A. W. and Jorgenson, J. W. (1995) Comprehensive three-dimensional separation of peptides using size-exclusion chromatography/reversed-phase liquid chromatography/optically gated capillary zone electrophoresis. *Anal. Chem.* **67**, 3456–3463.
62. Lewis, K. C., Opiteck, G. J., Jorgenson, J. W., and Sheeley, D. M. (1997) Comprehensive on-line RPLC-CZE-MS of peptides. *J. Am. Soc. Mass Spectrom.* **8,** 495–500.
63. Isaaq, H. J., Chan, K. C., Janini, G. M., and Muschik, G. M. (1999) A simple two-dimensional high performance liquid chromatography/high performance capillary electrophoresis set-up for the separation of complex mixtures. *Electrophoresis* **20**, 1533–1537.
64. Isaaq, H. J., Chan, K. C., Cheng, S. L., and Qingbo, L. (2001) Multidimensional high performance liquid chromatography-capillary electrophoresis separation of a protein digest: an update. *Electrophoresis* **22,** 1133–1135.

65. Kirkland, J. J., Henderson, J. W., De Stefano, J. J., van Straten, M. A., and Claessens, H. A. (1997) Stability of silica-based, endcapped columns with pH 7 and pH 11 mobile phases for reversed-phase high-performance liquid chromatography. *J. Chromatogr. A* **762,** 97–112.

66. Kirkland, J. J., van Straten, M. A., and Claessens, H. A. (1998) Reversed-phase high-performance liquid chromatography of basic compounds at pH 11 with silica-based packings. *J. Chromatogr. A.* **797,** 111–120.

67. Kirkland, J. J., Glajch, J. L., and Farlee, R. D. (1989) Synthesis and characterization of highly stable bonded phases for high-performance liquid chromatography column packings. *Anal. Chem.* **61,** 2–11.

68. Boyes, B. E. and Walker, D. G. (1995) Selectivity optimization of reversed-phase high-performance liquid chromatographic peptide and protein separations by varying bonded-phase functionality. *J. Chromatogr. A.* **691,** 337–347.

69. Mant, C. T., Parker, J. M. R., and Hodges, R. S. (1987) Size-exclusion HPLC of peptides: requirement for peptide standards to monitor non-ideal behavior. *J. Chromatogr.* **397,** 99–112.

70. Tripet, B., Howards, M. W., Jobling, M., Holmes, R. K., Holmes, K. V., and Hodges, R. S. (2004) Structural characterization of the SARS-coronavirus spike S fusion protein core. *J. Biol. Chem.* **279,** 20,836–20,849.

71. Chen, Y., Mehok, A. R., Mant, C. T., and Hodges, R. S. (2004) Optimum concentration of trifluoroacetic acid (TFA) for reversed-phase chromatography of peptides revisited. *J. Chromatogr. A* **1043,** 9–18.

72. Shibue, M., Mant, C. T., and Hodges, R. S. (2005) The perchlorate anion is more effective than trifluoroacetate anion as a ion-pairing reagent for reversed-phase chromatography of peptides. *J. Chromatogr. A* **1080,** 49–57.

73. Shibue, M., Mant, C. T., and Hodges, R. S. (2005) Effect of anionic ion-pairing reagent concentration (1 mM–60 mM) on reversed-phase chromatography elution behavior of peptides. *J. Chromatogr. A* **1080,** 58–67.

74. Antia, F. D. and Horváth, C. (1988) High-performance liquid chromatography at elevated temperatures: examination of conditions for the rapid separation of large molecules. *J. Chromatogr.* **435,** 1–15.

75. Li, J. W. and Carr, P. W. (1997) Evaluation of temperature effects on selectivity in RPLC separations using polybutadiene-coated zirconia. *Anal. Chem.* **69,** 2202–2206.

76. Chen, Y., Mant, C. T., and Hodges, R. S. (2003) Temperature selectivity effects in reversed-phase liquid chromatography due to conformation differences between helical and non-helical peptides. *J. Chromatogr. A* **1010,** 45–61.

77. Cornette, J. L., Cease, K. B., Margalit, H., Spouge, J. L., Berzofsky, J. A., and DeLis, C. D. (1987) Hydrophobicity scales and computational techniques for detecting amphipathic structures in proteins. *J. Mol. Biol.* **195,** 659–685.

78. Mant, C. T., Chen, Y., and Hodges, R. S. (2003) Temperature profiling of polypeptides in reversed-phase liquid chromatography: I. Monitoring of dimerization and unfolding of amphipathic α-helical peptides. *J. Chromatogr. A* **1009,** 29–43.

79. Rothemund, S., Beyerman, M., Krause, E., et al. (1995) Structure effects of double D-amino acid replacements: a NMR and CD study using amphipathic model helices. *Biochemistry* **34,** 12,954–12,962.

80. Chen, Y., Mant, C. T., and Hodges, R. S. (2002) Determination of stereochemistry stability coefficients of amino acid side-chains in an amphipathic α-helix. *J. Peptide Res.* **59,** 18–33.

81. Zhang, L., Benz, R., and Hancock, R. E. W. (1998) Influence of proline residues on the antibacterial and synergistic activities of α-helical peptides. *Biochemistry* **38,** 8102–8111.

82. Zhang, L., Falla, T., Wu, M., et al. (1999) Determinants of recombinant production of antimicrobial cationic peptides and creation of peptide variants in bacteria. *Biochem. Biophys. Res. Commun.* **247,** 674–680.

83. Chen, Y., Vasil, A. I., Rehaume, L., et al. (2006) Comparison of biophysical and biological properties of α-helical enantiomeric antimicrobial peptides. *Chem. Biol. Drug Design* **67,** 162–173.

84. Mills, J. B., Mant, C. T., and Hodges, R. S. (2006) One-step purification of recombinant proteins from a whole cell extract by reversed-phase high-performance liquid chromatography. *J. Chromatogr. A* **1133,** 248–253.

85. Hodges, R. S., Burke, T. W. L., and Mant, C. T. (1988) Preparative purification of peptides by reversed-phase chromatography: Sample displacement mode versus gradient elution mode. *J. Chromatogr.* **444,** 349–362.

86. Burke, T. W. L., Mant, C. T., and Hodges, R. S. (1988) A novel approach to reversed-phase preparative high-performance liquid chromatography of peptides. *J. Liq. Chromatogr.* **11,** 1229–1247.

87. Hodges, R. S., Burke, T. W. L., and Mant, C. T. (1991) Multi-column reversed-phase sample displacement chromatography of peptides. *J. Chromatogr.* **548,** 267–280.

88. Mant, C. T., Burke, T. W. L., Mendonca, A. J., and Hodges, R. S. (1992) Preparative reversed-phase sample displacement chromatography of peptides. *Proceedings of the 9th International Symposium on Preparative and Industrial Chromatography*, pp. 274–279.

89. Hodges, R. S., Burke, T. W. L., Mendonca, A. J., and Mant, C. T. (1993) Preparative reversed-phase sample displacement chromatography of peptides, in *Chromatography in Biotechnology* (Horváth, C. and Ettre, L. S., eds.), American Chemical Soc. Symposium Series 529, pp. 59–76.

90. Horváth, Cs., Frenz, J. H., and El Rossi, Z. (1983) Operating parameters in high-performance displacement chromatography. *J. Chromatogr.* **255,** 273–293.

91. Husband, D. L., Mant, C. T., and Hodges, R. S. (2000) Development of simultaneous purification methodology for multiple synthetic peptides by reversed-phase sample displacement chromatography. *J. Chromatogr. A* **893,** 81–94.

92. Veeraragavan, K., Bernier, A., and Braendli, E. (1991) Sample displacement mode chromatography: purification of proteins by use of a high-performance anion-exchange column. *J. Chromatogr.* **541**, 207–220.

93. Mehok, A. R., Mant, C. T., Gera, L., Stewart, J., and Hodges, R. S. (2002) Preparative reversed-phase chromatography of peptides: isocratic two-step elution system for high loads on analytical columns. *J. Chromatogr. A* **972,** 87–99.

2

Identification of Proteins Based on MS/MS Spectra and Location of Posttranslational Modifications

Kathryn L. Stone, Myron Crawford, Walter McMurray, Nancy Williams, and Kenneth R. Williams

Summary

Comparative protein profiling is a key approach to understanding the human and other proteomes. Systems-level profiling technologies, such as differential fluorescence two-dimensional gel electrophoresis (DIGE), often require the identification of the proteins that are contained within 50 or more spots per gel. A major focus of this chapter therefore is devoted to a general approach for high throughput protein identification that is based on liquid chromatography (LC)/tandem mass spectrometry (MS/MS) analysis of tryptic digests of individual proteins or mixtures of only a few proteins (i.e., as are usually obtained from individual DIGE spots), and that is also applicable to the analysis of complex protein extracts. Additionally, multiple techniques will be described for identifying sites of protein posttranslational modification, with emphasis on phosphorylation and Arg methylation.

Key Words: MS/MS; phosphorylation; methylation.

1. Introduction

The ultimate goal of proteomics is to comprehensively identify all proteins within a given tissue or cell type, their associated biological activities, posttranslational modifications, and protein–protein interactions, and to determine how the proteome is altered in response to a variable (i.e., treatment, disease). Mass spectrometry (MS)-based identification of proteins is typically carried out by first subjecting the sample to trypsin digestion followed by either high mass accuracy matrix-assisted laser desorption/ionization (MALDI) MS fingerprinting *(1)* or by tandem mass spectrometry (MS/MS) analysis *(2)* of individual peptides within the digest. Advantages to the MALDI-MS peptide

From: *Methods in Molecular Biology, vol. 386: Peptide Characterization and Application Protocols*
Edited by: G. Fields © Humana Press Inc., Totowa, NJ

mass database searching approach include its very low attomol/fmol sensitivity, short instrument run time, the relatively inexpensive instrumentation, and the comparative ease with which the data and the resulting database searches can be interpreted and understood. A major disadvantage of this approach, however, is that it is limited to highly purified samples that preferably contain only a single major protein (i.e., as might be expected for many spots derived from two-dimensional [2D] gel electrophoresis of partially purified cell fractions). While it may be possible to use peptide mass database searching to identify two to three proteins in a given sample, this approach cannot be used on the very complex mixtures that result from tryptic digestion of cell extracts. Generally, we find that the probability of identifying a protein by peptide mass database searching decreases as the number of peptide ions detected by MALDI-MS significantly exceeds the 50–100 peptides that might be obtained from one to two average-sized proteins. In MS/MS-based protein identification, which is typically performed on an electrospray/quadrupole time-of-flight (Q-TOF) or MALDI/tandem time-of-flight (TOF/TOF) instrument, the MS/MS spectrum obtained from *each* (approx 10–25 residue) peptide often contains sufficient information to enable its parent protein to be identified by database searching of the *uninterpreted* MS/MS spectrum. Hence, extremely complex protein digests are amenable to MS/MS based identification using any one of several different search algorithms *(3)*. One of the better programs, Mascot (http://www.matrix-science.com) is accessible to anyone over the Internet; thus, focus will be given to using this program for MS/MS identification of proteins. Often, if no match is found, and the quality of the MS/MS spectra is sufficiently good, it is possible to call a *de novo* sequence *(4,5)*. Another advantage of MS/MS-based protein identification is that the success rate with this approach is higher than with peptide mass database searching. In our experience from analyzing >3200 "unknown" proteins submitted by hundreds of investigators, we have observed a 43% success rate at identifying one or more proteins/sample from peptide mass database searching as compared to a 77% success rate with MS/MS-based protein identification.

Reversible protein phosphorylation is probably the most important mechanism used for intracellular signal transduction *(6)* and is involved in regulating cell-cycle progression, differentiation, transformation, development, peptide hormone response, and adaptation *(7–9)*. Because as many as one-third of mammalian proteins may be phoshorylated *(9)*, this modification is among the most important and widespread. Based on our experience, the majority of modifications occur at very low stoichiometry such that only a small fraction of the protein substrate is phosphorylated at any given peptide site. For this reason, it is often advantageous to tryptically digest an in vivo ^{32}P-labeled protein. The radiolabel tag serves as a tracer for locating the phosphopeptide. When the stoichiometry of phosphorylation is above approximately 10% at any given site, it is often possible to use mass spectrometric approaches to bring about an identification without the need for in vivo or in vitro [^{32}P]-labeling. One approach that works well under these conditions is comparative MALDI-MS fingerprinting of the trypsin-digested sample before and after alkaline phosphatase treatment. After identifying the mass of a peptide ion containing a site of phosphorylation (based on the mass shift after alkaline phosphatase treatment), a second aliquot of the sample is subjected to MS/MS analysis (or MS/MS can be done directly on MALDI-TOF/TOF instrument). There are many other approaches for locating sites of phosphorylation including β-elimination/biotin affinity tag modification of the phosphorylated site *(10–12)*. Given the low stoichiometry of many phosphorylation sites in proteins, procedures such as immobilized metal affinity chromatography (IMAC) *(13)* and affinity chromatography (Qiagen Phosphoprotein Purification Kit) that can enrich for phosphopeptides or phosphoproteins provide an attractive option prior to the use of MS or MS/MS-based approaches for structural analysis of sites of phosphorylation. Typically, we employ multiple, complementary approaches for locating sites of phosphorylation that often begin with in vivo or in vitro ^{32}P-labeling and trypsin digestion followed by preparative capillary high-performance liquid chromatography (HPLC). Identifications are then brought about by MALDI-MS mapping of isolated ^{32}P-labeled peptides (or of nonfractionated digests) before and after treatment with alkaline phosphatase, MS/MS analysis, and Edman chemical and radioisotope-based sequencing.

One of the ultimate goals of proteomics is to identify and quantify *all* protein *and* protein posttranslational changes between a control vs an experimental whole-cell protein extract. It is, however, extremely difficult to account for *all* tryptic peptide ions expected and/or observed following MALDI-MS or liquid chromatography (LC)/MS/MS analysis of a digest of even a single, average-sized, 75-kD protein. Incomplete cleavages, chymotryptic-like cleavages, in vivo and in vitro (e.g., occurring *during* gel electrophoresis) protein modifications, trypsin autolysis products, and keratin and other contaminating proteins all serve to frustrate a "complete" analysis of an isolated protein and a complete accounting of all observed ions. An alternative strategy that may be used in this case is comparative analysis of a native protein vs its counterpart that has been expressed in an organism such as *Escherichia coli* where proteins are subject to relatively few posttranslational modifications. With this approach, tryptic digests of the two proteins are compared to each other using MALDI-MS peptide mapping, LC/MS, microbore (preparative) 1D or 2D HPLC, or other similar approaches. With this approach peaks that are found only in the digest of the native or control protein usually represent sites of posttranslational modification which are then subjected to MS/MS and other structural analyses. When this comparative approach is feasible, we believe it provides one of the most facile means to quickly identify peptides that are posttranslationally modified. To increase the probability of finding "all" modifications in the native protein, this experiment can be repeated using one or more different proteases such as chymotrypsin or Staphylococcus protease. In the case of complex cell extracts, multidimensional protein identification (MudPIT) *(14)* provides one of the best approaches to identify all proteins and their posttranslational modifications that are *differentially* expressed in a control vs experimental sample. Although quantitative evaluation would require the use of isotopic labeling, qualitative comparisons can be based on LC/MS differences in total ion current.

2. Materials

2.1. LC/MS/MS

1. Q-TOF mass spectrometer (such as a Waters Q-Tof or Applied Biosystems (ABI) QSTAR instrument) equipped with a capillary HPLC system (such as a Waters capLC).
2. Buffer A: 98% water, 2% acetonitrile, 0.1% acetic acid.
3. Buffer B: 20% water, 80% acetonitrile, 0.1% acetic acid.
4. Waters Atlantis™ dC18, 3 µm particle size, 100 Å pore size, 100 µ × 150 mm NanoEase™ column.

2.2. Alkaline Phosphatase Digestion

1. 1X buffer: 100 m*M* NaCl, 50 m*M* Tris-HCl (pH 7.9), 10 m*M* MgCl$_2$, 1 m*M* dithiothreitol (DTT).
2. Alkaline phosphatase (New England Biolabs), 10 U/µL in 50 m*M* KCl, 10 m*M* Tris-HCl (pH 8.2), 1 m*M* MgCl$_2$, 0.1 m*M* ZnCl$_2$, 50% glycerol.
3. C18 ZipTip® (Millipore Corporation), 10-µL pipet tip with a 0.6-µL resin bed.
4. ZipTip wash buffer: 50% acetonitrile, 0.1% trifluoroacetic acid (TFA).
5. ZipTip equilibration buffer: 0.1% TFA.
6. ZipTip elution buffer: 50% acetonitrile, 0.1% aqueous formic acid.
7. Matrix solution: 5 mg/mL α-cyano-4-hydroxycinnamic acid matrix in 0.05%s aqueous TFA, 50% acetonitrile.

2.3. Comparative Peptide Mapping

1. Microbore HPLC system.
2. Vydac C18, 5 µm particle size, 300 Å pore size, 1 × 250 mm reversed-phase (RP) column.

3. Buffer A: 0.05% TFA, 100% water.
4. Buffer B: 80% acetonitrile, 20% water, 0.045% TFA.
5. Mass spectrometer: a MALDI or electrospray ionization (ESI) mass spectrometer can be used.

2.4. Radiosequencing

1. Edman degradation chemical sequencer (such as an Applied Biosystems [AB] Procise 494 cLC sequencer).
2. Sequelon™ membrane (Applied Biosystems).

3. Methods

The procedure for generating MS/MS data for protein identification and location of posttranslational modifications requires that the peptides of interest first be ionized and transferred in the gas phase into a mass spectrometer. Typically, either a MALDI or ESI source is used to generate and introduce the peptide ions into a MALDI-TOF/TOF, ion trap, or Q-TOF type mass spectrometer. Tryptic peptides (which often range from 10 to 25 residues in length) tend to ionize and fragment well using electrospray, forming primarily doubly and triply charged ions. Hence, trypsin is usually the enzyme of choice for MS/MS-based protein identification (*see* **Note 1**). In an ESI-Q-TOF instrument (such as that used in this study), the ions enter into the first quadrupole, where individual m/z ions are selected, based on criteria set up in the method, for transfer to the collision cell for collision-induced dissociation (CID). During CID, peptides undergo collisions with a pressurized neutral gas such as Ar or He. These collisions cause the peptide to break along the peptide backbone producing product/sequence ions. These ions are then mass measured by the TOF analyzer in a Q-TOF instrument or by an ion trap in the latter type instrument and their relative intensities are measured at the detector. The sequence ions observed are called a, b, c and x, y, z ions and are produced by cleavage at C-C, C-N, or N-C bonds, respectively. Generally, each doubly charged ion cleaves to produce two singly charged fragments. If the charge is retained on the N-terminal peptide fragment, the resulting ions are called a, b, or c ions. If the charge is retained on the C-terminal fragment, these are called x, y, or z ions (**Fig. 1**). Ions are further labeled with a subscript to indicate the number of amino acids in the fragment (*4,5*). The most commonly observed ions in an ESI-Q-TOF instrument are the b and y ions.

3.1. MS/MS Analysis

The procedure described below is designed for analyzing protein digests using LC/MS/MS on a Waters/Micromass Q-Tof equipped with a Waters

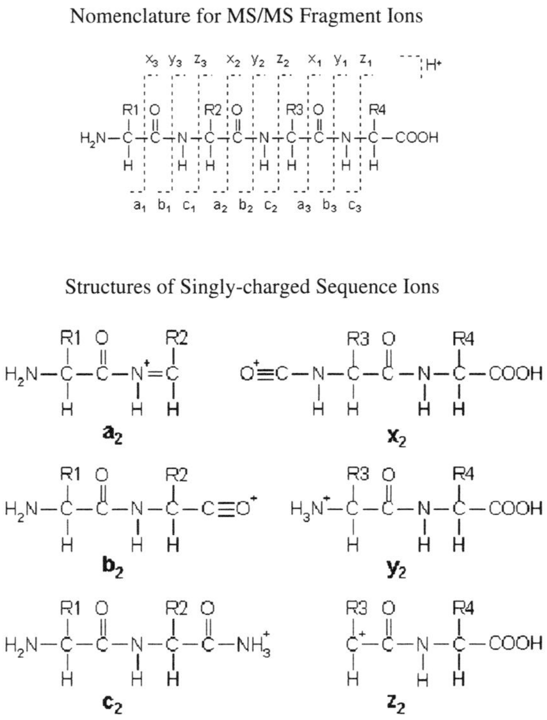

Nomenclature for MS/MS Fragment Ions

Structures of Singly-charged Sequence Ions

Fig. 1. Tandem mass spectrometry peptide fragment ions. Ions a, b, and c have the charge retained on the N-terminal fragment. Ions x, y, and z have the charge on the C-terminal fragment. From Mascot (http://www.matrix-science.com).

capLC. The Mascot Distiller and Mascot search program described can analyze MS/MS data from any vendor's mass spectrometer.

1. Five microliters of the sample dissolved in buffer A is directly injected onto a 75 or $100\,\mu \times 150\,mm$ Atlantis column running at 300 or 500 nL/min, respectively (*see* **Notes 2** and **3**).
2. Gradient (*see* **Note 4**):

Time	%B
0–23 min	5–37
23–55 min	37–75
55–65 min	75–95

3. Data-dependent acquisition is performed so that the mass spectrometer switches automatically from MS to MS/MS modes when the total ion current increases above the 1.5 counts/second threshold set point.
4. To obtain good fragmentation, a collision energy ramp is set for the different peptide mass ranges and charge states (**Table 1**) (*see* **Note 5**). Preference is given to doubly and triply charged species for fragmentation over singly charged ions which generally do not fragment as well as the higher charged ions.
5. All MS/MS are searched using the Mascot algorithm *(15)* for uninterpreted MS/MS spectra after using the Mascot Distiller program to generate Mascot compatible files. The Mascot Distiller program combines sequential MS/MS scans from profile data that have the same precursor ion. Charge states of +2 and +3 are preferentially located with a signal-to-noise ratio of 1.2 or greater and a peak list is generated for database searching.
6. Using the Mascot database search algorithm, a protein is considered to be identified when Mascot lists it as a significant match and two or more peptides are matched to the same protein.
7. The database searched is the NCBInr, which is chosen over genome-specific databases because a match to a protein from the correct species in the NCBInr

Table 1
Collision Energy in Volts used to Obtain
MS/MS Spectra for Peptides with the
Indicated Charge State and m/z

Collision energy ramp		
Charge state	m/z	Collision energy used
1	300	22
	566	29
	955	55
	1800	80
2	400	26
	653	28
	740	30
	820	32
	1200	55
	1800	65
3	435	30
	547	35
	605	38
	1800	50

database provides an added degree of certainty that the identified protein is indeed present in the sample. Additionally, when the genome for the organism being studied has not yet been completed and the sequence for the isolated peptide is not yet in *any* database, a significant match may still be found in the NCBInr database if the isolated peptide shares sufficient sequence identity to the corresponding peptide in a homologous protein found in another species (*see* **Note 6**).

8. The score determined by Mascot for an individual MS/MS peptide match is based on the probability (P) that the observed match between the experimental MS/MS spectra and the ions predicted from a database peptide sequence is a random event *(16)*. The match with the *lowest* probability is thus the best match. Whether this match is significant depends upon the size of the database. Because current databases contain very large numbers of entries, a significant match has a very low Mascot probability that would need to be expressed in scientific notation. Because this is inconvenient, the reported Mascot score is $-10Log_{10}(P)$, where P is the probability. This means that the best peptide match is the one with the *highest* score. If we use a significance threshold of <0.05, which is a widely used threshold, this means that the probability of the observed match resulting from chance is less than 1 in 20. With this threshold a significant peptide match typically has a Mascot score of 70 or higher. The Mascot *protein* score is derived from the individual peptide scores. The best indicator that the highest scoring protein is present in the sample is the presence of multiple, significant peptide matches to the same database entry. Typical parameters used for searching are partial Met oxidation and acrylamide modified Cys (*see* **Note 7**), a peptide tolerance of ±0.4 Da, MS/MS fragment tolerance of ±0.2 Da, peptide charges of $+2$ or $+3$, and a significance threshold of <0.05.

3.2. De Novo MS/MS Interpretation

If no protein is identified by an automated database search, then it may be possible to call either a sequence tag *(17)*, which is a partial sequence, or the entire sequence of the peptide from the MS/MS spectra. **Figure 2** contains the MS/MS fragmentation pattern of a tryptic peptide from a bovine serum albumin (BSA) digest. This spectrum is used as an example in the explanation below on interpreting MS/MS spectra.

1. Determine the precursor mass (722.34), the charge state [$+2$ (doubly charged)], and then the mass of the intact peptide (*see* **Note 8**).
2. Locate an area of the spectra that appears to be "clean" but contains fragment ions (*see* **Note 9**).
3. Pick the most abundant ion, and subtract the neighboring most abundant ion. **Table 2** contains the masses of the different amino acids in a peptide. Gly (57 Da) is the smallest amino acid, while Trp is the largest (186 Da). Hence, the fragment ion differences will fall between these masses providing that no posttranslational modifications are present.

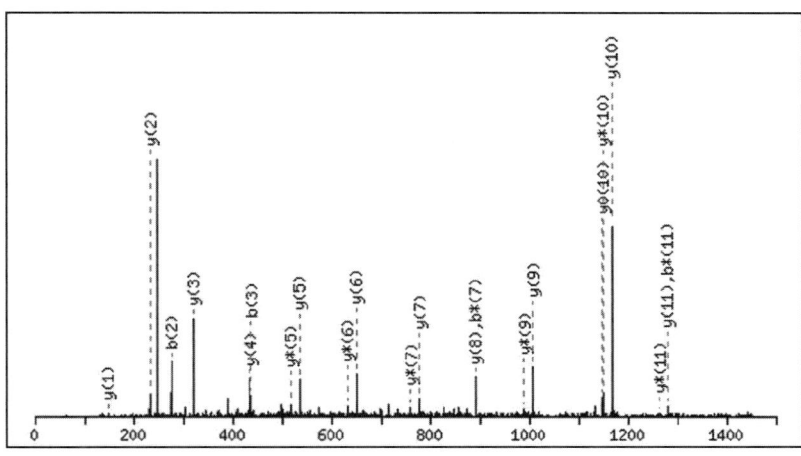

Peptide from BSA tryptic digest

Average Mass = 1443.5547, Monoisotopic Mass = 1442.6346
Residues: 1–12
N-Terminus = H, C-Terminus = OH

a	136.08	249.16	409.19	524.22	638.26	766.32	881.35	982.39	1095.48	1182.51	1269.54	-
b	164.07	**277.16**	**437.19**	552.21	666.26	794.31	909.34	1010.39	1123.47	1210.51	1297.54	-
c"	181.10	294.18	454.21	569.24	683.28	811.34	926.37	1027.42	1140.50	1227.53	1314.56	-
i	136.08	86.10	133.04	88.04	87.06	101.07	88.04	74.06	86.10	60.04	60.04	101.11
	1	2	3	4	5	6	7	8	9	10	11	12
	Tyr	Ile	CAM	Asp	Asn	Gln	Asp	Thr	Ile	Ser	Ser	Lys
	12	11	10	9	8	7	6	5	4	3	2	1
x	-	1306.56	1193.47	1033.44	918.42	804.37	676.32	561.29	460.24	347.16	260.12	173.09
y"	-	**1280.58**	**1167.50**	**1007.46**	**892.44**	**778.39**	**650.34**	**535.31**	**434.26**	**321.18**	**234.15**	**147.11**
z	-	1263.55	1150.47	990.44	875.41	761.37	633.31	518.28	417.23	304.15	217.12	130.09

Fig. 2. Tandem mass spectrometry (MS/MS) spectra obtained on a tryptic peptide from bovine serum albumin (BSA) using a Waters Q-Tof API. 250 fmols of a BSA tryptic digest was subjected to liquid chromatography/MS/MS analysis as described in Methods. The spectra were acquired automatically using data-dependent MS/MS acquisition.

4. If the mass difference between the two selected ions is not equal to the mass of an amino acid, then continue to the next closest ion until one is located whose mass difference from the most abundant ion is equal to the mass of an amino acid (*see* **Note 10**).

5. Next, try to extend the sequence by picking the ion next to the second ion and subtracting their masses until another fragment ion is located.

6. Continue this process until the C-terminus of the peptide is identified. The C-terminal amino acid will have a mass that is equal to the sum of one of the amino acid masses listed in **Table 1** plus the C-terminal group, which adds 19 Da

Table 2
Listing of the Amino Acids and Their Corresponding Monoisotopic Masses Within a Peptide and Their Immonium Ions

| Amino acid | 1-letter code | Mass values for amino acids in peptides | |
		Monoisotopic mass, integer value	Immonium ion
Glycine	G	57 Da	30
Alanine	A	71 Da	44.1
Serine	S	87 Da	60
Proline	P	97 Da	70.1
Threonine	T	101 Da	74.1
Cysteine	C	103 Da	76
Isoleucine	I	113 Da	86.1
Leucine	L	113 Da	86.1
Asparagine	N	114 Da	87.1
Aspartic acid	D	115 Da	88
Glutamine	Q	128 Da	101.1
Lysine	K	128 Da	101.1
Glutamic acid	E	129 Da	102.1
Methionine	M	131 Da	104.1
Histidine	H	137 Da	110.1
Phenylalanine	F	147 Da	120.1
Arginine	R	156 Da	129.1
Tyrosine	Y	163 Da	136.1
Tryptophan	W	186 Da	159.1

($H_2O + H^+$) to the amino acid mass. Hence, if the peptide ends in Lys, the C-terminal mass would be 147, and if in Arg, 175. In the y ion series, the C-terminal amino acid is called y_1.

7. Once several amino acids have been called from the y ions, the sequence can be confirmed by identifying the corresponding b ions. The b ion series will be seen primarily in the low mass region and corresponds to the amino terminus of the peptide. In a Q-TOF instrument, the b ions generally are not observable above the precursor mass. As with the y ions, the mass of the C-terminal amino acid in the b ion series will be equal to an amino acid residue mass listed in **Table 2** plus 19 Da. The N-terminal amino acid fragment in this series is called b_1.

8. The b_2 ion typically loses CO or -28 Da to form an a_2 ion. Hence, a strong b_2/a_2 pair is often observed and can be used to begin calling the b ion series.

9. **Table 2** also contains immonium ions for each amino acid. Immonium ions are internal fragments of the amino acid that can be used to confirm the presence, but not the location, of that amino acid in the sequence.

10. In most MS/MS analyses it is not possible to differentiate between the isomeric (Ile/Leu) and isobaric (Phe/oxidized Met and Gln/Lys) amino acids. However, if a tryptic digestion was performed, and a mass difference of 128 Da is observed within the peptide sequence, this is most likely a Gln.

11. Other helpful aids are as follows:

Thr/Ser: can show a loss of OH (-17 Da) or H_2O (-18 Da).
Met: can oxidize to Met sulfoxide (Met $+ 16$ Da $= 147$ Da).
Trp: several combinations of amino acids also add up to the 186 mass of Trp.
Gln: an amino terminal Gln can cyclize to pyroglutamic acid (Gln -17 Da).
Acetylation: amino terminal acetylation adds 42 Da to the mass of the peptide.

12. If the sequence has been called correctly, all "pieces" or fragment ions will fit. If an error has been made, then often the mass difference remaining at the N- or C-terminus will be inconsistent with the peptide mass.

13. If only a partial sequence has been called (i.e., a sequence tag), then this sequence will have to be database searched in both the forward and reverse directions. This is necessary because unless the C-terminus of the peptide has been identified, it is impossible to know if the fragment ion series called is a b or y ion series.

14. Programs that are commonly used for searching manually called peptide sequences are BLAST (http://www.ncbi.nlm.nih.gov/BLAST) and MS-PATTERN (http://prospector.ucsf.edu/ucsfhtml4.0/mspattern.htm).

3.3. Location of Sites of Phosphorylation

Whenever feasible, our general approach to identifying sites of protein phosphorylation begins with an in gel trypsin digest of the Coomassie Blue stained, [^{32}P]-labeled phosphorylated protein. When >5000 cpm are present in the gel band, we have succeeded in identifying the site(s) of phosphorylation in approximately 75% of the samples analyzed by the following approach.

1. An in gel tryptic digest of the ^{32}P-labeled protein is subjected to preparative, microbore RP-HPLC with the entire gradient being collected with peak detection into 1.5-mL capless Eppendorf tubes.

2. *All* of the resulting Eppendorf tubes are Cerenkov-counted to quickly locate those fractions which contain phosphorylated peptides (*see* **Note 11**).

3. MALDI-MS analysis is carried out on approx 3% of each of the radioactive HPLC fractions containing a tryptic phosphopeptide (*see* **Note 12**).

4. Up to 50% of each radioactive fraction that contains >1500 Cerenkov cpm is coupled to a Sequelon solid support per the manufacturer's directions and is then subjected to radioactive Edman sequencing (*see* **Note 13**).

5. The ABI 494 cLC protein/peptide sequencer is modified with a fraction collector attached through the ATZ port. No flask cycles or HPLC gradients are used. A modified cartridge Begin cycle is used that has an extended 100% TFA wash to remove any unbound peptide that remains on the Sequelon support. The usual cartridge chemistry cycle is modified to extract each anilinothiazolinone (ATZ) amino acid after cleavage in 100% TFA for transfer to the fraction collector instead of to the reaction flask. An argon flush through the cartridge to the fraction collector, followed by a rinse of the line to the fraction collector, and another argon flush ensure collection of any residual [^{32}P] ATZ cpm from the initial transfer. Collected fractions are dried, scintillation fluid is added, and each cycle is subjected to scintillation counting. This data identifies the cycle(s) at which the cpm are eluted and thereby which residue number(s) in the peptide are phosphorylated.

6. Normal Edman or MS/MS sequencing can be obtained on the other half of the HPLC fraction to identify the [^{32}P] peptide. Because the identity of the phosphorylated protein is usually known, the identity of the phosphorylated peptide(s) can often be inferred by comparing its mass, which is determined in **step 3** above by MALDI-MS, to the masses of the predicted tryptic peptides. In these instances the radioactive, solid-phase Edman sequencing as well as normal Edman or MS/MS sequencing provide confirmatory data (*see* **Notes 14** and **15**).

3.4. Alkaline Phosphatase Digestion/MALDI-MS Approach to Locating Sites of Protein Phosphorylation

Alkaline phosphatase digestion of an aliquot of a tryptic digest prior to comparative MALDI-MS analysis of the treated and nontreated tryptic digests can often be used to locate site(s) of phosphorylation. This approach is most suitable for highly purified protein samples.

1. An aliquot of an in gel tryptic digest carried out as described in Stone and Williams *(18)* is brought to dryness in a Speedvac and then re-dissolved in 5 μL of 1X alkaline phosphatase digestion buffer.

2. One microliter or 10 U of alkaline phosphatase is mixed with the tryptic digest aliquot.

3. Digestion proceeds at 37°C for 2 h.

4. Sample desalting is carried out by using a C18 ZipTip.

   a. The ZipTip is washed three times with 10 μL of the ZipTip wash buffer.
   b. Equilibration of the ZipTip to prepare for peptide binding is carried out by washing six times with 10 μL of the equilibration buffer.
   c. Add 6 μL of the equilibration buffer to the sample.
   d. The sample digest is loaded onto the ZipTip by drawing the sample up and expelling it ten times.
   e. Wash the ZipTip six times with 10 μl of the equilibration buffer.
   f. Elute the peptides with 3 μl of the elution buffer.

g. Repeat this procedure with an equal amount of the control tryptic digest that has not been treated with alkaline phosphatase.

5. The ZipTip-eluted alkaline phosphatase digested tryptic digest and the control tryptic digest are each mixed with 1 μL of the matrix solution and loaded on a MALDI-MS target plate.

6. Comparative MALDI-MS mapping is carried out as shown in **Fig. 4** (*see* **Note 16**).

7. After locating the phosphorylated tryptic peptide, MS/MS sequencing is carried out by MALDI-MS/MS, nanospray MS/MS or LC/MS/MS using an include list for the peptide of interest.

3.5. Location of Other Sites of Protein Posttranslational Modification

Locating *all* sites of protein posttranslational modification is a difficult task that generally requires MS analysis and accounting of all peptides that result from tryptic and/or other protease cleavage of the protein of interest. Whenever feasible, it is advantageous to first obtain a high mass accuracy ESI or MALDI-MS analysis of the native protein which (depending upon the mass accuracy that can be achieved and the MW of the protein) may be able to determine whether or not the protein is modified and if so, the total MW increase resulting from these modification(s). Often, it is of more interest and easier to determine *differential* protein posttranslational modification(s) between an experimental and control sample. In this case comparative MALDI-MS and/or RP-LC/MS on one or more proteolytic digests of the two samples may succeed in identifying the peptides that are differentially modified, providing that the stoichiometry of modification is reasonably high (i.e., >10%). Because relatively few protein posttranslational modifications occur in *E. coli*, comparative reverse phase (RP)-HPLC, MALDI-MS, and/or RP-LC/MS of tryptic and other protease digests of the native vs recombinant protein expressed in *E. coli* can potentially identify numerous sites and types of post-translational modifications. In the case of preparative RP-HPLC, aliquots of any peptide peaks that are unique to either the native or recombinant protein would be subjected to MALDI-MS, Edman, and/or nanospray MS/MS sequencing to assist in identifying the modification(s) and its location(s) in the protein (*see* **Note 17**).

In some cases (e.g., when the protein of interest is being isolated from tissue culture cells), it may be possible to radiolabel the posttranslational modification, which provides an enormous benefit in terms of being able to isolate RP-HPLC fractions containing the modified peptides and also in terms of being able to track the recovery of these peptides. Hence, labeling with [³H] S-adenosyl Met can be used to tag sites of protein Arg methylation. In this instance an overall approach may be used that is similar to that outlined in the protein phosphorylation section above. After trypsin or other protease digestion and RP-HPLC collection, it would be necessary (unless an in-line ^{3}H-detector is available) to remove 10% of each fraction, mix it with a scintillation fluid, and then use a scintillation counter to locate fractions containing the methylated peptides. Once the radio-chemically tagged peptide(s) is/are located, MALDI-MS would be performed along with radiochemical Edman sequencing and/or nanospray MS/MS analysis.

Acknowledgments

This work was funded with Federal support from National Heart, Blood, and Lung Institute (NHLBI)/National Institutes of Health (NIH) contract N01-HV-28186 and National Institute on Drug Abuse (NIDA)/NIH grant 1 P30 DA018343-01.

4. Notes

1. Trypsin is a very specific protease cleaving at the C-terminal side of Lys and Arg. This specificity is particularly useful for de novo sequencing, because the C-terminal amino acid can be specifically located as Lys (at 147 Da) or Arg (at 175 Da).
2. The Waters capLC is capable of running reproducibly at 500 nl/min and can do a direct injection of 5 μL of sample without concentration on a trap column. For other capillary HPLC systems, such as the LC Packings Ultimate, we run at 20 μL/min with stream splitting performed by the LC Packings Switchos, to 500 nL/min. In this instance the sample is preconcentrated on a Waters Symmetry® NanoEase™ Trapping column.
3. Based on our experience, the Waters Atlantis NanoEase columns have an average lifetime of more than 600 runs per column. A 100-μ inner diamter (ID) column eluted at 500 nL/min was found to be considerably more robust than the 75-μ ID columns eluted at 190 nL/min. The 100-μ ID columns also allow the capLC to operate at 500 nL/min, which eliminates problems associated with stream splitting.
4. For very complex protein mixtures, it can be advantageous to extend the capLC run time. This will help to better separate the peptides so that additional MS/MS spectra will be obtained.
5. Singly, doubly, and triply charged peptides, as well as different length peptides differ in terms of the collision energies needed to obtain optimum fragmentation. By using a collision energy ramp in the method, it is thus possible to increase the quality of the resulting MS/MS spectra.
6. A few other databases are available at the Matrix Science web site. Specific genomes can be searched using Mascot if an in-house license is obtained.
7. Typically, proteins are isolated in 1- or 2D gels prior to tryptic digestion. During gel electrophoresis, cysteines can become modified with acrylamide which converts

them to Cys-S-propionamide *(19)*. Thus, it is important to consider this modification when doing a database search.

8. The precursor mass is often found in the MS/MS spectrum header, or it can readily be located in the Mascot search results, under "Observed." Sometimes the charge state can be determined by dividing the precursor mass by the highest mass fragment ion observed in the MS/MS spectrum. Assuming that the mass spectrometer has sufficient resolution to resolve the parent ion isotopes, the best approach to determine the charge state is by examining the LC/MS spectrum. Isotopes that differ by 1 amu indicate the parent ion is singly charged; by 0.5 amu, doubly charged; by 0.3 amu, triply charged, and so on. The peptide mass is calculated by subtracting the proton from the precursor ion, then multiplying that value by the charge state to determine the mass M of the peptide. The protonated mass would be the $M + H$ species. In this example, $722.34 - 1.01 = 721.33$; $721.33 \times 2 = 1442.66$ (M), and $1442.66 + H = 1443.67$ $(M + H)$.

9. Typically, the ions observed in the mass range above the precursor ion mass are y ions. This is generally a good place in the spectrum to begin to determine a *de novo* peptide sequence.

10. On a Q-TOF instrument the mass difference between two fragment ions will differ from that of the correct amino acid by less than ± 0.1 Da.

11. Because Cerenkov counting does not require the addition of scintillation fluid, each entire fraction may be counted without loss of any sample. Cerenkov radiation results from energetic β-particles (electrons) from ^{32}P passing through a transparent medium of high refractive index (e.g., water). The resulting bluish-white light ("Cerenkov" light) may be detected using the ^3H-channel on most scintillation counters.

12. When the stoichiometry of phosphorylation is high, MALDI-MS analysis of either the tryptic digest mixture (without alkaline phosphatase treatment) or the RP-HPLC fractions containing the phosphorylated peptides in both linear and reflectron modes can often be used to locate the phosphopeptide mass. In "linear" MALDI-MS, the phosphorylated peptide will have a mass that is +80 Da compared to that of any predicted tryptic peptide. In "reflectron" MALDI-MS, the phosphorylated peptide will be unique in that it often will show a characteristic fragmentation product resulting from loss of phosphate during postsource decay (PSD). Peptides that contain phosphoserine or phosphothreonine often will undergo detectable loss of phosphate resulting in a characteristic peak due to loss of 98 Da $[MH - H_3PO_4]$ from the protonated molecular ion. In the reflectron mode the $[MH - H_3PO_4]$ peak is observed in a Waters MALDI-L/R at about $MH - 90$ Da. This mass is not seen at the expected $MH - 98$ Da because the reflector is calibrated for ions with full accelerating energy, which this ion does not have. In addition, this new mass will show substantial loss of resolution (or peak broadening). Phosphotyrosine, phosphoserine, and phosphothreonine will all show a second fragment ion at a characteristic loss of

80 Da [MH − HPO₃]. This ion is less abundant and may not be readily observed (http://www.abrf.org/ABRFNews/1995/December1995/dec95maldi.html). By comparing the linear to the reflectron MALDI-MS spectra it is thus possible to tentatively assign which peptide mass corresponds to that for the phosphorylated peptide. This identification is then confirmed and the location of the site of phosphorylation is determined by MS/MS or radioactive Edman sequencing as described above.

13. Phosphoamino acids and their ATZ and phenylthiohydantoin derivatives are insoluble in the organic solvents used to extract these derivatives on the ABI 494 cLC protein/peptide sequencer and thus are not observed in a normal Edman sequencing run. They are, however, soluble in 100% TFA. For this reason, during radioactive Edman sequencing the phosphorylated peptides are covalently attached to the ABI/Perseptive Biosystems Sequelon membrane as per the manufacturer's directions so that the ATZ derivative may be extracted at each cycle with TFA without loss of the remaining peptide.

14. **Figure 3** contains the MS/MS spectrum (m/z 450 to 1550 only) obtained on a [³²P] labeled phosphopeptide that had been isolated by microbore RP-HPLC following

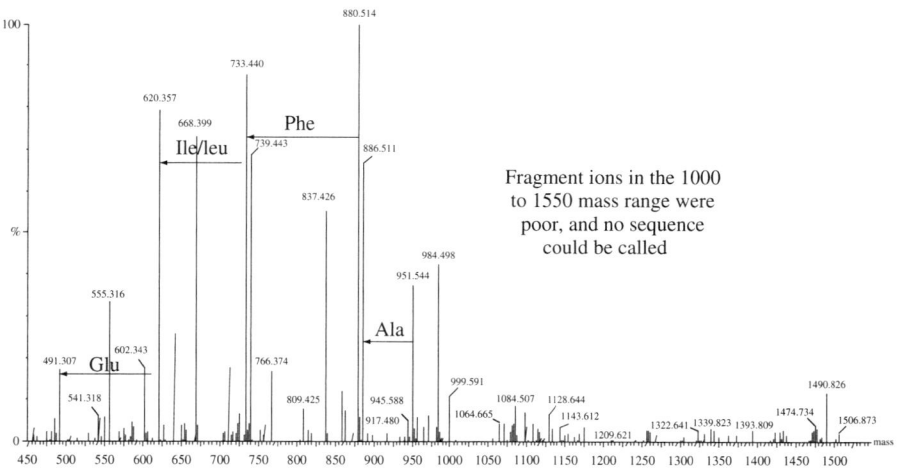

Fig. 3. Tandem mass spectrometry (MS/MS) spectra of a tryptic phosphopeptide. The [³²P]-radiolabeled high-performance liquid chromatography fraction was subjected to both radioactive Edman sequencing and liquid chromatography/MS/MS analysis. The spectra contains the mass range of m/z 450 to 1550 for the collision-induced dissociation analysis of the phosphorylated peptide. As indicated above, the fragmentation observed in the mass range extending from m/z 1000 to 1550 was poor, with no sequence ions observed. Hence, as a result of poor fragmentation, it was not possible to locate the site of phosphorylation based on the MS/MS spectra.

tryptic digestion. Radiosequencing was performed on 50% of the sample and this analysis identified residue 5 as containing the phosphoamino acid. The MS/MS spectra, which were obtained on the remaining sample, contained a good y ion series that enabled the sequence to be manually called for residues 6–14. One Ser was located within these residues but it was not phosphorylated. The fragment ions for residues 1–5 were poor and could not be identified from the MS/MS spectra. Because the peptide has Ser at residues 3 and 5, it was not possible to determine the site of phosphorylation from the MS/MS data. However, by combining the radioactive Edman sequencing results, which demonstrated that residue 5 was phosphorylated, with the MS/MS identification of the peptide, the Ser at residue 5 was determined to be the site of phosphorylation.

15. In those instances where the stoichiometry of phosphorylation at any given site is low (i.e., less than 10%), the analysis can be carried out either on phosphorylated protein that has been separated from its non-phosphorylated counterpart via 2D gel electrophoresis or it may be possible to indirectly infer the site of phosphorylation. Because a phosphorylated tryptic peptide generally elutes from RP-HPLC slightly in front (i.e., perhaps a minute earlier using a 90 min acetonitrile gradient *[17]*) of its nonphosphorylated counterpart, it may be possible to infer the identification of the phosphorylated peptide by MS, MS/MS, and/or conventional Edman sequencing of its nonphosphorylated counterpart. In this instance, the identity of the phosphorylated peptide and the site of modification can be confirmed by a number of approaches including radiochemical Edman sequencing of an aliquot of the ^{32}P-phosphate labeled fraction which has been coupled to Sequelon membrane (as described above and which thus identifies the Edman cycle at which cpm are released), in vitro mutagenesis of the putative site of phosphorylation, co-chromatography of the ^{32}P-phosphorylated peptide with a synthetic version of the putative phosphorylated peptide, and/or by carrying out a digest with another enzyme such as chymotrypsin and verifying that the ^{32}P-labeled tryptic and chymotryptic peptides (both of which were identified indirectly via analysis of their non-phosphorylated counterparts) indeed overlap and include a possible site of phosphorylation.

16. Phosphopeptides can be identified by comparing the MALDI-MS spectra before and after treatment of the tryptic digest with alkaline phosphatase and then looking for the characteristic 80-amu loss due to loss of -HPO$_3$. This approach is demonstrated in **Fig. 4** where the mass at 1769.8 is missing from the alkaline phosphatase-treated sample suggesting that it derives from a phosphopeptide (albeit there is not a clear peak at the expected m/z of 1689.8). However, the characteristic loss of MH − 90 Da from Ser and Thr phosphorylated peptides observed in reflectron mode in MALDI-MS, which is discussed in **Note 12**, is also demonstrated in **Fig. 4**. Hence, in the lower panel there is a peak in the control sample at the expected mass of 1769.8 − 90 amu = 1679.8 that results from PSD and that shows a very characteristic loss of resolution. This finding further supports the conclusion that the 1769.8 mass does indeed result from a tryptic peptide containing one site

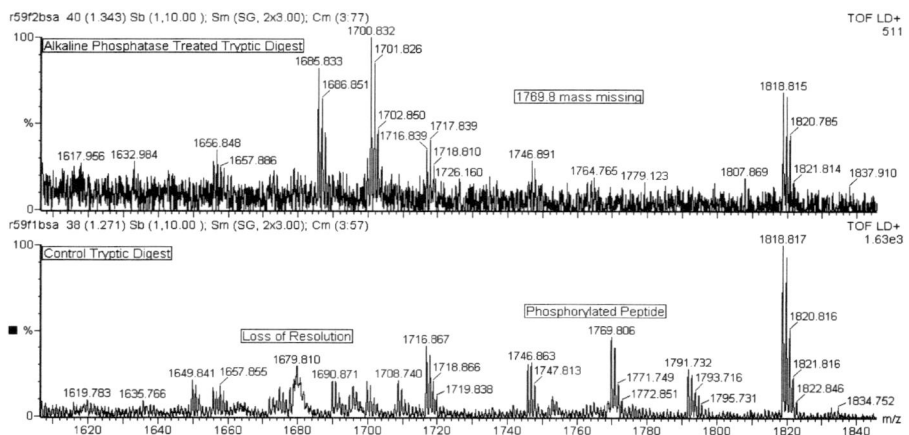

Fig. 4. Reflectron matrix-assisted laser desorption/ionization (MALDI)-mass spectrometry (MS) analysis of the tryptic digest of a 132-kDa phosphorylated protein
before (bottom) and after (top) alkaline phosphatase treatment. The top panel shows
the MALDI-MS spectrum for the mass region of interest after alkaline phosphatase
treatment; the bottom panel shows the corresponding spectrum prior to phosphatase
treatment (*see* **Note 16**).

of phosphorylation. This digest of the phosphorylated protein was then subjected
to C-18 ZipTip de-salting and nanospray MS/MS analysis on a Waters/Micromass
Q-Tof API mass spectrometer. The doubly charged species at m/z of 885.4 (corresponding to the singly charged MALDI-MS mass of 1769.8) was subjected to
CID. Manual interpretation of the resulting MS/MS spectrum located the site of
phosphorylation to position 11 in the phosphopeptide.

17.   **Figure 5** illustrates the use of comparative RP-HPLC tryptic maps to locate sites of
protein Arg methylation. The HPLC profile from the tryptic digest of native HeLa cell
A1 hnRNP protein (**Fig. 5A**), which contains four dimethylarginines, is compared to
the corresponding profile produced by the tryptic digest of the nonmethylated recombinant A1 protein isolated from *E. coli* (**Fig. 5B**). The mass of the peaks in the HPLC
profile unique to either the native or recombinant A1 were determined by MALDI-MS
and each of these peaks was subsequently subjected to Edman sequencing and MS/MS
analysis. Together, these data resulted in locating four sites of Arg dimethylation in
the A1 hnRNP protein. Because trypsin does not cleave at dimethylated arginines,
this resulted in the isolation of the corresponding overlapping tryptic peptides from
the native hnRNP A1 protein and thus explains the very large differences apparent
in **Fig. 5**. Hence, the presence of dimethylarginines at residues 205, 217, and 224
explains why three tryptic peptides in fractions 34, 32, and 62, respectively, in
Fig. 5B (that would be predicted to result from cleavage after arginines 205, 217,

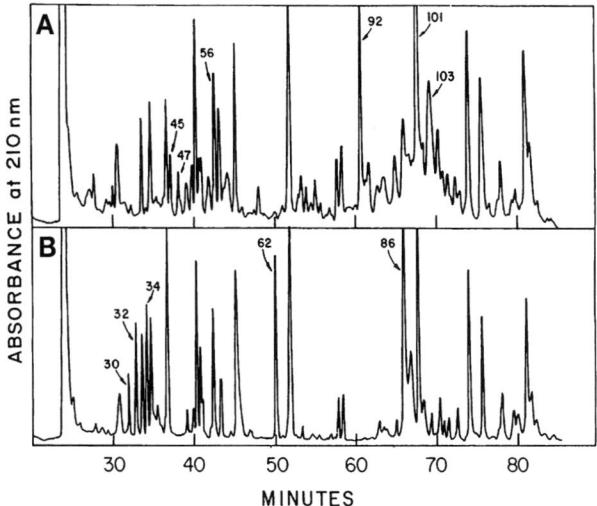

Fig. 5. Comparative high-performance liquid chromatography tryptic peptide maps of A1 hnRNP protein from HeLa cells (**A**) and recombinant A1 expressed in *Escherichia coli* (**B**) (*see* **Note 17** and **ref. *19***).

and 224, respectively) are missing from the HeLa A1 digest in **Fig. 5A**. Similarly, the presence of these three dimethylarginines explains why peak 92 in the HeLa A1 digest, which contains the tryptic peptide spanning residues 196–231, is missing from the *E. coli* A1 digest (*see* Kim et al. *[20]*). Obviously, had chymotrypsin or endopeptidase Lys-C been used instead of trypsin, the differences observed upon comparative RP-HPLC would have been far less pronounced. Although LC-MS/MS analysis could have been used instead of preparative HPLC (which was followed by MALDI-MS, MALDI PSD MS/MS, and Edman sequencing of aliquots of each fraction of interest), some of the overlapping (methylated) tryptic peptides in the A1 hnRNP digest (e.g., peak 92 corresponding to residues 196–231 which provided only a poor PSD MS/MS spectrum) probably would have been too long to have been amenable to MS/MS (CID) analysis. Instead, the preparative HPLC provided the opportunity to subject aliquots of the methylated tryptic peptides to Edman sequencing.

References

1. Henzel, W. J., Watanabe, C., and Stults, J. T. (2003) Protein identification: the origins of peptide mass fingerprinting. *J. Am. Soc. Mass Spectrom.* **14**, 931–942.
2. Yates, J. R., Eng, J. K., McCormack, A. L., and Schieltz, D. (1995) Method to correlate tandem mass spectra of modified peptides to amino acid sequences in the protein database. *Anal. Chem.* **67**, 1426–1436.

3. Gulcicek, E. E., Colangelo, C. M., McMurray, W., Stone, K., Wu, T., Zhao, H., Spratt, H., Kurosky, A., Wu, B., and Williams, K. (2006). Proteomics and the Analysis of Proteomic Data: An Overview of Current Protein-Profiling Technologies, in *Current Protocols in Bioinformatics* (Baxevanis, A. D., Davison, D. B., Page, R. D. M., Petsko, G. A., Stein, L. D., and Stormo, G. D., eds.), 1311–13,131, John Wiley & Sons Inc, Hoboken, NJ.

4. Biemann, K. and Martin, S. A. (1987) Mass spectrometric determination of the amino acid sequence of peptides and proteins. *Mass Spectrom. Rev.* **6**, 1–76.

5. Roepstorff, P. and Fohlman, J. (1984) Proposal for a common nomenclature for sequence ions in mass spectra of peptides. *Biomed. Mass Spectrom.* **11**, 601.

6. Hubbard, M. J. and Cohen, P. (1993) On target with a new mechanism for the regulation of protein phosphorylation. *Trends Biochem. Sci.* **18**, 172–177.

7. Cohen, P. (1982) the role of protein phosphorylation in neural and hormonal control of cellular activity. *Nature* **296,** 613–620.

8. Cohen, P. (1992) Signal integration at the level of protein kinases, protein phosphatases and their substrates. *Trends Biochem. Sci.* **17**, 408–413.

9. Pawson, T. and Scott, J. D. (1997) Signaling through scaffold, anchoring, and adaptor proteins. *Science* **278**, 2075–2080.

10. Goshe, M. B., Conrads, T. P., Panisko, E. A., Angell, N. H., Veenstra, T. D., and Smith, R. D. (2001) Phosphoprotein isotope-coded affinity tag approach for isolating and quantitating phosphopeptides in proteome-wide analyses. *Anal. Chem.* **73**, 2578–2586.

11. Oda, Y., Nagasu, T., and Chait, B. T. (2001) Enrichment analysis of phosphorylated proteins as a tool for probing the phosphoproteome. *Nat. Biotechnol.* **19**, 379–382.

12. McLachlin, D. T. and Chait, B. T., (2003) Improved β-elimination-based affinity purification strategy for enrichment of phosphopeptides. *Anal. Chem.* **75**, 6826–6836.

13. Ficarro, S. B., McCleland, M. L., Stukenberg, P. T., et al. (2002) Phosphoproteome analysis by mass spectrometry and its application to Saccharomyces cerevisiae. *Nat. Biotechnol.* **20**, 301–305.

14. Wolters, D. A., Washburn, M. P., and Yates, J. R., (2001) An automated multidimensional protein identification technology for shotgun proteomics. *Anal. Chem.* **73**, 5683–5690.

15. Hirosawa, M., Hoshida, M., Ishikawa, M., and Toya, T. (1993) MASCOT: multiple alignment system for protein sequences based on three-way dynamic programming. *Comput. Appl. Biosci.* **9**, 161–167.

16. Perkins, D., Pappin, D., Creasy, D., and Cottrell, J. (1999) Probability-based protein identification by searching sequence databases using mass spectrometry data. *Electrophoresis* **20,** 3551–3567.

17. Mann, M. and Wilm, M. (1994) Error-tolerant identification of peptides in sequence databases by peptide sequence tags. *Anal. Chem.* **66**, 4390–4399.

18. Stone, K. L. and Williams, K. R. (2006) Enzymatic Digestion of Proteins in Gels for Mass Spectrometric Identification and Structural Analysis, in *Current Protocols in Protein Science* (Coligan, J. E., Dunn, B. M., Ploegh, H. L., Speicher, D. W., and Wingfield, P.T., eds.) John Wiley & Sons, New York: in press.
19. Brune, D. C. (1992) Alkylation of cysteine with acrylamide for protein sequence analysis. *Anal. Biochem.* **207**, 285–290.
20. Kim, S., Merrill, B. M., Rajpurohit, R., et al. (1997) Identification of N^G-methylarginine residues in human heterogeneous RNP protein A1: Phe/Gly-Gly-Gly-Arg-Gly-Gly-Gly/Phe is a preferred recognition motif. *Biochemistry* **36**, 5185–5192.

3

Determination of Phosphorylated and *O*-Glycosylated Sites by Chemical Targeting (CTID) at Ambient Temperature

Gary M. Hathaway

Summary

In the analytical approach called chemically targeted identification (CTID), peptides containing phosphorylated or glycosylated serine and threonine underwent β-elimination to produce an unsaturated double bond. Nucleophilic addition of 2-aminoethanethiol to this bond occurred, yielding aminoethylcysteine. Thus, sites containing posttranslational modifications were made susceptible to lysine endopeptidase. Structural information could then be obtained by mass analysis of the proteolytic products. The method was demonstrated by the analysis of β-casein tryptic digest peptides and an *O*-glycosylated peptide. Contrary to an earlier report, the glycopeptide was found to react with essentially the same kinetics as phosphopeptides. Conversion of all five phosphoserines in residues 15, 17, 18, 19, and 35 in N-terminal tryptic phosphopeptides from bovine β-casein were followed by monitoring the time course of the addition reaction. The chemistry proceeded rapidly at room temperature with a half-reaction time of 15 min. No side reaction products were observed. However, care had to be taken to minimize all counterions, which either precipitate barium or neutralize the base. In the case of 2-aminoethanethiol, excess Ba(OH)$_2$ was needed to offset the effect of the hydrochloride. Alternatively, pre-incubation with base followed by nucleophilic addition was found to work satisfactorily. The use of water-soluble thiol allowed the procedure to be carried out in the solid phase, with a micro pipet greatly facilitating sample cleanup.

Key Words: Phosphorylation; *O*-glycosylation; β-elimination; mass spectrometry; post-translational.

1. Introduction

Sequence analysis of peptides by mass spectrometry often relies on induced fragmentation by collision of the ion with a neutral gas. Although frequently employed and remarkably useful, the method has its limitations. For example,

From: *Methods in Molecular Biology, vol. 386: Peptide Characterization and Application Protocols*
Edited by: G. Fields © Humana Press Inc., Totowa, NJ

length and composition affect both the completeness and uniformity of the fragmentation, and thus the amount of sequence information, that can be obtained. In the case of phosphopeptides or glycosylated peptides, their propensity toward neutral loss of the side chain of the modified amino acid residue yields little more than the precursor ion *(1–3)*. As a means for analyzing this type of peptide, our laboratory proposed introducing a lytic site chemically and then enzymatically cleaving the peptide *(4)*. We have called this method chemically targeted identification (CTID) *(5)*. The technique involved β-eliminating phosphate or *O*-glycosyl side chains from modified serine or threonine residues (*see* **Fig. 1**) and introducing by nucleophilic addition 2-aminoethanethiol (AET). This resulted in a lysine analog, cleavable by lysine endopeptidase *(6,7)* (*see* **Fig. 2**). However, the relatively harsh conditions used for β-elimination at 37–50°C *(4,5,8–10)* have recently been called into question

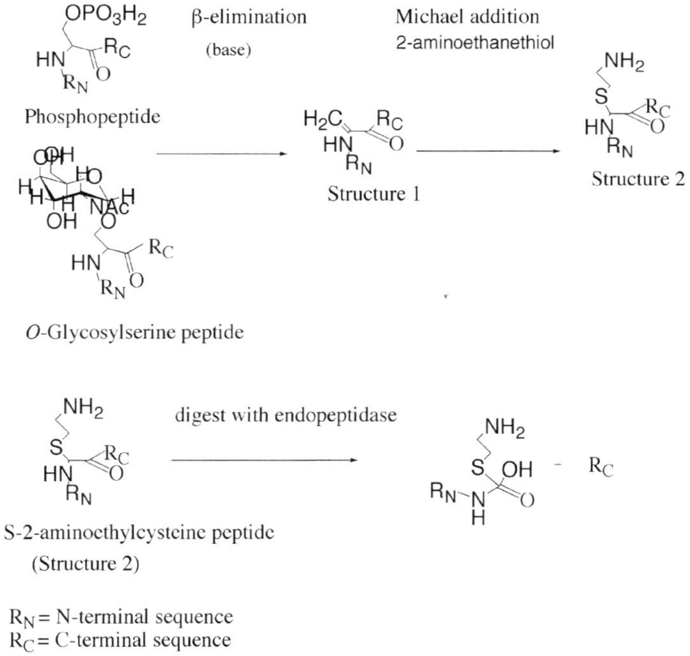

Fig. 1. Reaction scheme for chemical conversion of phosphorylated or *O*-glycosylated peptides to protease-sensitive sites. In the base-dependent reaction, phosphate or sugar is removed, leaving an unsaturated bond. The residue, dehydroalanine (in the case of serine) or its β-methyl analog (in the case of threonine) can then serve as electrophile in an addition reaction with 2-aminoethanethiol.

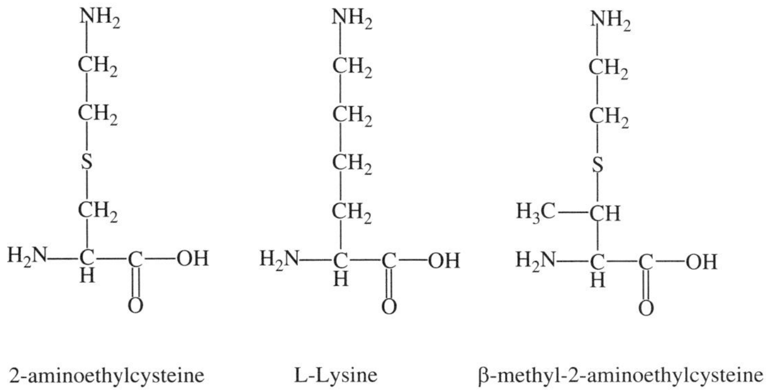

2-aminoethylcysteine   L-Lysine   β-methyl-2-aminoethylcysteine

Fig. 2. Three homologous residues that serve in site-directed proteolysis by lysine endopeptidase (*see* **refs. (6,7)**).

(11). Additionally, most of these methods required the use of some organic solvent, presumably to solvate the thiol adduct *(8–10)*. This led us to investigate an alternate approach and conditions for carrying out β-elimination and nucleophilic addition at room temperature in an aqueous environment which greatly facilitated sample desalting by micro pipet reversed-phase and even enabled carrying out the chemistry on peptides bound to the solid reversed-phase *(12)*.

2. Materials
2.1. β-Elimination and Michael Addition

1. Barium hydroxide octahydrate [Ba(OH)$_2$] and AET are purchased from Sigma-Aldrich (St. Louis, MO).
2. Micro pipet C18 "ZipTips" and the Milli-Q system for producing 18 megohm, doubly deionized water are obtained from Millipore Corp. (Bedford, MA).
3. Trifluoroacetic acid (TFA) and acetonitrile (ACN) are from Applied Biosystems Inc. (Foster City, CA). Formic acid is from J.T. Baker (Phillipsburg, NJ).
4. AET is prepared fresh as a 0.2 *M* (22.72 mg/mL) solution in deionized water. The pH 5.0 solution is used without adjustment.

2.2. Phosphopeptides and Glycosylated Peptide

1. Bovine β-casein, urea, ammonium bicarbonate, and Amberlite MB-1 are purchased from Sigma-Aldrich.
2. The *O*-glycosylated peptide Phe-Ala-Ala-(*O*-GlcNAc-Ser)-Asn-Tyr-Pro-Ala-Leu (1.156 kDa) is a kind gift from Dr. Mona Shagaholi (California Institute of Technology).

2.3. MALDI-TOF Mass Spectrometry

1. The matrix 2,5-dihydroxybenzoic acid (DHB) (98%) is purchased from Sigma-Aldrich. Ammonium hydroxide was from J. T. Baker.
2. Matrix is dissolved to 50 mg/mL in 20% ACN/0.05% TFA/20 mM ammonium phosphate *(13)* (*see* **Notes 5** and **6**).
3. Ammonium phosphate, prepared by mixing ammonium hydroxide and phosphoric acid (Fisher Scientific, Fair Lawn, NJ) to pH 3.0, is stored at −30°C as a 0.1 M solution.

2.4. Casein Phosphopeptides

1. Modified trypsin is from Promega Biochemicals (Madison, WI) and lysine endopeptidase (Lys-C) is purchased from WAKO (Richmond, VA).
2. A 100-μL volume of 8 M urea is deionized by vortexing 15 s with mixed-bed resin (the tip of a spatula) and filtering through a spin filter (0.2 micron, Amicon, Beverly, MA).
3. Tris-HCl is from Sigma-Aldrich.

3. Methods

While it is convenient to carry out both β-elimination and addition reactions in a single reaction mixture, when using the barium hydroxide base certain precautions must be undertaken (*see* **Note 1**). Because the divalent metal readily forms highly insoluble salts with many counterions such as chlorides, sulfates, phosphates, and acetates, these ions must be avoided. In this report, the hydrochloride salt of 2-aminoethanethiol is used because it is a solid and soluble in aqueous solvents. For optimal results, the base has to be maintained above a critical concentration. Initially, the concentration for AET is kept high to speed the reaction. However, this means that for 0.1 M AET a minimum of 50 mM base is required to overcome the effect of the hydrochloride (see **Fig. 3**). To avoid this complication, experiments are carried out in which the β-elimination reaction preceded nucleophilic addition *(14)*.

3.1. Preparation of Ba(OH)$_2$

1. Twice the Ba(OH)$_2$ to give a saturated solution at room temperature (126 mg/mL) is added to doubly deionized water with gentle swirling *(15)*.
2. The saturated solution is overlaid with argon, tightly capped and used for subsequent experiments for a week (*see* **Notes 1–3**).
3. Prior to use, aliquots are briefly centrifuged to pellet excess reagent and precipitates (mostly carbonates).
4. A simple gravimetric analysis is used to estimate Ba(OH)$_2$ stock solution concentration. Equal volumes of saturated supernatant and 0.5 M ammonium bicarbonate

are transferred to preweighed microcentrifuge tubes, mixed, and centrifuged. The pellets are washed once with deionized water and re-centrifuged. Precipitates from triplicate samples are dried in a SpeedVac centrifuge and their weights recorded. Results are averaged and found to be in good agreement ($\pm 1\%$). Measured concentration is 180 mM. This figure is in good agreement with published data on Ba(OH)$_2$ solubility (177 mM) in cold solution *(15)*. As an alternate technique, titration with acid has also been reported *(14)*.

3.2. β-Elimination and Michael Addition

Two methods are used which differ by the order of addition of reagents and chemicals. A third method performs the elimination and addition reactions

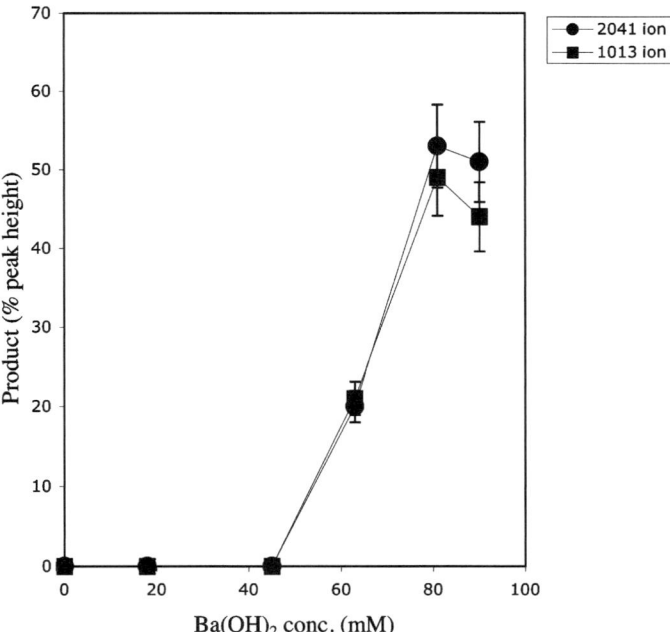

Fig. 3. Effect of chloride ion on β-elimination of a mixture of the (circles) 2.06-kDa β-casein phosphopeptide and (squares) (*O*-GlcNAc)peptide. Calculated mass for the glycosylated peptide is 1155.5 and 1011.5 for the modified peptide, a net loss of 144.1. Calculated mass for the singly phosphorylated casein peptide is 2061.8 and 2040.9 for the modified peptide, a net loss of 21. Equal volumes of 0.2 M 2-aminoethanethiol (AET) and Ba(OH)$_2$ at varying concentration are mixed with the dried peptides. After 60 min, the reaction is terminated by adding formic acid. No reaction occurred when the reaction concentration of Ba(OH)$_2$ is below 45 mM. This suggested complexation with chloride ion present as the counterion to AET prevented β-elimination.

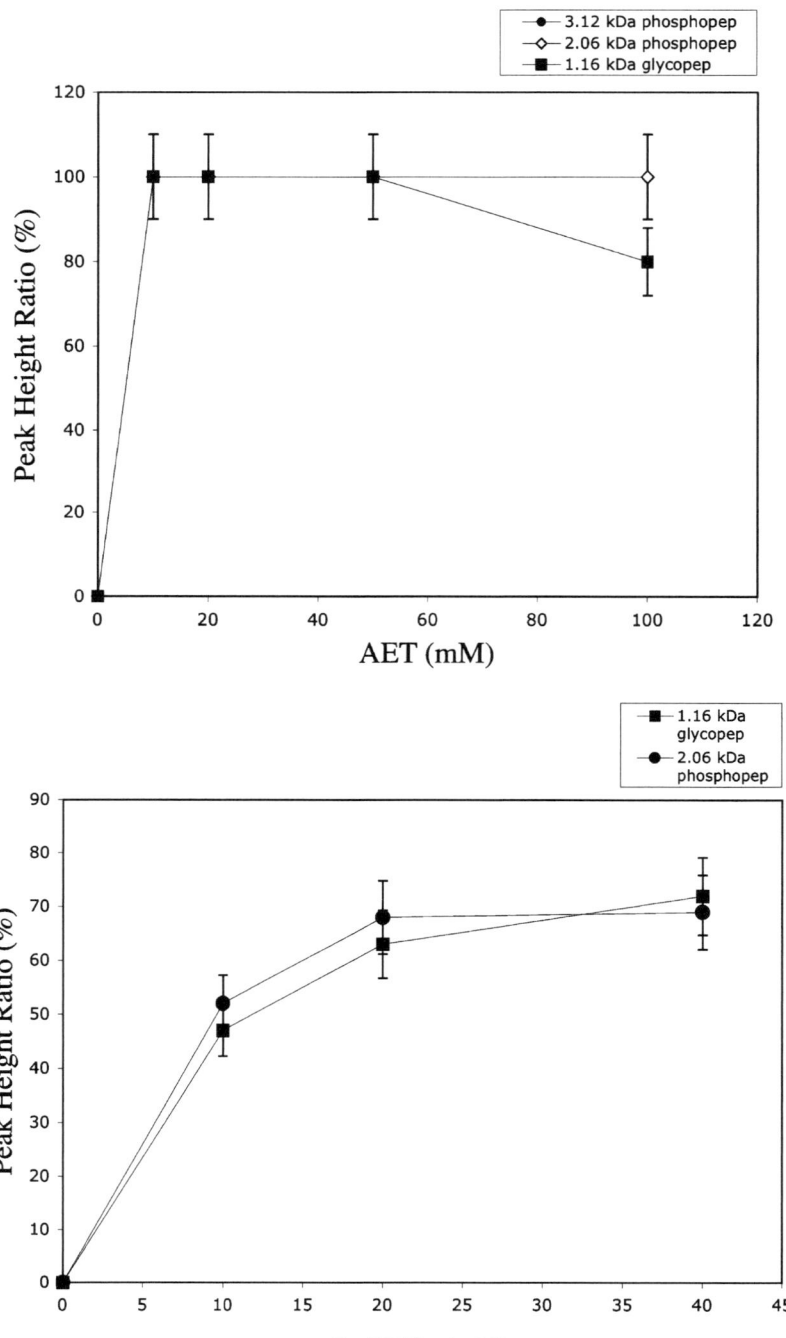

with peptide bound to C18 reversed-phase. All experiments are conducted at ambient temperature (25°C measured with a total immersion thermometer).

1. Method 1: peptides are dried in the vacuum centrifuge. Equal volumes of 0.2 M AET and Ba(OH)$_2$ are added (*see* **Fig. 3**). The maximum reaction concentration achievable by method 1 is 90 mM Ba(OH)$_2$. The data show that no reaction occurs until base exceeds 45 mM, i.e., half the concentration of AET.
2. Method 2: Ba(OH)$_2$ is added in the absence of thiol, and β-elimination is allowed to proceed. Then AET is added and nucleophilic addition is allowed to proceed (*see* **Fig. 4**).
3. Samples (2 μL) from the reactions are diluted with 8 μL of 10% formic acid, and a further 10- or 25-fold with matrix, before spotting. Alternatively, samples are diluted with 8 μl of 10% formic acid and desalted by C18 reversed-phase ZipTips. In either case, 0.5-μL aliquots of the treated samples are placed on the prespotted positions and analyzed.
4. Method 3: peptide is bound to a micro pipet C18 ZipTip in 0.005% TFA. Elimination is initiated by flushing 10X with 90 mM Ba(OH)$_2$. The micro pipet is enclosed in a 1.5-mL centrifuge tube and in contact with a few microliters of liquid below and above the C18 bed. After 30 min at 25°C, the micro pipet is flushed 10X with 0.1 M AET, which had been adjusted to pH 8.0 with a small amount of Ba(OH)$_2$. After 60 min, the micro pipet is rinsed 10X with 0.1% TFA and the peptide eluted with 3 μL of a 45% ACN/5% formic acid solution (**Fig. 5**) (**Table 3**).

Fig. 4. Reaction of a mixture of peptides (3.12-kDa and 2.06-kDa phosphopeptides and the glycosylated peptide) as a function of 2-aminoethanethiol (AET) or Ba(OH)$_2$ concentration. Upper panel: Ba(OH)$_2$ is added to 90 mM and AET to 0, 10, 20, and 50 mM. Samples are allowed to react for 60 min at 25°C. The reaction is terminated and samples treated as described in **Subheading 3.2., step 3**. Products are expressed as their peak height ratios as defined in **Subheading 3.2., step 5**. Complete conversion of the peptides, including the glycosylated peptide, is obtained at the lowest concentration of AET. The mass spectrometry spectrum for no addition of AET shows peaks at 935.5, 1964.0, and 2731.6, corresponding to the dehydroalanine products of the glycosylated peptide and two phosphopeptides, respectively. Lower panel: mixture of the (circles) 2.06 kDa phosphopeptide and the (squares) glycosylated peptide is preincubated with base at the indicated concentration for 30 min. Eighty-five micromolar AET [minimally twofold molar excess over Ba(OH)$_2$ to stop the elimination reaction] is added and incubation continued for another 30 min. No ions corresponding to the addition of water to dehydroalanine are observed, and essentially no difference in reactivity is found for the two types of posttranslational modifications.

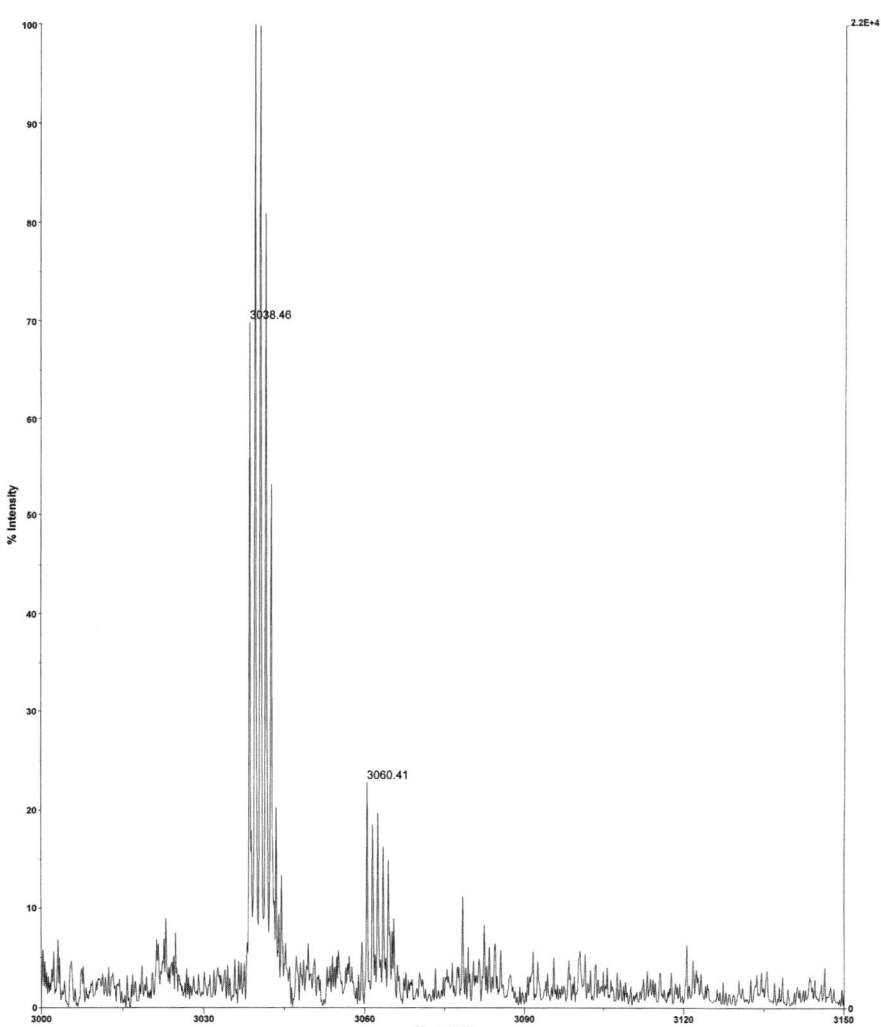

Fig. 5. Solid-phase reaction of the 3.12-kDa tetraphosphopeptide. The peptide (4 pmol) is bound to a micro pipet as detailed in **Subheading 3.2., step 4**. As shown by the data, 100% elimination occurs. Approximately 10% of an ion with m/z 2961 is also observed. This ion is believed to be the peptide with all phosphates eliminated but with the addition of three 2-aminoethanethiol residues and a single dehydroalanine.

Table 1
Sequence and Accurate Mass for β-Casein Tryptic Phosphopeptides and Their Aminoethylcysteine Derivatives

Peptide	Mass	
Sequence	(calculated)	(observed)
RELEELNVPGEIVEpS*LpSpSpSEESITR	3121.3	3121.2
RELEELNVPGEIVEpSLpSpSXEESITR	3100.3	3100.3
RELEELNVPGEIVEpSLpSXXEESITR	3079.4	3079.6
RELEELNVPGEIVEpSLXXXEESITR	3058.4	3058.7
RELEELNVPGEIVEXLXXXEESITR	3037.5	3037.7
FQpSEEQQQTEDELQDK	2060.8	2060.8
FQXEEQQQTEDELQDK	2039.8	2040.1

Mass analysis was by matrix-asisted laser desorption/ionization time-of-flight with 2,5-dihydroxybenzoic acid, as described under **Subheading 3.4**.
*pS = phosphoserine, X = S-2-aminoethylcysteine.
The sequence of β-casein is
RELEELNVPGEIVEpSLpSpSpSEESITRINKKIEKFQpSEEQQQTEDELQDKIHPFAQTQSLV
YPFPGPIPNSLPQNIPPLTTPVVVPPFLQPEVMGVSKVKEAMAPKHKEMPFPKYPVEPFTE
SQSLTLTDVENLHLPLPLLQSWMHQPHQPLPPTVMFPPQSVLSLSQSKVLPVPQKAVPYP
QRDMPIQAFLLYQEPVLGPVRGPFPIIV

5. Reaction time courses are followed by plotting relative peak height ratios vs time. Relative peak–height ratio is defined as the peak height of the protonated ions and their metal ligated counterparts divided by the sum of the corresponding product and reactant peak heights and expressed as percent (**Fig. 6**). The method is used for the data presented in **Figs. 3, 4**, and **6**.

3.3. Digestion by Lysine Endopeptidase

Lys-C is purchased from WAKO Chemicals (Richmond, VA). The protease is dissolved in deionized water, and stored dry as 0.5-μg aliquots at −30°C. Peptides prepared for CTID as described in **Subheading 3.2., step 1** are desalted and dried in the SpeedVac. Lys-C (0.1 μg) in 20 μL of 20 m*M* Tris-HCl, pH 9.2, is added. After overnight digestion at 25°C, the reaction is terminated by adding 1 μL of 10% formic acid. Aliquots (1.0 μL) are diluted with 4 μL DHB and analyzed (*see* **Fig. 7**).

Table 2
Mass Analysis of the Chemically Targeted Identification Products of β-Casein phosphopeptides

	(M + H)		
	(calc)	(obs)	(residues)
Tetraphosphopeptide residues 1–25			
RELEELNVPGEIVEpS*LpSpSpSEESITR	3122.27	3122.3	
RELEELNVPGEIVEXLXpSpS	2365.12	2365.2	15,17,18,19
RELEELNVPGEIVEXLXpSX	2344.12	2344.2	15,17,18,19
RELEELNVPGEIVEXLpSpS	2219.08	2219.1	15,17,18
RELEELNVPGEIVEXLXpS	2198.08	2198.1	15,17,18
RELEELNVPGEIVEpSLX	2051.97	2052.1	15,17
RELEELNVPGEIVEX	1771.88	1772.0	15
EESITR	734.37	734.4	19
Phosphopeptide residues 33–48			
FQpSEEQQQTEDELQDK	2061.82	2062.1	
FQXEEQQQTEDELQDK	2057.1	2057.2	
EEQQQTEDELQDK	1619.70	1619.7	35

Tryptic phosphopeptides were converted to their 2-aminoethylcysteine analogs for 2 h at 25°C. The mixed reaction products were desalted and digested with lysine endopeptidase overnight at room temperature in 20 mM Tris-HCl, pH 9.2. Mass analysis was by matrix-asisted laser desorption/ionization time-of-flight with 2,5-dihydroxybenzoic acid, as described under **Subheading 3.4**.
  * pS = phosphoserine, X = S-2-aminoethylcysteine

Table 3
Chemically Targeted Identification Performed on a pH-Stable, Micro Pipet C18 Tip

Control	Applied	Recovery	Recovery
	(pmol)	(pmol)	(%)
No treatment control	14	N/A	100
Applied control	14	12.4	89
Experimental	14	13.1	94

The β-casein 3.120-kDa tetraphosphopeptide was applied to a micro pipet C18 tip as outlined in **Subheading 3.2., step 3**. The "no treatment" control was loaded directly to a sequencing disc and analyzed by N-terminal sequence analysis. Yield was calculated from a plot of log cycle yield vs cycle number assuming a sequence efficiency factor of 0.6. The applied control was bound to the pipet tip and treated with 2-aminoethanethiol, but without exposure to Ba(OH)$_2$.

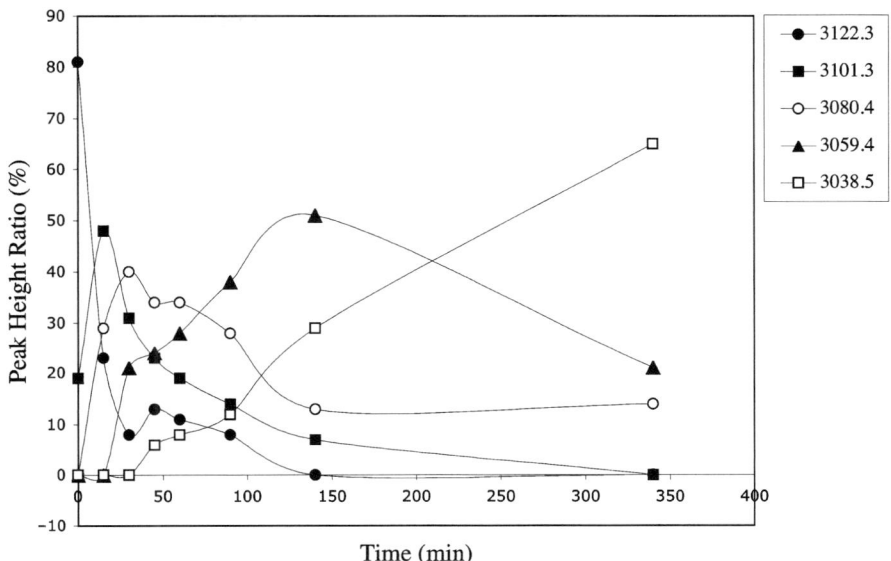

Fig. 6. Time course for the conversion of the multiply phosphorylated 3.12-kDa tetraphosphopeptide. The mass of the anticipated product ions are given in **Table 1** and as their m/z values in the inset to the figure. The method described in **Subheading 3.2., step 1** is used. The time course is followed by terminating the reaction at various times, diluting 1:250 with matrix and analyzing the products by matrix-assisted laser desorption/ionization time-of-flight mass spectrometry. The samples are analyzed in triplicate, averaged, and expressed as peak height ratios. The reacting peptide ion signal (solid circles) fell over the course of the experiment while the intermediate products are observed to rise and fall. The time for half reaction of the peptide is 15 min. These data are consistent with four phosphorylated residues within the peptide and with no significant difference in chemical reactivity among the multiple sites.

3.4. Analysis by MALDI-TOF Mass Spectrometry

1. Experimental measurement is performed with a Voyager De.str operated in reflector mode (Applied Biosystems, Inc., Foster City, CA). Monoisotopic values are recorded.
2. A 0.5-μL aliquot of matrix is spotted to the sample slide and dried in a stream of warm air at 30°C.
3. A 0.5-μL aliquot of sample is spotted to the dry matrix spot. Matrix and sample were dried in a warm (30°C) air stream.

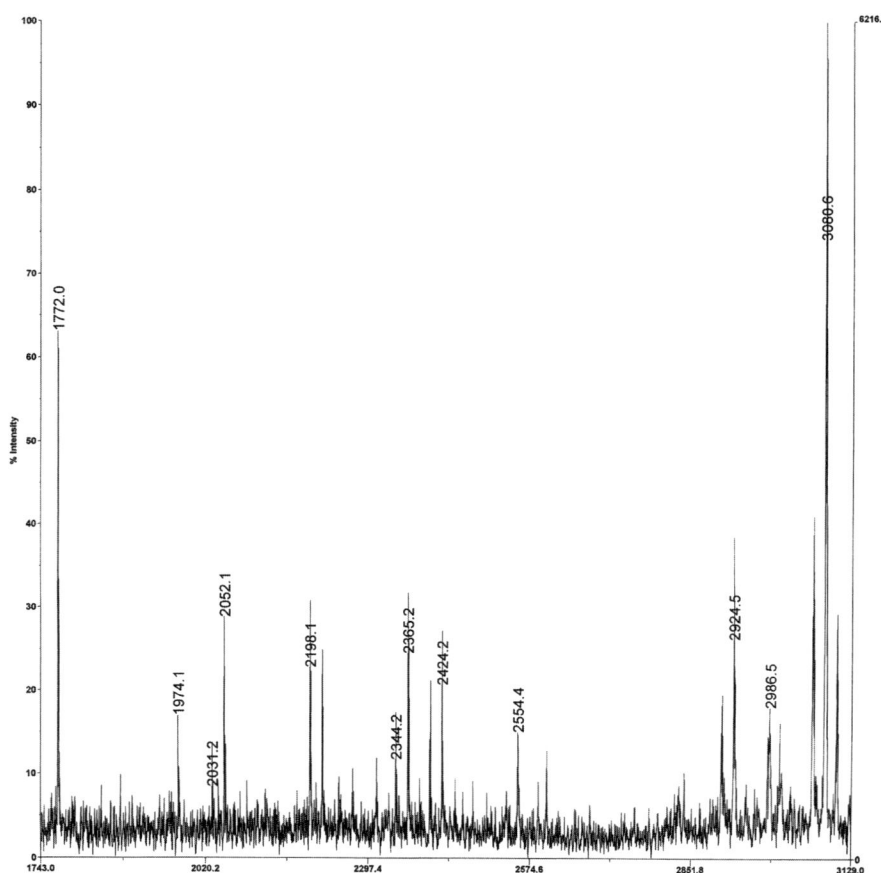

Fig. 7. Digestion of the reacted tetraphosphopeptide by Lys-C. The product mixture from the 60 min time point from the experiment presented in **Fig. 6** is digested with Lys-C, desalted and analyzed. Ions are recorded at 734.4, 1772.0, 2052.1, 2198.1, 2219.1, 2344.2, and 2365.2. These values are in good experimental agreement with the calculated singly charged ions for cleavage on the C-terminal side of serine residues 15, 17, 18, and 19 (*see* **Table 2**).

3.5. Preparation of Casein Phosphopeptides

1. 20 mg of β-casein is dissolved in 20 μL of deionized 8 *M* urea and incubated at 37°C for 20 min. The sample is diluted fourfold with 0.1 *M* ammonium bicarbonate, pH 8.0, which contained 20 μg trypsin. After overnight digestion at room temperature, TFA is added to a concentration of 1%.

2. A 2-μL aliquot is injected for high-performance liquid chromatography purification. Fractions are analyzed by mass spectrometry and those containing the 2.06-kDa phosphopeptide (FQpSEEQQQTEDELQDK) and 3.12-kDa tetraphosphopeptide (RELEELNVPGEIVEpSLpSpSpSEESITR) are stored at 4°C (*see* **Table 1**).

4. Notes

1. When using the procedure described in **Subheading 3.2., step 1,** it is absolutely necessary to avoid any buffer or counterion that neutralized the effect of the base. Chlorides, carbonates, phosphates, and acetates must be removed from the mixtures of peptides. C18 reversed-phase micro pipets are used and found to be a convenient and practical way to desalt peptides described in **Subheading 3.2., steps 1** and **2**. The micro pipets described in **Subheading 3.2., step 4** are stable to pH > 13.0 as specified by the manufacturer (http://www.hejie.com.cn/download/ziptip.pdf). Examination of the micro pipets under the light microscope after 2 h treatment with up to 180 m*M* Ba(OH)$_2$ shows no degradation of the C18 reversed-phase bed.
2. A second source of error is the possible reaction or β-elimination of cysteine containing peptides. Although the casein used in this work contained no cysteine, alkylation with iodoacetamide or other suitable thiol blocking agents have been proposed *(14)*.
3. Ba(OH)$_2$ will react with CO$_2$ and should be kept in tightly closed containers overlaid with an inert gas such as argon.
4. The CTID method will not work with phosphotyrosine residues because of its inability to form an energetically favorable double bond, nor will it work with *N*-glycosylated peptides.
5. Conversion of phosphopeptides to their aminoethylcysteine derivatives results in a net loss of mass 21. Care must be taken to measure the change, as the sodiated form of the modified ion differs from the unmodified peptide by unit mass.
6. DHB was found to be the matrix of choice to observe phosphopeptide ions. When α-cyano-4-hydroxycinnamic acid or sinapic acid is substituted, the signals from the phosphopeptides greatly diminished or disappeared completely. Therefore, when recording peak height ratios, it is essential to use DHB.
7. The method for calculating peak height ratios used here was compared with the use of an internal reference ion *(16,17)*. For the data presented in **Fig. 6**, a nonphosphorylated peptide in the preparation is used to check the validity of the relative peak ratio method. Comparable results are obtained (data not shown).

References

1. Carr, S. A., Huddlestone, M. J., and Annan, R. S. (1996) Selective detection and sequencing of phosphopeptides at the femtomole level by mass spectrometry. *Anal. Biochem.* **239**, 180–192.

2. Annan, R. S. and Carr, S. A. (1996) Phosphopeptide analysis by matrix-assisted, laser desorption time-of-flight mass spectrometry, *Anal. Chem.* **68**, 3413–3421.

3. Liao, P. C., Leykam, J., Andrews, P. C., Gage, D. A., and Allison, J. (1994) An approach to locate phosphorylation sites in a phosphoprotein: mass mapping by combining specific enzymatic degradation with matrix-assisted, laser desorption time-of-flight mass spectrometry. *Anal. Biochem.* **219**, 9–20.

4. Rusnak, F., Zhou, J., and Hathaway, G. M. (2002) Identification of phosphorylated and glycosylated sites in peptides by chemically targeted proteolysis. *J. Biomol. Tech.* **13**, 228–237.

5. Zhou, J., Rusnak, F., and Hathaway, G. (2003) A method for the identification of post-translationally modified peptides by chemical targeting. The 51st meeting of The American Society for Mass Spectrometry, Montreal, Canada, poster 361.

6. Cole, R. D. (1967) S-aminoethylation. *Meth. Enzymol.* **11**, 315–317.

7. Masaki, T., Takiya, T., Tsunasawa, S., Kuwahara, S., Sakiyama, F., and Soehma, M. (1994) Hydrolysis of S-2-aminoethylcysteinyl peptide bond by achromobacter protease I. *Biosci. Biotech. Biochem.* **58(1)**, 215–216.

8. Meyer, H. E., Hoffmann-Posorske, E., Korbe. H., and Heilmeyeer, L. M., Jr. (1986) Sequence analysis of phosphoserines-containing peptides. Modification for picomolar sensitivity. *FEBS Lett.* **204**, 61–66.

9. Molloy, M. P. and Andrews, P. C. (2000) Phosphopeptide derivatization signatures to identify serine and threonine phosphorylated peptides by mass spectrometry. *Anal. Chem.* **73**, 5387–5394.

10. Knight, K., Schilling, B., Row, R. H., Kenski, D. M., Gibson, B. W., and Shokat, K. M. (2003) Phosphospecific proteolysis for mapping sites of protein phosphorylation. *Nat. Biotech.* **21**, 1047–1054.

11. Li, W., Backlund, P. S., Boykins, R. A., Wang, G., and Chen, H-C. (2003) Susceptibility of the hydroxyl groups in serine and threonine to β-elimination/Michael addition under commonly used moderately high-temperature conditions. *Anal. Biochem.* **323**, 94–102.

12. Nika, H., Hawke, D. H., and Kobayashi, R. (2003) Derivatization on reversed-phase supports for enhanced detection of phosphorylated peptides. The 51st meeting of The American Society for Mass Spectrometry, Montreal, Canada, poster 282.

13. Smirnov, I. P., Zhu, X., Taylor, T., et al. (2004) Suppression of α-cyano-4-hydroxycinnamic acid matrix clusters and reduction of chemical noise in MALDI-TOF mass spectrometry. *Anal. Chem.* **76**, 2958–2965.

14. Byford, M. F. (1991) Rapid and selective modification of phosphoserine residues catalysed by Ba^{2+} ions for their detection during peptide microsequencing. *Biochem. J.* **280**, 261–265.

15. Weast, R.C. (ed.) (1975) *Handbook of Chemistry and Physics*, 55th edition. CRC, Cleveland, OH: B–71.

16. Jespersen, S., Niessen, W. M. A., Tjaden, U. R., and van der Greef, J. (1995) Quantitative bioanalysis using matrix-assisted laser desorption/ionization mass spectrometry. *J Mass Spectrom.* **30**, 357–364.
17. Houston, C. T., Taylor, W. D., Widlanski, T. S., and Reilly, J. P. (2000) Investigation of enzyme kinetics using quench-flow techniques with MALDI-ToF mass spectrometry. *Anal. Chem.* **72**, 3311–3319.

4

Synthesis, Biosynthesis, and Characterization of Transmembrane Domains of a G Protein–Coupled Receptor

Fred Naider

Summary

Peptide fragments have been widely used in biophysical studies on specific regions of integral membrane proteins. Because of their inherent insoluble nature and tendency to aggregate the preparation of such model peptides is challenging. We have developed synthetic and biosynthetic approaches to prepare peptides containing single and multiple domains of a G protein–coupled receptor. Both the synthetic and biosynthetic products can be isolated by reversed-phase high-performance liquid chromatography to near homogeneity. The biosynthetic product, a fusion protein, is processed by CNBr cleavage to yield the target peptide in various isotopic forms. The final peptides are studied by circular dichroism spectroscopy to determine their secondary structure under a variety of conditions.

Key Words: G protein–coupled receptor; membrane proteins; circular dichroism spectroscopy; α-factor receptor; biosynthesis of membrane proteins.

1. Introduction

As a result of the abundance and lack of sufficient study of membrane proteins, there is significant interest in their structure and function. This interest necessitates the synthesis and biophysical analysis of such molecules. However, the inherent insolubility of membrane peptides and their tendency to self-aggregate renders their preparation and isolation quite challenging. Nevertheless, the need for membrane peptides as surrogates for regions of transporters and receptors, and as analogs of naturally occurring peptide antibiotics continues to grow. In this chapter, the author discusses methods used in his

From: *Methods in Molecular Biology, vol. 386: Peptide Characterization and Application Protocols*
Edited by: G. Fields © Humana Press Inc., Totowa, NJ

laboratory to synthesize, biosynthesize, purify, and characterize peptides corresponding to regions of a G protein–coupled receptor. Although the laboratory has used circular dichroism, infrared, and nuclear magnetic resonance (NMR) spectroscopies to determine the secondary structure of these peptides, only experiments with circular dichroism spectroscopy will be described.

1.1. G Protein—Coupled Receptors

G protein–coupled receptors (GPCRs) represent one of the largest families of proteins in higher cells. There are approx 1000 examples of these ubiquitous molecules in eukaryotes *(1)*. They function in all modalities of signal transduction including light reception, taste and pain perception, growth regulation, immunity, and general metabolism. Their ligands vary from biogenic amines to amino acids, from small to medium and large-sized polypeptides. They are a major target of the pharmaceutical industry, with between 30% and 50% of drugs acting as agonists or antagonists of GPCRs *(2)*. Despite the diversity and versatility of this class of membrane proteins they have a common over all topology. Hydropathy analysis predicts that members of this superfamily of proteins have seven transmembrane helices connected by three extracellular and three intracellular loops. The amine terminus is on the extracellular side of the membrane while the carboxy terminus resides in the cytosol (**Fig. 1**). In contrast to the ubiquitous GPCRs and their critical biological functions, only one high- resolution structure of a GPCR has been reported, that of bovine rhodopsin *(3)*.

The α-factor receptor (Ste2p) of the yeast *Saccharomyces cerevisiae* is used as a model system to develop methodologies for studying GPCR structure and biochemistry. This receptor is involved in the sexual conjugation of yeast haploid strains *MAT***a** and *MAT*α and recognizes a 13-residue mating pheromone—the α-factor. The receptor belongs to Class D of the GPCR superfamily and has been subjected to extensive investigation in numerous laboratories *(4)*. The author's laboratory has synthesized or biosynthesized many regions of this receptor and subjected these peptide surrogates to a variety of biophysical analyses *(5–13)*.

1.2. Synthesis of Membrane Peptides (30–60 Residues): Peptide Design and Synthesis

The peptides synthesized in the author's laboratory correspond to transmembrane domains of the α-factor receptor of *S. cerevisiae* (Ste2p) which are predicted by hydropathy analysis and experimentally verified by several methods *(14,15)*. Each polypeptide is designed to include the entire

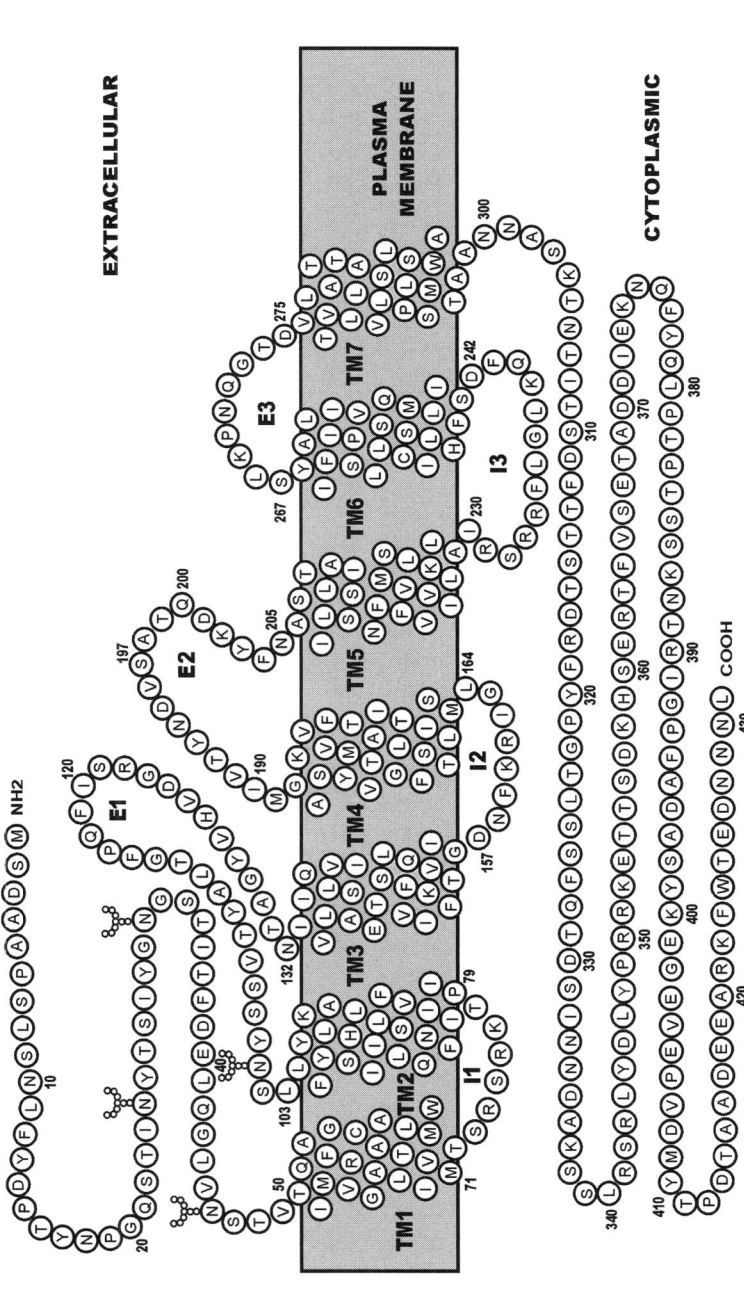

Fig. 1. Cartoon of the α–factor receptor (Ste2p). Domains are indicated by the following: E: extracellular loops; I: intracellular loops; TM: transmembrane. The four Asn residues represent glycosylation positions.

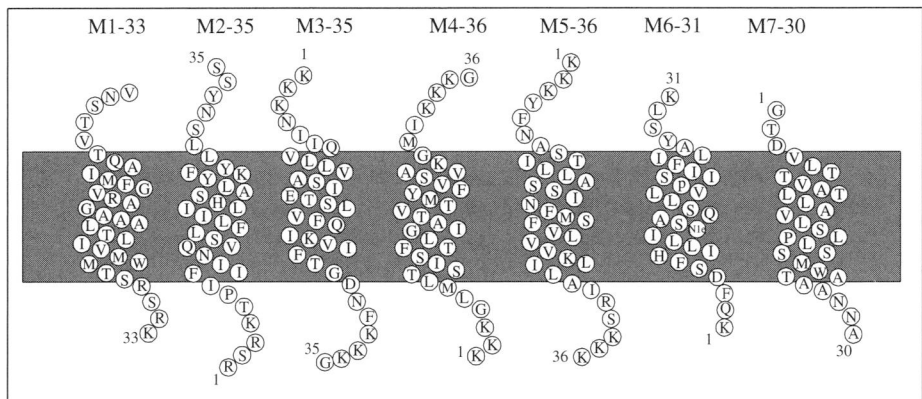

Fig. 2. Cartoon illustrating synthetic peptides corresponding to the transmembrane domains (TMDs) of Ste2p. In some cases, Lys (K) residues flank the naturally occurring TMDs. Taken from **ref. *10***.

transmembrane helix plus some portion of the flanking loop sequences (**Fig. 2**). On the basis of the author's experience, the inclusion of residues in the inter-helix loops on both sides of a given transmembrane helix is essential to attain solubilities needed for isolation by high-performance liquid chromatography (HPLC) and subsequent biophysical studies. For some transmembrane peptides, the addition of natural loop residues does not suffice and several artificial Lys residues with a positively charged side chain were added to both termini. Wild-type Ste2p contains two Cys residues that are not essential for receptor function. These residues are changed to Ala or Leu in our synthetic peptides to avoid problems of Cys oxidation. Met is also changed to Nle or Leu for similar reasons.

All syntheses are carried out on an Applied Biosystems automated peptides synthesizer model #433A. The steps described can be adapted to any synthesizer or can be carried out manually.

2. Materials

2.1. Synthesis of Membrane Peptides

1. Wang resin (Advanced ChemTech, Louisville, KY).
2. Most Fmoc protected amino acids are purchased from Advanced ChemTech (Louisville, KY). Fmoc-His(Trt) is from Calbiochem-Novabiochem Corp. (San Diego, CA) and Bachem Inc. (Torrance, CA).
3. *O*-Benzotriazoleyl-*N,N,N′,N′*-tetramethyluronium hexafluorophosphate (HBTU) (Advanced ChemTech, Louisville, KY).

4. 1-Hydroxybenzotriazole (HOBt) (Advanced ChemTech, Louisville, KY).
5. *N*-methylpyrrolidinone (NMP) (Advanced ChemTech, Louisville, KY).
6. 1-Hydroxy-7-azabenzotriazole (HOAt) (PerSeptive Biosystems, Framingham, MA).
7. *N*, *N'*-diisopropylethylamine (DIEA) (Aldrich Chemical Co., Milwaukee, WI).
8. Dicyclohexylcarbodiimide (DCC) (Aldrich Chemical Co., Milwaukee, WI).
9. Trifluoroacetic acid (TFA) (Aldrich Chemical Co., Milwaukee, WI).
10. Thioanisole (Aldrich Chemical Co., Milwaukee, WI).
11. 1,2-Ethanedithiol (EDT) (Aldrich Chemical Co., Milwaukee, WI).
12. *N*, *N'*-dimethyl-aminopyridine (DMAP) and all other reagents (Aldrich Chemical Co., Milwaukee, WI).
13. Solvents used for synthesis and purification are purchased from VWR Scientific (Piscataway, NJ) and Fisher Scientific (Springfield, NJ).

2.2. Biosynthesis of Membrane Peptides

2.2.1. Vectors

1. The plasmid pMD602 containing the *STE2* gene with mutations Met218Leu and Met250Ala is derived using site-directed mutagenesis from the parent plasmid pMD194 obtained from Mark Dumont, University of Rochester *(16)*.
2. The cloning vector pRSET B is purchased from Invitrogen (Carlsbad, CA).
3. The plasmid pMMHa is obtained as a gift from Peter Kim, Massachusetts Institute of Technology *(17)*.

2.2.2. Strains

1. The original yeast strain transformed with the plasmid DNA harboring wild-type and mutant Ste2p receptors is A232 (MATa *ste2-Δcryl^R ade2-1 his4-580 lys2$_{oc}$ tyr1$_{oc}$ SUP4-3^{ts} leu2 ura3 bar1-1 FUS1::p[FUS1-lacZ TRP1]*). A232 is obtained as a courtesy from M. Dumont *(16)*.
2. *Escherichia coli* expression strains BL21(DE3) and BL21(DE3)pLysS are purchased as competent cells from Promega.
3. *E. coli* propagation cells DH5α and XL-1 Blue are purchased from Gibco BRL Life Technologies (Grand Island, NY) and Stratagene (La Jolla, CA), respectively.

2.2.3. Chemicals

1. Polymerase chain reaction reagents (Promega, Madison, WI).
2. dNTP (dATP, dGTP, dCTP, and dTTP) (Promega, Madison, WI).
3. Taq polymerase (Promega, Madison, WI).
4. Restriction enzymes (Promega, Madison, WI).
5. Ligase (Promega, Madison, WI).
6. Agarose and low melting point agarose (GIBCO BRL, Grand Island, NY).
7. The anti-Xpress mouse monoclonal immunoglobulin (Ig)G$_1$ antibody (Invitrogen, Carlsbad, CA).

8. Isopropyl-β-D-thiogalactopyranoside (IPTG) (Invitrogen, Carlsbad, CA).
9. M13 phage that contain T7 RNA polymerase (Invitrogen, Carlsbad, CA).
10. Immobilized metal affinity column for protein purification (ProBond resin that contains $NiCl_2$) (Invitrogen, Carlsbad, CA).
11. Guanidinium hydrochloride (GuHCl) (Sigma, St. Louis, MO).
12. The detergents sodium deoxycholic acid, igepal CA60, sodium *N*-laurylsarcosine, and sodium dodecyl sulfate (SDS) (Sigma, St. Louis, MO).
13. Buffer chemicals (Trizma-HCl, sodium phosphate, sodium chloride, potassium chloride, magnesium chloride) (Sigma, St. Louis, MO).
14. Media ingredients such as Luria Broth base and yeast extract (Gibco BRL Life Technologies).
15. Peptone (Difco Laboratories, Detroit, MI).
16. Agar (Difco Laboratories, Detroit, MI).
17. Tryptone (Sigma, St. Louis, MO).
18. SDS-polyacrylamide gel electrophoresis (PAGE) reagents (Coomassie brilliant blue G-250, acrylamide, glycerol, ammonium persulfate, *N, N, N′, N′*-tetramethylethy - lenediamine [TEMED], and β-mercaptoethanol) (Sigma, St. Louis, MO).
19. Luminol reagents for the development of immunoblots of His tags (Amersham Life Science, Buckinghamshire, UK).
20. Secondary IgG mouse antibody horseradish peroxidase labeled (Sigma, St. Louis, MO).
21. Nitrocellulose membrane, 0.45-μm pore size (Sigma, St. Louis, MO).
22. India His Probe (Pierce, Rockford, IL) (*see* **Note 1**).
23. Phospholipids used to prepare the lipid vesicles: 1,2-dimyristoyl-*sn*-glycero-3-phosphocholine (DMPC) and 1,2-dimyristoyl-*sn*-glycero-3-[phospho-*rac*-(1-glycerol)] (DMPG) (Avanti Polar Lipids, Alabaster, AL).

3. Methods

3.1. Synthesis of Membrane Peptides

1. α-Fmoc amino acid loaded Wang resin is weighed into the reaction vessel.
2. Syntheses are carried out on a 0.1-mmol scale (*see* **Note 2**).
3. The Fmoc group is used for protection of all N-α groups, and Boc, Trt, Trt, *t*Bu, *t*Bu, *t*Bu, Pmc were employed for protection of Lys, Gln, His, Ser, Tyr, Asp, and Arg, respectively.
4. The Fmoc group is cleaved using 20% piperidine in NMP.
5. α-Fmoc amino acids with appropriate side-chain protection are coupled to the free amine using HBTU/HOBT activation (*see* **Note 3**) in NMP.
6. A 10-fold molar excess of α-Fmoc protected amino acid is used in the coupling step.
7. Each amino acid is routinely double coupled.
8. Unreacted chains are capped with acetic anhydride.

9. The synthesizer monitors the completeness of the coupling at the Fmoc deprotection step by measuring conductivity.
10. In a normal synthesis there is a gradual decrease in conductivity upon deprotection as the length of the peptide increases.
11. After completion of chain assembly, the N-terminal Fmoc group is removed using 20% piperidine in NMP.
12. The resin is washed with NMP and methylene chloride and the peptide-resin dried to constant weight in a vacuum dessicator.
13. The peptide resin is weighed and the weight gain on synthesis recorded.
14. The side-chain protecting groups are removed and the peptide released from the resin using a cleavage cocktail consisting of TFA/water and various scavengers.
15. The resin is treated with a scavenger-solution containing 0.75 g phenol, 0.5 ml thioanisole, 0.25 ml EDT, 0.5 mL water, and 10 mL TFA if the peptide contained Arg or Met.
16. 9.5 mL TFA, 0.25 mL EDT, and 0.25 mL water are used when the peptide contained neither Arg nor Met but contained Trt or Trp.
17. The cleavage reaction is carried out for 1–2 h at room temperature.
18. The cleavage cocktail is evaporated using a rotary evaporator at a vacuum of about 0.1 mmHg and a temperature below 35°C.
19. The residue is poured into anhydrous ether to precipitate the crude peptide.
20. The precipitate is allowed to stand for several hours at room temperature.
21. The precipitated peptide is filtered through a medium pore-sized sintered glass funnel, washed with cold ether, and dried in a vacuum dessicator.
22. The resulting peptide is weighed to determine the crude yield.
23. The crude peptide is ready for HPLC characterization.

3.2. Biosynthesis of Membrane Peptides

The chemical synthesis of membrane peptides is limited by the physical properties of these molecules. Because of their propensity to self-associate, final products often contain numerous impurities. Above a chain length of 30–40 residues, removal of the truncated peptides and deletion sequences that constitute the impurities is often beyond the resolution of HPLC. Furthermore, the high-resolution NMR analysis of membrane peptides requires ^{15}N- and/or ^{13}C/^{15}N-labeled compounds. Chemical synthesis of these molecules is prohibitively expensive.

As an alternative to synthesis, the author's laboratory is using biosynthesis to prepare membrane peptides. Direct biosynthesis of fragments of GPCRs has been unsuccessful. The products appear to be toxic to the bacteria used for expression. Reasonable success has been achieved with the synthesis of fusion proteins containing the target sequence followed by CNBr release of the desired peptide.

3.2.1. Preparation of Media and Buffers

1. Liquid Luria-Bertani (LB) medium is prepared by mixing 10 g of tryptone, 5 g of yeast extract, and 10 g of NaCl in 1 L of water (*see* **Note 4**).
2. Plates of LB medium are prepared in the same way as the liquid with the addition of 15 g of agar.
3. SOB medium for glycerol stocks: 20 g of tryptone, 5 g of yeast extract, 0.5 g of NaCl, and 186.0 mg of KCl are mixed in 1 L of water (*see* **Note 5**).
4. SOC medium for transformation is prepared exactly as the SOB medium with the addition of 20 mM glucose.
5. Minimal medium: mix 200 mL of 5X M9 salts (15 g of KH_2PO_4, 34 g of Na_2HPO_4, 2.5 g of NaCl, 5 g of NH_4Cl in 1 L of water, pH 7.2), 8 mL of 50% glucose (filtered-sterilized), 1 mL of 2 M $MgSO_4$, 0.2 mL of 0.5 M $CaCl_2$, and 788.3 mL of deionized water. ^{15}N-minimal media and $^{13}C/^{15}N$-minimal media is prepared in the same way except for the 5X M9 salts, which contain $^{15}NH_4Cl$ instead of NH_4Cl. Uniformly labeled ^{13}C glucose is used for the $^{13}C/^{15}N$-minimal media.
6. All media except for minimal medium are brought to pH 7.0. 5X M9 salts are adjusted to pH 7.2, except for the glucose solutions, which are filter-sterilized, the other media components are sterilized by autoclaving. Sterilized glucose is then added to the cooled sterilized media.
7. After autoclaving, filter-sterilized antibiotics are added to the cool media in a ratio of 1 μL antibiotic to 1 mL medium. Stock solutions of the following antibiotics are used: ampicillin (200 mg/mL), chloramphenicol (35 mg/mL).
8. Tris-buffered saline + Tween 20 (TBST): TBS + 0.05% Tween-20 (w/v).
9. Phosphate-buffered saline (PBS) contains 10 mM sodium phosphate, 138 mM NaCl, and 2.7 mM KCl, pH 7.4.
10. TE buffer: 8.3 mM Tris-HCl and 40 μM ethylenediamine tetraacetic acid (EDTA).
11. Guanidinium buffer: 6 M GuHCl, 20 mM sodium phosphate, 500 mM sodium chloride, pH 7.8.
12. Solvents used for HPLC purification are 2-propanol:1-butanol (2:1) (used for double domain) or acetonitrile containing 0.1% TFA.
13. Buffers used to resuspend lipid vesicles are 0.1 mM potassium phosphate, pH 4.0, and 10 mM ammonium acetate, pH 4.0.

The protocol under **Subheading 3.2.2.** describes the construction of the pSW02 plasmid that expresses the TrpΔLE-M6 fusion protein. Different plasmids coding for the expression of other single domain, multiple domain, and double transmembrane domain regions of Ste2p have been constructed in a similar manner (*12,13*).

3.2.2. Construction of Expression Plasmids

1. The plasmid pMMHa (*17*) is a pET-derived plasmid that also contains the gene for β-lactamase. This plasmid contains the T7 RNA polymerase binding site and the

N-terminus leader peptide TrpΔLE derived from the tryptophan operon *(17,18)*. The expression vector also contains multiple restriction enzyme cloning sites, including *Bam*H I and *Hind* III. The vector codes for the TrpΔLE protein fused to bovine pancreatic trypsin inhibitor and is used as the parent vector to construct fusions of transmembrane domains to the TrpΔLE carrier protein (**Fig. 3**).

2. Polymerase chain reaction is used to amplify the M6 DNA sequence containing the mutation M250A from the pMD602 plasmid (*see* **Note 6**). The primers used are designed to introduce the respective *Hind* III and *Bam*H I restriction sites to the 5′ and 3′ ends of the M6 DNA sequence.

3. The PCR product is purified using the QIAquick PCR Purification Kit (Qiagen Inc.) and analyzed on a 2% agarose gel.

4. The *Hind* III and *Bam*H I restriction enzymes are then used to digest the PCR product as well as the pMMHa vector (*see* **Note 6**). The reaction is allowed to run overnight at 37°C.

5. The digested PCR product and pMMHa vector are purified using the QIAquick PCR Purification Kit and analyzed on a 2% agarose gel.

6. Ligation of the digested M6 PCR product to the digested pMMHa vector is carried out using T4 ligase (*see* **Note 6**). The reaction mixture is left at 4°C overnight. This creates the pSW02 plasmid.

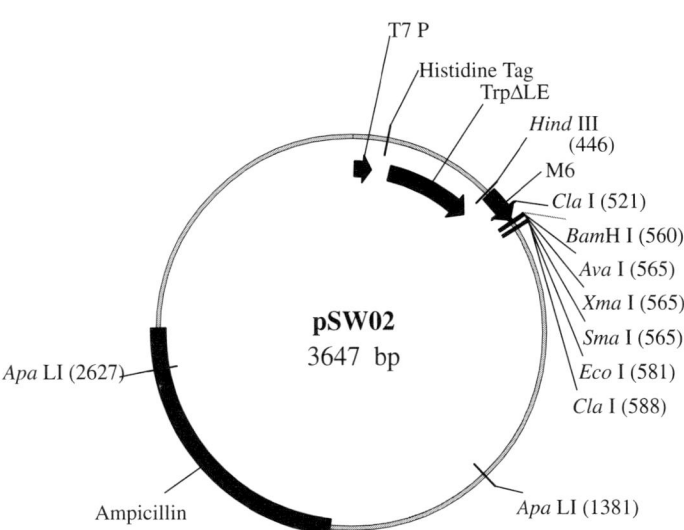

Fig. 3. Expression plasmid for fusion proteins. pSW02 is derived from pMHHa *(17)* using the procedures described in the text. The plasmid codes for a fusion protein TrpΔLE-M6 that contains the sixth transmembrane domain of Ste2p.

7. The pSW02 plasmid is transformed into DH5α cells by adding 2–5 µL of plasmid to 50 µL of cells.
8. The cells are placed in an ice bath for 10 min and heat-shocked at 42°C for 1 min.
9. The cells are placed in the ice for 2 min, 400 µL of SOC medium is added, and the vessel is incubated at 37°C and 225 rpm for 1 h.
10. 50–100 µL of the transformation mixture is spread on a LB plate containing the appropriate antibiotics.
11. A single colony from the LB plate is grown in 15 mL LB media containing ampicillin (200 µg/mL) until late log phase (OD$_{600}$ ~ 0.7–0.8).
12. The pSW02 plasmid is isolated from the culture using the Wizard Plus SV Minipreps DNA purification system (Promega).
13. The purified pSW02 plasmid is digested with the *Bam*H I and *Hind* III restriction enzymes and analyzed on a 2% agarose gel to determine whether the size of the insert corresponded to that expected for M6 DNA.
14. DNA sequencing of the pSW02 plasmid is performed to confirm the correct construct.
15. Glycerol stocks of the pSW02 plasmid are made by transforming 2–5 µL of the plasmid into 50 µL of DH5α cells.
16. A single colony from the LB plate is grown in 15 mL LB media containing ampicillin (200 µg/mL) until late log phase (OD$_{600}$ ~ 0.7–0.8).
17. 850 µL of cells is combined with 150 µL sterile glycerol and quickly frozen in dry ice plus ethanol. The frozen stocks are stored at −70°C.

3.2.3. Expression of Fusion Proteins

1. To achieve expression, 5 µL of the pSW02 plasmid is transformed into 50 µL of BL21(DE3)pLysS *E. coli* cells using the above procedure. The BL21(DE3)pLysS cells contain the pLysS plasmid, which confers resistance to chloramphenicol.
2. A single colony from an LB plate is inoculated into 5 mL of LB media containing ampicillin (200 µg/mL) and chloramphenicol (35 µg/mL) and allowed to grow overnight.
3. Two milliliters of the overnight culture are then diluted into 250 mL of the same media and cells are allowed to grow to late log phase (OD$_{600}$ ~ 0.7–0.8).
4. Cells are induced with 1 mM IPTG, incubated for 6 h, and harvested by centrifugation. The BL21(DE3)pLysS cells contain the DE3 lysogen, which encodes the T7 RNA polymerase under control of the IPTG-inducible lacUV5 promoter. Upon induction, T7 RNA polymerase is expressed and converts almost all of the host cells resources to transcribe the gene of interest.
5. Inclusion bodies are isolated as below and the expression of target proteins is followed by SDS-PAGE. Induction of expression of a fusion protein representing the sixth transmembrane domain is illustrated in **Fig. 4** (*see* **Note 7**).

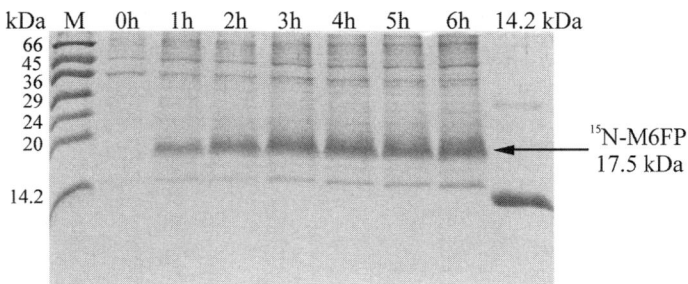

Fig. 4. Sodium dodecyl sulfate-polyacrylamide gel electrophoresis analysis of induction of the expression of ^{15}N-TrpΔLE-M6. Inclusion bodies were isolated after cells were induced with 1 mM isopropyl-β-D-thiogalactopyranoside for the indicated times. The gel was stained with Coomassie Blue.

3.2.4. Isolation of Inclusion Bodies

1. Expression cells are harvested by centrifugation at 6160 g for 20 min at 4°C.
2. The cell pellet is resuspended in lysis solution (see above) at a ratio of 5 mL per gram of wet cells.
3. Cell suspensions are sonicated for 5 to 10 min and centrifuged at 25400 g for 15 min at 4°C.
4. The resulting pellet is resuspended and sonicated in 3 mL of lysis buffer (see above).
5. The solution is centrifuged using the same conditions as above.
6. The washing step is repeated with 3 mL of inclusion bodies washing buffer and 3 mL of distilled water.
7. The final inclusion bodies preparation is resuspended by sonication in 3–4 mL of 6 M guanidinium buffer and stored at −20°C.

3.3. SDS-PAGE and Western Blot Procedures

The procedure used in preparing gels for SDS-PAGE and for running these gels follows instructions provided by Sigma *(19)*.

3.3.1. Reagents

1. A 48% acrylamide solution: dissolve 48 g of acrylamide and 1.5 g N, N'-methylene-bis-acrylamide in distilled water up to a final volume of 100 mL (*see* **Note 8**). The solution is stored in an amber bottle at 4°C.
2. The gel buffer: dissolve 36.34 g Trizma Base and 0.30 g of SDS in 60 mL of distilled water. The pH is adjusted to 8.45 with 1 N HCl and the final volume is adjusted to 100 mL with distilled water. This buffer is stored at 4°C.

3. A 20% SDS stock solution is prepared by dissolving 10 g of SDS in 50 mL water by gently warming until dissolution is complete. Store at 4°C.

4. 1 *M* Tris-HCl: dissolve 12.1 g of Trizma Base in 80 mL of distilled water, adjusting the pH to 6.8 with 1 *N* HCl, and adjusting the volume to 100 mL with distilled water. Store at 4°C.

5. Sample buffer: mix 4 mL of 20% SDS solution, 2.4 mL glycerol, 0.4 mL 2-mercaptoethanol, and 1.0 mL of 1 *M* Tris-HCl together and adjust the volume to 20 mL with distilled water. A few flakes of Coomassie Brilliant Blue G are added and the solution is stored at 4°C. Gentle heating might be needed before use.

6. 10X anode buffer: dissolve 121.1 g of Trizma Base in distilled water to a final volume of 500 mL. The pH is adjusted to 8.9 with 1 *N* HCl. Store at 4°C.

7. 10X cathode buffer: dissolve 121.1 g Trizma Base, 179.2 g Tricine, and 10 g SDS in distilled water to a final volume of 1 L. Store at 4°C.

8. 10% ammonium persulfate (APS) solution: dissolve 100 mg ammonium persulfate in 1 mL of distilled water. Store at 4°C.

9. Staining solution: dissolve 0.5 g Coomassie Brilliant Blue G in 500 mL of 40% methanol/10% acetic acid/50% distilled water.

10. Destaining solution: 40% methanol/10% acetic acid/50% distilled water. The following solutions are prepared following a protocol supplied by Pierce with their SuperSignal® West HistProbe™ Kit.

11. Transfer buffer: dissolve 6 g Trizma Base, 28.8 g glycine, and 2 g SDS in 400 mL of methanol. Distilled water is added to a final volume of 2 L and the pH adjusted to 8.3 with 1 *N* HCl.

12. Blocking buffer: dissolve 10 mg bovine serum albumin (BSA) in 1 mL TBST.

13. HisProbe*TM*-HRP solution: dissolve 2 mg HisProbe*TM*-HRP in 0.5 mL autoclaved water. Store as 2-μL aliquots at −20°C.

14. HisProbe*TM*-HRP working solution: dilute the HisProbe*TM*-horseradish peroxidase (HRP) solution 1:5000 in TBST. SuperSignal West Pico substrate working solution is prepared by mixing equal volumes of SuperSignal West Pico Luminol/Enhancer Solution and SuperSignal West Pico Stable Peroxide Solution.

3.3.2. Preparing Gels for SDS-PAGE

1. Set up the gel apparatus (*see* **Note 9**). Place two separators on the large glass plate and place the small glass plate on top of them. Place the glass sandwich in its holder. Check for leaks!

2. Combine the above solutions to make a 16% acrylamide gel consisting of three sections: A stacking gel, a spacer gel, and a separating gel. The solutions are combined as per **Table 1**.

3. After the solutions are prepared, the APS and TEMED are added to the separating gel solution. The resulting solution is mixed and 3.25 mL of the separating gel solution carefully pipetted between the glass sandwich plates.

Table 1
Solutions for Sodium Dodecyl Sulfate-Polyacrylamide Gel Electrophoresis

	Stacking gel	Spacer gel	Separating gel
Acrylamide	0.25 mL	0.76 mL	2.5 mL
Gel buffer	0.775 mL	1.25 mL	2.5 mL
Glycerol	–	–	0.8 mL
Water	2.1 mL	1.74 mL	1.7 mL
APS	25 μL	12.5 μL	25 μL
TEMED	3.5 μL	3 μL	2.5 μL

All the reagents are combined with the exception of the ammonium persulfate (APS) and N, N, N', N'-tetramethylethylenediamine (TEMED).

4. The APS and TEMED are then immediately added to the spacer gel solution and 700 μL is carefully loaded between the plates using a pipet.
5. Water is added to the remaining space between the plates until it reaches the top of the glass sandwich. After about 20 min, the above solutions polymerize to form a solid gel. At this time, the water is emptied from between the plates and a comb (which is supplied by the manufacturer) is inserted to provide distinct loading channels in the stacking gel.
6. The APS and TEMED are added to the stacking gel solution and the solution mixed by swirling.
7. The stacking gel solution is loaded between the plates until it overflows. After about 20 min, this solution polymerizes to a solid.
8. The comb is then removed and the wells thoroughly washed with distilled water.
9. Gels prepared as above can be used immediately or stored at 4°C for up to a week. To store the gel, fill the wells with distilled water and wrap it in a wet paper towel. Place the gel in a plastic bag, seal, and store at 4°C.

3.3.3. Running SDS-PAGE

1. Two gels are connected to a voltage source, creating a chamber between them.
2. 10X cathode buffer is diluted to 1X with distilled water and used to fill the chamber created by the two gels. Check for leaks!
3. The 10X anode buffer is diluted to 1X with distilled water and the main chamber holding the gels filled half way with this buffer.
4. The samples being analyzed are dissolved in sample buffer and sonicated.
5. The sample is boiled for 2–3 min and 10–20 μL loaded into the wells in the stacking gel.
6. 20 μL of the appropriate molecular weight markers are loaded into the gel.

7. The gels are placed into the main chamber and the apparatus connected to the voltage box.
8. Gels are electrophoresed at 20 mA for 1–2 h.
9. Once the sample dye reaches the level of the anode buffer the voltage can be increased to 40 mA (*see* **Note 10**).
10. The gels are run until the dye runs out of the gel (5–6 h).
11. The gels are stained with Coomassie blue or transferred to a nitrocellulose membrane for Western blotting.

3.3.4. Staining SDS-PAGE Gels

1. Once the sample dye runs out of the gels, the gels are removed from their holder and the wells cut off with a metal spatula.
2. The gels are placed in the staining solution overnight with gentle rocking at room temperature.
3. In the morning, the gels are removed from staining solution and rinsed with distilled water.
4. The gels are washed with destaining solution for about a half hour with gentle rocking at room temperature.
5. The destaining wash is repeated until the desired contrast between peptide or protein bands and the gel background is obtained.
6. Dry the gels.

3.3.5. Western Blotting

3.3.5.1. TRANSFER OF PEPTIDES/PROTEINS TO NITROCELLULOSE

1. Once the sample runs out of the gels, the gels are removed from their holder and the wells cut off with a metal spatula.
2. The gels are placed in the Western blot sandwich apparatus in the transfer buffer.
3. The gel is placed on the negative side of the sandwich apparatus and the nitrocellulose membrane placed over it (*see* **Note 11**).
4. The sandwich apparatus is closed and placed in its chamber along with a stir bar.
5. The chamber is filled with transfer buffer and connected to a voltage box.
6. The protein is transferred from the SDS-PAGE gel to the nitrocellulose membrane at a voltage of 40 V (approx 167 mA) for about 1 h. This should be done at 4°C with constant stirring.
7. After 1 h the voltage was increased to 220 mA (approx 46 V) for 1 h. This should also be done at 4°C with constant stirring.
8. The nitrocellulose membrane is removed from the apparatus and blocked overnight with 10 mL of blocking buffer at 4°C and gentle shaking.

3.3.5.2. PROBING

The following steps are all performed at room temperature with gentle shaking.

1. In the morning the nitrocellulose membrane is washed twice with 15 mL TBST for 10 min each time.
2. The membrane is then incubated in a 10 mL solution of the HisProbe-HRP Working Solution for 1 h.
3. After 1 h, the nitrocellulose membrane is washed four times with 15 mL TBST for 10 min each time.
4. The nitrocellulose membrane is then incubated in 10 mL of SuperSignal West Pico Substrate Working Solution for 5 min in a dark room.
5. Excess liquid is removed from the membrane with a tissue and wrapped in plastic wrap.
6. The wrapped membrane is placed over a piece of film in a film cassette.
7. The film is exposed anywhere from 10 s to 5 min. Optimal exposure time must be determined.
8. Develop the film.

3.4. Purification of Fusion Protein

The fusion constructs generated by the above procedures contain the TrpΔLE protein at the N-terminus and the target peptide at the C-terminus. A histidine tag precedes the TrpΔLE leader protein. In initial studies, attempts were made to isolate the fusion protein using immobilized metal affinity chromatography (IMAC). Although this procedure worked, the final protein was only 80% to 90% homogeneous as judged by HPLC. Therefore, we needed further purification using semi-preparative HPLC. Subsequently we found that the guanidinium hydrochloride suspensions of the fusion proteins could be directly injected onto an HPLC column and highly homogeneous (>95%) fusion protein isolated in good yield. Currently, we have eliminated the IMAC step and only use HPLC to isolate the fusion proteins.

1. The guanidinium hydrochloride suspension of the inclusion bodies (2–3 mL) is injected directly onto an HPLC column.
2. HPLC purification is carried out using a Hewlett Packard-Agilent instrument equipped with a gradient-based solvent system.
3. Chromatograms were monitored at $\lambda = 220$ nm.
4. Fusion protein purification is carried out using a Vydac 259VHP82215 preparative reversed-phase polymer column (22 mm × 150 mm; 8 μm; 300 Å) with a water jacket

at 50°C and a water-acetonitrile (0.1% TFA) gradient from 30% to 60% acetonitrile in 80 min at a flow rate of 4 mL/min (*see* **Note 12**).

5. Fractions are analyzed using a Vydac 259VHP54 reversed-phase polymer column (4.6 mm × 150 mm, 5 µm, 300 Å) at 50°C and a water-acetonitrile (0.1% TFA) gradient from 30% to 60% acetonitrile in 20 min at a flow rate of 1 mL/min.

6. Final fusion proteins are over 95% homogeneous as judged by reversed-phase (RP)-HPLC and SDS-PAGE.

7. Yields of fusion protein vary with the exact construct and average 10 to 100 mg/L of culture in rich medium.

8. Expression in minimal media used for ^{15}N- or $^{13}C/^{15}N$-labeled peptides lowers the yields in some cases and has little affect in others (*see* **Note 13**).

3.5. Cyanogen Bromide Release of Target Membrane Peptide

1. The lyophilized pure fusion protein is cleaved with 1 M CNBr.
2. Pure fusion proteins are dissolved in 50% or 70% TFA.
3. 2–3 mg of fusion protein is used per mL of cleavage mixture.
4. CNBr is dissolved separately in the TFA medium.
5. A 500- to 5000-fold molar excess of CNBr is used.
6. The final concentration of CNBr is 1 M.
7. The CNBr solution is prepared by adding the required volume of 70% TFA to CNBr crystals and vortexing until dissolution is complete.
8. The CNBr solution is then added to the peptide solution.
9. The cleavage is carried out in the dark at room temperature.
10. Immediately upon addition of the CNBr to the peptide, 5 µL of the reaction solution is injected into the HPLC to follow the cleavage.
11. Cleavage of a single transmembrane fusion peptide requires 24 h whereas optimal cleavage of a 73-residue multi-domain peptide occurs after 4.5 h (*see* **Note 14**).
12. To stop the cleavage reaction, the reaction mixture is frozen in dry ice and lyophilized.

3.6. HPLC Purification of Target Membrane Peptide

The HPLC purification of membrane peptides is carried out using instrumentation and columns that are widely used in the purification of peptides. The major challenge in the purification of membrane peptides is dissolving the peptide prior to purification and preventing both aggregation and precipitation in the HPLC. In the case of certain transmembrane peptides for Ste2p, it was virtually impossible to dissolve the peptide unless Lys residues were added at both the amino and carboxyl termini. Crude peptides from either the CNBr cleavage or the solid-phase synthesis were dissolved in strong solvents including mixtures of $CH_3CN:H_2O:CF_3COOH$ (30:40:30) or neat dimethyl sulfoxide (DMSO). Best results were achieved when high concentrations of peptide were obtained (approx 10 mg/mL), and therefore low volumes of these peptide solutions were injected. In many cases, only 2 mg of crude peptide could be injected at a time. In cases in which the peptide was more soluble, as much as 10 or 15 mg could be injected. It was observed that the best results were achieved when separations were carried out at 50°C to 65°C. In many cases, the quality of the crude peptide appears quite poor (**Fig. 5**) as a result of the tendency of hydrophobic peptides to aggregate during synthesis, leading to

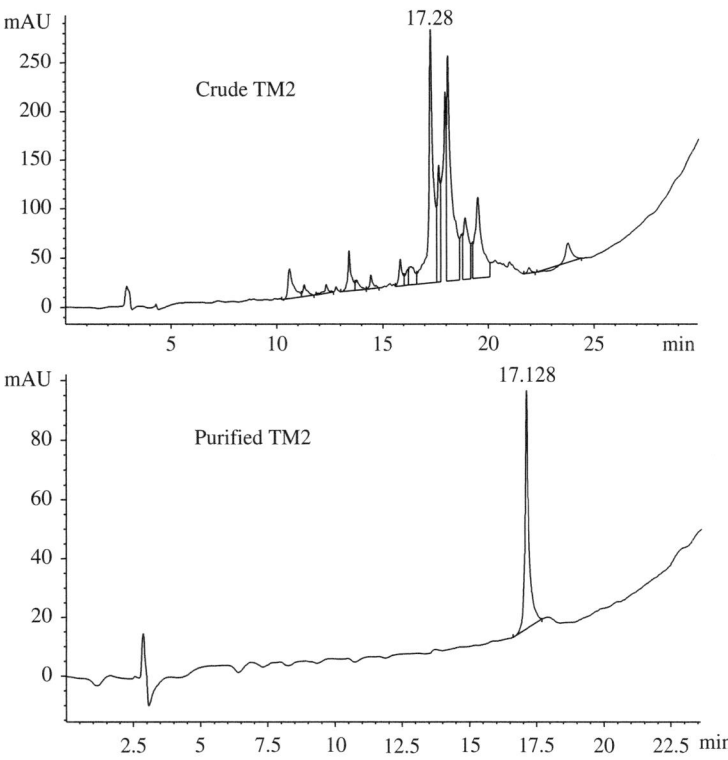

Fig. 5. High-performance liquid chromatography spectra of synthetic transmembrane peptides. Data from M2-35 is presented as a representative example. (*See* **Fig. 2** for structure.) **A**, crude peptide after cleavage from resin; **B**, purified peptide. Vydac 259VHP54 polymer column (4.6 × 250 mm) at 80°C. Gradient: 90% A-100% B in 30 min, where A was water (0.1% TFA), B was acetonitrile (0.1% trifluoroacetic acid), detection at λ = 220 nm. Taken from **ref. 7**.

incomplete coupling and to aggregation on the HPLC column during analysis and purification. Nevertheless, high quality final product can be obtained.

1. HPLC purification is carried out using a Hewlett Packard-Agilent instrument equipped with a gradient-based solvent system.
2. Chromatograms are monitored at $\lambda = 220$ nm.
3. Crude synthetic peptides or CNBr cleaved peptides are purified using a Waters μBondpak™ preparative C18 reversed-phase column (19 mm × 300 mm, 10 μm, 125 Å) with a water jacket at 50°C and a water-acetonitrile (0.1% TFA) gradient from 40% to 80% acetonitrile in 80 min at a flow rate of 4 mL/min (*see* **Note 15**).
4. Two to ten mg of the peptide mixture is dissolved in 200 μL to 1 mL of an $CH_3CN:H_2O:CF_3COOH$ mixture. Attempts are made to minimize the amount of organic solvent and to try to maintain its level below the starting acetonitrile concentration of the gradient.
5. If the solubility in the $CH_3CN:H_2O:CF_3COOH$ is not sufficient to obtain high concentrations, 2,2,2-trifluoroethanol (TFE) is added as a co-solvent or neat DMSO is used.
6. The peptide solution is filtered through a Gelman Acrodisc 13CR PTFE 0.45 μm filter prior to injection (*see* **Note 16**).
7. Fractions from the preparative column are analyzed using a Waters Delta Pak C18 reversed-phase column (3.9 mm × 150 mm, 5 μm, 100 Å) at 50°C and water-acetonitrile (0.1% TFA) gradient from 30% to 60% acetonitrile in 20 min at a flow rate of 1 mL/min.
8. All peptides are purified to over 95% homogeneity as judged by RP-HPLC.
9. The ultimate recovery of the peptides is usually on the order of 10% to 20%.
10. The final products are assessed by electrospray ionization (ESI)-mass spectrometry (MS).
11. In order to retain column efficiency, it is useful to integrate a daily "regeneration" by ending the day with an injection of DMSO (approx 2–5 mL) to wash insoluble material and aggregated peptide from the column.

3.7. Circular Dichroism Spectroscopy of Membrane Peptides

Circular dichroism (CD) spectroscopy is a technique that is highly sensitive to the conformational state of peptide chains, exhibiting distinct spectral patterns for α-helices, β-sheets, turns, and disordered peptides *(20,21)*. The phenomenon is based on the different extinction coefficients that optically active molecules exhibit for left-handed and right-handed circularly polarized light. Extensive theoretical studies have resulted in an understanding of the fundamental electronic and electromagnetic interactions that are involved in the phenomenon. For the purpose of investigating membrane peptides, key issues involve sample preparation because for the most part membrane peptides corresponding to regions of GPCRs are either helical or disordered. There is little evidence of significant contributions from β-sheet structures to the secondary structure of these proteins. Nevertheless, the author's laboratory and others have found that in certain solvents, membrane peptides tend to aggregate and exhibit CD patterns characteristic of sheets. One must be concerned whether this is artifactual or reflects the biophysical tendencies of a specific domain of a protein.

3.7.1. Characteristic CD Spectra of Polypeptides

Figure 6 presents characteristic CD spectra for helical, sheet, and disordered polypeptide chains. In the case of the α-helix a minimum is noted at $\lambda = 222$ nm another minimum at $\lambda = 208$ nm and a maximum at about $\lambda = 195$ nm. The minimum at $\lambda = 222$ nm is assigned to the $n \rightarrow \pi^*$ transition of the amide chromophore. This transition is normally forbidden on quantum mechanical grounds; however, in the dissymmetric environment of the helix it becomes allowed. The $\pi \rightarrow \pi^*$ transition is split by exciton interactions into the $\lambda = 208$ nm minimum and the $\lambda = 195$ nm maximum. CD spectra of α-helices are similar both qualitatively and quantitatively irrespective of the amino acid composition or sequence. In rare cases, peptides rich in aromatic residues show unusual CD curves. However, this would not be expected to significantly affect the spectra of membrane peptides as aromatic residues are usually excluded from transmembrane domains. The helicity of a membrane peptide is highly dependent on the solvent used for the investigation and the sequence of the peptide. For amphiphilic peptides, the structure is often random in aqueous buffer and becomes helical in membrane mimetic solvents such as TFE or in detergent micelles. For peptides corresponding to transmembrane regions of GPCRs, the water solubility is so low that CD spectra cannot be measured. In membrane mimetic solvents, helical CD patterns are often observed.

The fraction helix (f_h) can be estimated using equations developed by Fasman and coworkers (*20*) and later modified by Chen and coworkers

$$f_h = ([\theta] - [\theta]^0)/([\theta]^{100} - [\theta]^0) \tag{1}$$

where $[\theta]$ is the experimentally observed mean residue ellipticity, $[\theta]^0$ and $[\theta]^{100}$ correspond to 0 and 100% helical content at either $\lambda = 208$ nm or $\lambda = 222$ nm. Normally we use $\lambda = 222$ nm values of 2,000 and 30,000 deg•cm²/dmol for $[\theta]^0$ and $[\theta]^{100}$, respectively (*22,23*) (*see* **Note 17**).

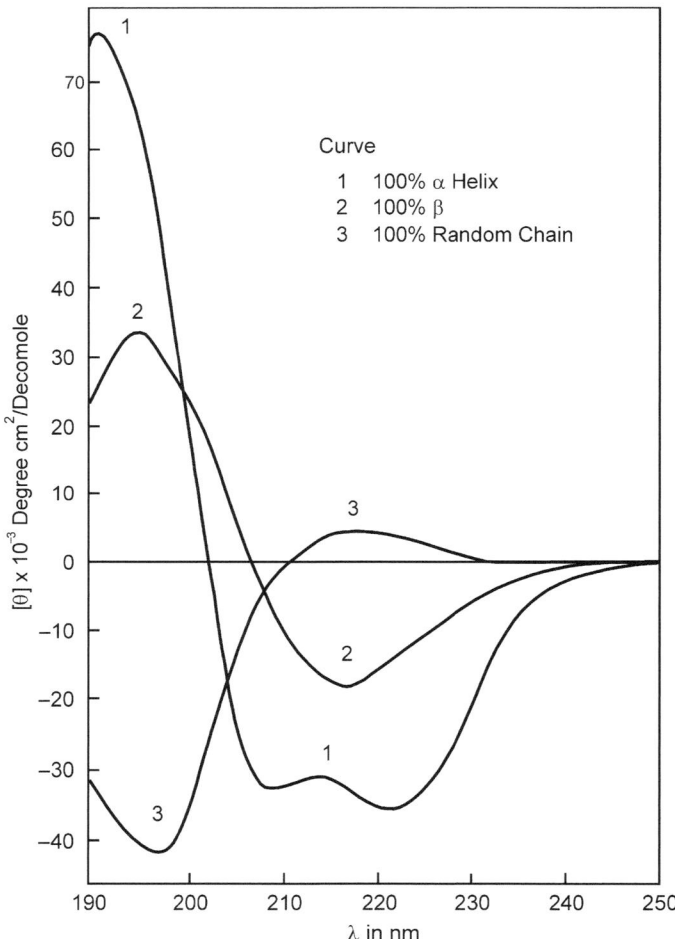

Fig. 6. Circular dichroism (CD) spectra of characteristic secondary structures found in proteins. The figure illustrates the CD patterns for α-helical, β-sheet, and disordered structures based on poly-L-Lys. Taken from **ref. 20**.

3.7.2. Measuring CD Spectra of Membrane Peptides

A major concern in the quantitation of helicity from CD studies on membrane peptides is determining the peptide concentration accurately. Our approach is to prepare stock solutions in high concentrations of TFE and determine the exact concentration of these stock solutions using knowledge of the extinction coefficient. These stock solutions are then used to prepare all other solutions used in CD measurements.

1. Weigh membrane peptide sample into a clean vial using a microbalance with a precision of 0.01 mg.
2. Using a micropipetter, add the desired volume of TFE.
3. Measure the ultraviolet (UV) spectrum.
4. Pipette a known volume of solution into a tube and conduct quantitative amino acid analysis.
5. Calculate the extinction coefficient (ε) for the peptide based on UV and amino acid analysis results. Beers law ($A = \varepsilon CL$) is used in this calculation, where A is the measured absorbance at a given wavelength, C is the concentration calculated from the amino acid analysis, and L is the path length. In some cases, the ε at $\lambda = 280$ nm is calculated using an extinction coefficient of 5350 $M^{-1}cm^{-1}$ for a Trp residue and 1340 $M^{-1}cm^{-1}$ for a Tyr residue.
6. The TFE stock solution is used to prepare sample solutions at (a) different TFE/H_2O ratios or for measurements in (b) detergent or (c) vesicles.
7. For measurements in TFE/H_2O mixtures, the required volume of stock solution, TFE, and water is added to a vial.
8. For measurements in the presence of detergent micelles, the peptide stock solution is added to a vial and lyophilized overnight. Then, 200 μL of 100 mM detergent 10 mM phosphate buffer is added to the lyophilized peptide.
9. The peptide/detergent mixture is sonicated at 50°C for 15 min.
10. Final peptide concentrations of 50 to 400 μM with detergent/peptide ratios from 20/1 to 400/1 are prepared in this manner.
11. For measurements in the presence of synthetic lipid vesicles, the TFE/H_2O stock solutions of the membrane are added to 2 mg of DMPC/DMPG (4:1) in 1 mL of $CHCl_3$.
12. The resulting solution is dried under N_2 flow. Residual traces of organic solvent are removed by placing the dried film under vacuum overnight. Then the peptide/lipid mixture is resuspended in 500 μL of 10 mM phosphate buffer, pH 6.4. The suspension is sonicated at 50°C for 60 min. The final peptide concentration is 50 μM and the lipid to peptide ratio is 1:117.
13. The CD spectra of the peptides are recorded on an AVIV model 62-DS CD instrument (AVIV Associates, Lakewood, NJ).
14. Quartz cuvets with pathlengths of 1 mm and 0.2 mm are used for peptides in TFE solution and DMPC/DMPG (4:1) vesicles with concentration of 50 μM. Cuvets with light path lengths of 0.2 and 0.1 mm are used for peptides in PPG and DPC detergents (4 mg/mL) with concentration of 50 μM to 400 μM.
15. All spectra are the average of three to five scans between $\lambda = 260$–280 nm and $\lambda = 185$ nm at an interval of 1 nm with a 3- to 5-s integration time at each wavelength. The bandwidth for each measurement is set to 2 nm. Blanks are collected on all solvent media and subtracted from the spectra containing the protein.
16. CD intensities are expressed as mean residue ellipticities (deg cm^2 dmol^{-1}).

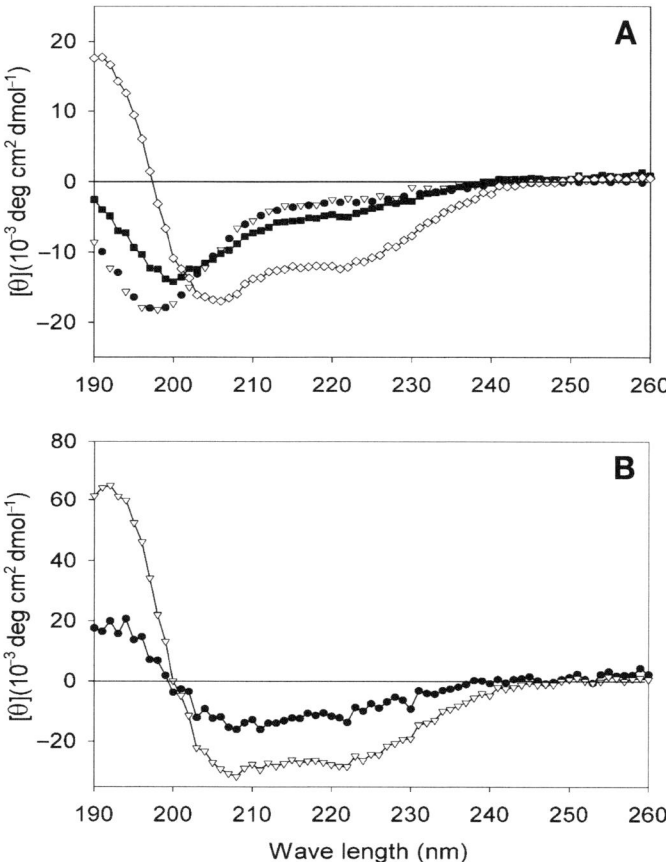

Fig. 7. Circular dichroism spectra as mean residue ellipticity vs wavelength for T40, M7-12-T40, and M7-24-T40 in DMPC/DMPG (4:1) vesicles suspended in phosphate buffer (0.1 m*M*, pH 6.3). (**A**) T40 in vesicles (closed circles) and phosphate buffer (inverted open triangles); M7-12-T40 in vesicles (closed squares); M7-24-T40 in vesicles (open diamonds). (**B**) Calculated difference spectra for M7-12 (closed circles) and M7-24 (inverted open triangles) based on the spectra of M7-12-T40, M7-24-T40 and T40 (*see* **Subheading 3.**, eq. 2). Taken from **ref. *11***.

17. For multidomain peptides containing portions of the cytosolic tail (T40), it is assumed that the transmembrane and tail portions behave independently and the mean residue ellipticity of E3-M7-24 is estimated using the following equation:

$$[\theta]_{\text{E3-M7-24}} = \{([\theta]_{\text{E3-M7-24-T40}} \times 73) - ([\theta]_{\text{T40}} \times 40)\}/33 \qquad (2)$$

18. Mathematical subtraction of the CD spectra of T40 from the E3-M7-24-T40 peptide spectra is carried out using the calculated molar ellipticity values of each spectrum.
19. Finally, the mean residue ellipticity ($[\theta]$) of E3-M7-24 is obtained by dividing the molar ellipticity by the appropriate number of residues.
20. A typical CD spectrum for a synthetic 64-residue peptide containing a transmembrane domain and 40 residues of the cytosolic tail of Ste2p is shown in **Fig. 7**.

Acknowledgments

Many talented and dedicated students have participated in the development of the methods outlined in this chapter. They are: Enrique Arevalo, Boris Arshava, Michael Breslav, Patricia Cano-Sanchez, Fa-Xiang Ding, Gary Eng, Jacqueline Englander, Racha Estephan, Sanjay Khare, Shufing Liu, Jennifer Madeo, Abdulla Reddy, Joseph Russo, David Schreiber, Beatrice Severino, V.V. Sureshbabu, and Haibo Xie. I am indebted to all of them. I also owe much to my lifelong friend and collaborator Professor Jeffrey M. Becker who introduced me to *Saccharomyces cerevisiae* and who has stimulated my laboratory to conduct analyses of G protein–coupled receptors. I also wish to thank Racha Estephan, Jacqueline Englander, Leah Cohen, Boris Arshava, and Jeff Becker for carefully reading this chapter and for their helpful suggestions. All of these studies have been supported from by a grant from the National Institute of General Medical Sciences (GM 22086).

4. Notes

1. At present, all of the reagents necessary to detect a His tag in a protein are available from Pierce as the Super Signal West HisProbe Kit.
2. In some cases, syntheses may be carried out on a 0.05-mmol scale. This results in a significant savings in the amount of amino acid and solvents that are used in the synthesis. In the case of the 433A synthesizer, this requires adjustment of the volumes and concentrations of all reagents. If the standard volumes are used, the concentration of amino acids and coupling reagents will be diluter than normal, leading to lower coupling efficiencies.
3. Some sequences proved to be highly resistant to complete coupling. This was particularly a problem with stretches of β-branched amino acid residues such as Ile, Val, and Thr. In these cases, other solvents, such as DMSO, or stronger coupling reagents, such as HATU and HOAt, were sometimes used in the coupling reaction.
4. The author's laboratory currently uses a prepared mix from Sigma to make LB medium.
5. Glycerol stocks may also be prepared in LB medium.

6. The protocols for PCR, restriction enzyme digestion, and ligation are supplied by the provider of these reagents.

7. The levels of expression vary significantly depending on the domain expressed and the vector used for expression. For the Ste2p domains, better expression is observed for a 73-residue multidomain peptide *(24)* than for 33-residue peptides corresponding to the sixth transmembrane domain. The expression of the 73-residue peptide is also higher than that of a double transmembrane domain fusion peptide containing the fifth transmembrane domain, the third intracellular loop, and the sixth transmembrane domain. Better expression was also found using TrpΔLE rather than thioredoxin fusions *(24)*.

8. Gentle heating is often needed before use, as this solution solidifies on storing at 4°C.

9. The author uses an electrophoresis device purchased from BioRAD. This device runs a minimum of two gels simultaneously, and the gel holders create a space that the author's laboratory refers to as the cathode chamber between the two gels. The cathode buffer is poured into this chamber. The gel holder is then placed into an electrophoresis chamber that is hooked up to the power supply. The buffer in this chamber is the anode buffer.

10. The current specified is for two gels. When four gels are run simultaneously, the current is doubled.

11. Make sure there are no air bubbles by rolling a pipet over the nitrocellulose. Do not put pressure on the membrane! This should all be done while in the transfer buffer.

12. The gradient used in the separation of the fusion protein varies with the target domain that was being synthesized. In the case of very hydrophobic domains, e.g., fusion proteins containing two transmembrane domains, better results are obtained with isopropanol/butanol/water gradients than with acetonitrile-water gradients. The exact organic component used in the eluent must be empirically optimized depending on the transmembrane domain being targeted.

13. The level of expression is also dependent on the fusion protein being expressed in the medium. Higher expression is obtained in rich as compared to minimal medium for the 73-residue peptide. Little difference in expression levels in rich and minimal medium is observed for M6 fusion proteins. Other laboratories have found drops in expression yields in minimal medium and have improved expression yields by including 5% rich medium in the minimal medium *(25)*.

14. The percent acid used affects the rate of cleavage, but not the final purity of a 73-residue multi-domain Ste2p peptide that the author prepared. Cleavage is slightly slower in 50% TFA than in 70% TFA and it is easier to control the reaction.

15. Transmembrane domain peptides often aggregate badly on reversed-phase columns and the pattern obtained for the HPLC chromatogram sometimes reflects this aggregation. Typically, attempts are made to recover the major peak. However, detailed studies show that in some cases, other peaks exhibited molecular weights expected

for the product. When beginning a study on a new system, if multiple peaks are observed, it is worthwhile to change the eluent composition and vary the temperature. In some cases, superior results are obtained with mobile phases containing organic alcohols (methanol, isopropanol, *n*-butanol) instead of acetonitrile. Column jackets for both analytical and preparative HPLC columns are available.

16. Hydrophobic peptides can be lost during the filtration and controls must be run to test for this possibility. Filters should be washed with strong solvents and scrutinized for the presence of the peptide.

17. There are many algorithms that have been developed to deconvolute CD spectra into contributing secondary structures. Most of these were developed for globular proteins and the best of these reflect contributions from helices, sheets, turns, and "random" regions of the protein. For transmembrane peptides in bilayers, it is highly likely that the residues are either helical or disordered. Therefore, the author favors the simple approach of eq. 1.

References

1. Fredriksson, R., Lagerstrom, M. C., Lundin, L. G., and Schioth, H. B. (2003) The G-protein-coupled receptors in the human genome form five main families. Phylogenetic analysis, paralogon groups, and fingerprints. *Mol. Pharmacol.* **63**, 1256–1272.

2. Drews, J. (2000) Drug discovery: a historical perspective. *Science* **287**, 1960–1964.

3. Palczewski, K., Kumasaka, T., Hori, T., et al. (2000) Crystal structure of rhodopsin: A G protein-coupled receptor. *Science* **289**, 739–745.

4. Dohlman, H. G. and Thorner J. W. (2001) Regulation of G protein-initiated signal transduction in yeast: paradigms and principles. *Annu. Rev. Biochem.* **70**, 703–754.

5. Reddy, A. P., Tallon, M. A., Becker, J. M., and Naider, F. (1994) Biophysical studies on fragments of the α-factor receptor protein. *Biopolymers* **34**, 679–689.

6. Arshava, B., Liu, S. F., Jiang, H., Breslav, M., Becker, J. M., and Naider, F. (1998) Structure of segments of a G protein-coupled receptor: CD and NMR analysis of the *Saccharomyces cerevisiae* tridecapeptide pheromone receptor. *Biopolymers* **46**, 343–357.

7. Xie, H., Ding, F. X., Schreiber, D., et al. (2000) Synthesis and biophysical analysis of transmembrane domains of a *Saccharomyces cerevisiae* G protein-coupled receptor. *Biochemistry* **39**, 15,462–15,474.

8. Naider, F., Arshava, B., Ding, F. X., Arevalo, E., and Becker, J. M. (2001) Peptide fragments as models to study the structure of a G-protein coupled receptor: the α-factor receptor of *Saccharomyces cerevisiae*. *Biopolymers* **60**, 334–350.

9. Ding, F., Xie, H., Arshava, B., Becker, J. M., and Naider, F. (2001) ATR-FTIR study of the structure and orientation of transmembrane domains of the *Saccharomyces cerevisiae* α-mating factor receptor in phospholipids. *Biochemistry* **40**, 8945–8954.

10. Arshava, B., Taran, I., Xie, H., Becker, J. M., and Naider, F. (2002) High resolution NMR analysis of the seven transmembrane domains of a heptahelical receptor in organic-aqueous medium. *Biopolymers* **64**, 161–176.
11. Naider, F., Ding, F. X., VerBerkmoes, N. C., Arshava, B., and Becker, J. M. (2003) Synthesis and biophysical characterization of a multi-domain peptide from a *Saccharomyces cerevisiae* G protein-coupled receptor. *J. Biol. Chem.* **278**, 52, 537–52,545.
12. Arevalo, E., Estephan, R., Madeo, J., et al. (2003) Biosynthesis and biophysical analysis of domains of a yeast G protein-coupled receptor. *Biopolymers* **71**, 516–531.
13. Naider, F., Estephan, R., Englander, J., et al. (2004) Sexual conjugation in yeast: A paradigm to study G-protein-coupled receptor domain structure. *Biopolymers* **76**, 119–128.
14. Cartwright, C. P. and Tipper D. J. (1991) *In vivo* topological analysis of Ste2, a yeast plasma membrane protein, by using beta-lactamase gene fusions. *Mol. Cell. Biol.* **11**, 2620–2628.
15. Lin J. C., Parrish W., Eilers M., Smith S. O., and Konopka J. B. (2003) Aromatic residues at the extracellular ends of transmembrane domains 5 and 6 promote ligand activation of the G protein-coupled alpha-factor receptor. *Biochemistry* **42**, 293–301.
16. Martin, N. P., Leavitt, L. M., Sommers, C. M., and Dumont, M. E. (1999) Assembly of G protein-coupled receptors from fragments: identification of functional receptors with discontinuities in each of the loops connecting transmembrane segments. *Biochemistry* **38**, 682–695.
17. Staley, J. and Kim, P. (1994) Formation of a native-like subdomain in a partially folded intermediate of bovine pancreatic trypsin inhibitor. *Protein Science* **3**, 1822–1832.
18. Miozzari, G. and Yanofsky, C. (1978) Translation of the leader region of the *Escherichia coli* tryptophan operon. *J. Bacteriol.* **133**, 1457–1466.
19. Schagger, H. and von Jagow, G. (1987) Tricine-sodium dodecyl sulfate-polyacrylamide gel electrophoresis for the separation of proteins in the range from 1 to 100 kDa. *Anal. Biochem.* **166**, 368–379.
20. Greenfield N. and Fasman G. D. (1969) Computed circular dichroism spectra for the evaluation of protein conformation. *Biochemistry* **8**, 4108–4116.
21. Fasman, G. D. (ed.) (1996) *Circular Dichroism and the Conformational Analysis of Biomolecules.* Plenum, New York.
22. Chen, Y. H., Yang, J. T., and Chau, K. H. (1974) Determination of the helix and beta form of proteins in aqueous solution by circular dichroism. *Biochemistry* **13**, 3350–3359.
23. Wu, C. S., Ikeda, K., and Yang, J. T. (1981) Ordered conformation of polypeptides and proteins in acidic dodecyl sulfate solution. *Biochemistry* **20**, 566–570.

24. Estephan, R., Englander, J., Arshava, B., Samples, K. L., Becker, J. M., and Naider, F. (2005) Biosynthesis and NMR analysis of a 73-residue domain of a Saccharomyces cerevisiae G protein-coupled receptor. Biochemistry **44**, 11,795–11,810.
25. Lindhout, D. A., Thiessen, A., Schieve, D., and Sykes, B. D. (2003) High-yield expression of isotopically labeled peptides for use in NMR studies. *Protein Science* **12**, 1786–1791.

II

APPLICATIONS

5

Application of Topologically Constrained Mini-Proteins as Ligands, Substrates, and Inhibitors

Janelle L. Lauer-Fields, Dmitriy Minond, Keith Brew, and Gregg B. Fields

Summary

Protein–protein interactions are governed by a variety of structural features. The sequence specificities of such interactions are usually easier to establish than the "topological specificities," whereby interactions may be classified based on recognition of distinct three-dimensional structural motifs. Approaches to explore topological specificities have been based primarily on assembly of mini-proteins with well defined secondary, tertiary, and/or quarternary structures. The present chapter focuses on three approaches for constructing topologically well defined mini-proteins: template-assembled synthetic proteins (TASPs), disulfide-stabilized structures, and peptide-amphiphiles (PAs). Specific examples are given for applying each approach to explore topologically-dependent protein–protein interactions. TASPs are utilized to identify a metastatic melanoma receptor that binds to the $\alpha 1(IV)1263–1277$ region of basement membrane (type IV) collagen. A disulfide-stabilized structure incorporating a sarafotoxin (SRT) 6b model was examined as a matrix metalloproteinase (MMP)-3 inhibitor. PAs were developed as (a) fluorogenic triple-helical or polyPro II substrates for MMPs and aggrecanase members of the a disintegrin and metalloproteinase with thrombospondin motifs (ADAMTS) family and (b) glycosylated and nonglycosylated ligands for metastatic melanoma cells. Topologically constrained mini-proteins have proved to be quite versatile, helping to define critical primary, secondary, and tertiary structural elements that modulate enzyme and receptor functions.

Key Words: Triple-helix; peptide-amphiphile; matrix metalloproteinase; collagen; integrin.

1. Topology and Templates

Protein–protein interactions are governed by a variety of structural features. The sequence specificities of such interactions are usually easier to establish than the "topological specificities," whereby interactions may be classified

From: *Methods in Molecular Biology, vol. 386: Peptide Characterization and Application Protocols*
Edited by: G. Fields © Humana Press Inc., Totowa, NJ

based on recognition of distinct three-dimensional structural motifs. Approaches to explore topological specificities have been based primarily on assembly of mini-proteins with well defined secondary, tertiary, and/or quarternary structures. The *de novo* design of mini-proteins with distinct topologies has been pursued for several decades (*1–3* and references cited therein). The examination of biological functionality of such constructs has a more recent history. The present chapter focuses on three approaches for constructing topologically well defined mini-proteins: template-assembled synthetic proteins (TASPs), disulfide-stabilized structures, and peptide-amphiphiles (PAs). The application of such mini-proteins for evaluating topological specificities of receptors and proteases is also described.

1.1. Template-Assembled Synthetic Proteins

The use of a template-assisted approach to create topologically defined small proteins minimizes the entropic barriers to making multistranded species. A large variety of domains have been created using this concept, including 2, 3, and 4 α-helix bundles, ββα zinc fingers, and triple-helices (*4–14*). In recent years, much progress has been made in template-assisted synthetic methodologies, including orthogonal protection schemes, chemoselective ligation, and the design of novel and increasingly flexible templates. It is now possible to design peptides to form specialized structures such as ion channels (*14*) or receptor binding sites (*9,15–18*) while controlling their assembly into clearly defined higher-ordered structures containing two or more strands. Many templates have been utilized to create these structures. The main categories of template strategies are those that involve a scaffold to which the peptide strands are covalently coupled or the use of a metal ion as a noncovalent scaffold on which to build the molecule. The array of covalent templates includes (1) multi-Lys branching schemes (*5,9,10,14,19–24*), (2) cis,cis-1,3,5-trimethylcyclohexane-1,3,5-tricarboxylic acid (Kemp triacid [KTA]) (*25–31*), (3) tris(2-aminoethyl)amine (TREN) (*32*), (4) *N*-(benzyloxycarbonyl)-tris(carboxyethoxymethyl)aminomethane [Z-TRIS(33)$_3$] (*34*), (5) cyclotriveratrylene (CTV) (*35*), or (6) carbohydrates (*36,37*). Alternatively, the metal ion templates have used Ca^{++}, Ru(II), Ni(II), Fe(II), Cd(II), or Hg(II) in conjunction with amino acids or N-terminal pyridyl or bipyridyl functionalities to create multistrand linkages (*4,6,7,13,38–40*). The ultimate benefit to each of these templates is the incredible flexibility in designing the overall structure. Not only can the valency of the protein be modulated, but also the orientation of each peptide strand can be exquisitely controlled.

One can create parallel or antiparallel structures of homo- or heteromultimeric sequences. Additionally, the desired secondary structure of each peptide strand can be enhanced by bringing the strands together on the template *(4–8,11,13)*. For example, the presence of γ-carboxyglutamate in positions i, $i+4$, $i+7$, and $i+11$ in the neuroactive conopeptide conantokin-G promotes the formation of 2 α-helices arranged in an antiparallel fashion in the presence of Ca^{2+}, whereas the individual chains show minimal α-helical character in the absence of Ca^{2+} *(13)*. This arrangement appears to be very specific, as the secondary and tertiary structures are not formed in the presence of Mg^{2+}, Zn^{2+}, or Mn^{2+}, nor are they formed in the absence of all divalent cations. Interestingly, this association occurs and is stable under extremely mild conditions, which contrasts with the harsh conditions utilized for some of the metal complexes *(6,7,41)*.

Solid-phase assembly of triple-helical collagen-model peptides using a Lys-Lys *C*-terminal branch requires three different protecting group strategies (**Fig. 1**): N^{α}-amino protection (**A**), Lys side-chain protection (**B**), which must be stable to the N^{α}-amino group removal conditions, and C^{α}-carboxyl protection (**linker**), which must be stable to the N^{α}-amino and Lys side-chain protecting group removal conditions *(10,42)*. Branching is achieved by synthesizing **A**-[Lys(**B**)]$_2$-Tyr(**C**)-Gly-**linker** resin and deprotecting the N^{α}- and N^{ε}-amino groups. Tyr is incorporated prior to branching to provide a convenient chromophore for eventual concentration determination. Incorporation of **A**-Ahx (where Ahx is 6-aminohexanoic acid) onto all three *N*-termini provided a flexible spacer, as demonstrated by Roth and Heidemann *(19)*. All THP solid-phase methods are based on 9-fluorenylmethoxycarbonyl (Fmoc) N^{α}-amino group protection. For example, the type IV collagen derived α1 (IV)1263–1277 branch was assembled by a three-dimensionally orthogonal strategy, where **B** was *tert*-butyloxycarbonyl (Boc) and **linker** was the allyl-based 4-trityloxy-Z-but-2-enyloxyacetic acid [cleaved by (Ph$_3$P)$_4$Pd catalyzed nucleophilic transfer] *(9)*. A (Gly-Pro-Hyp)$_8$ branch has also been assembled by a three-dimensionally orthogonal strategy, where **B** was allyloxycarbonyl (Aloc) [cleaved by (Ph$_3$P)$_4$Pd catalyzed nucleophilic transfer] and **linker** was 4-hydroxymethylphenoxy (HMP; cleaved by trifluoroacetic acid) *(10)*. A third three-dimensionally orthogonal branch strategy was developed, where **B** was 1-(4,4-dimethyl-2,6-dioxocyclohex-1-ylidene)-ethyl (Dde) (cleaved by hydrazine/DMF, 1:49) and **linker** was HMP *(10)*. The Fmoc/Dde combination is now the most preferred, as exemplified by the synthesis of a topological model of collagen described below.

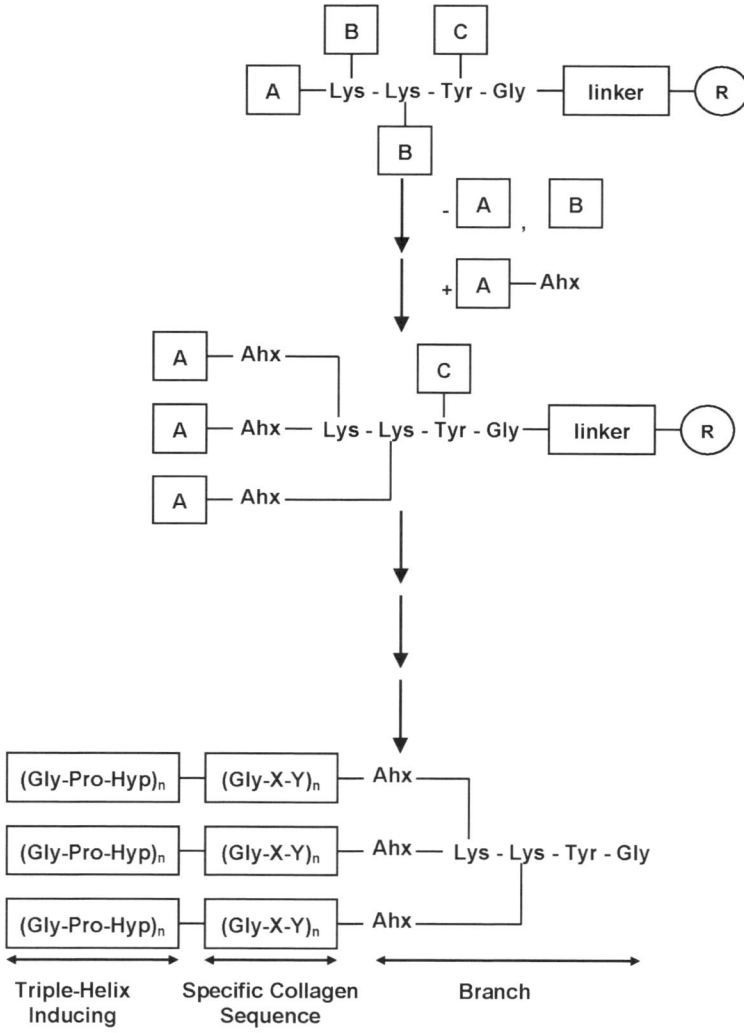

Fig. 1. General scheme for synthesis of branched, triple-helical peptides. Ahx is 6-aminohexanoic acid.

1.2. Disulfide-Stabilized Structures

One of the limitations in creating peptides with defined topological designs is the inherent thermodynamic instability of smaller molecules. This problem can be overcome by the use of disulfide bonds, which serve to shift the equilibrium from unfolded to folded species. Traditional *de novo* peptide design has focused on finding peptide sequences that have a threshold propensity to form defined structures. Alternatively, scientists have utilized naturally occurring biologically active structures as the starting point for synthetic modification. The variety of structurally stable peptide or protein scaffolds that are further stabilized by disulfide bonds is diverse in nature. Many of these scaffolds have been utilized as the basis for molecular design. Two basic categories of cystine-stabilized structures are created by one to two disulfides aiding the stability of simple structures such as α-helical hairpins or β-hairpin structures *(43,44)*. In both cases, a protein with the desired topological design was utilized as a scaffold. As expected, the desired secondary and tertiary structures are not stable on their own, but the addition of the properly placed cystine bridges allowed the desired α-helical or β-hairpin structures to be produced by 10–40 amino acids. This can be extended to more complicated structures such as the ββα motif, α-helical bundles, mixed poly Pro/α-helices, or triple-stranded β-sheets *(45)*.

Slightly more complex structures include the cystine-stabilized α-helical motif (CSH) and the cystine-stabilized β-sheet motif (CSB) *(46)*. The CSH motif is found in the endothelin family, the spider toxin apamin, insect defensins and toxins, plant γ-thionins, and neurotoxins *(46)*. One interesting example of a CSH motif is an 18-residue peptide based on the apamin scaffold. The substitution of five amino acids from the native apamin sequence produced an active enzyme capable of decarboxylating oxaloacetate at rates comparable to the best synthetic decarboxylates *(45)*.

Another common class of cystine-stabilized structures is the cystine knot (reviewed in **ref. 47**). This class can be further broken down into three families: (1) growth factor cystine knots such as nerve growth factor, transforming growth factor β-2, and platelet derived growth factor BB; (2) inhibitor cystine knots, such as trypsin inhibitors, conotoxins, and plant toxins; and (3) cyclic cystine knots, such as those present in plant cyclotides *(47)*. The cystine knot is characterized by an embedded ring formed by two disulfide bridges and the peptide backbone connecting them, surrounded by a third disulfide bond. Family members are 26–48 residues in length and retain such biological activities as ion channel blockade and hemolysis, as well as antiviral, antibacterial, and growth factor behavior *(47)*. The variety of biological activities as well as the flexibility of these scaffolds allows a myriad of possibilities for drug design.

If the initial scaffolds are naturally occurring, it is often found that direct oxidation using redox pairs, such as reduced and oxidized glutathione or β-mercaptoethanol and 2-hydroxyethylsulfide, produce correctly folded peptides in greater yields than multistep oxidations used in combination with orthogonal protection schemes *(43,48,49)* (Lauer-Fields et al., unpublished results).

Interestingly, the addition of organic solvents such as methanol, ethanol, or isopropanol will often aid in the folding process *(49)* (Lauer-Fields et al., unpublished results). Two possible mechanisms exist. If the correctly folded oxidized peptide is more hydrophobic than the improperly folded peptide or the fully reduced form, the organic solvent can preferentially solublize the folded species or reduce the formation of nonfolded aggregates *(49)*. Alternatively, if the properly folded species contains moderate α-helical content, solvents that aid in the formation of α-helices can increase the likelihood of properly folded construct *(50)* (Lauer-Fields et al., unpublished results). Isopropanol, trifluoroethanol, and hexafluoroisopropanol satisfy this criterion *(51)*.

The implications for future topological design work are enormous. The power within these scaffolds lies in the fact that relatively few amino acids, aside from the cysteines, are necessary for stabilizing the proper fold. As such, the scaffolds can be used to display limited combinatorial libraries of peptides for biological screening. Alternatively, one can rely on small scale multiple syntheses of singly substituted peptides to produce a panel for screening purposes. Robust folding kinetics as well as facile procedures, not relying on orthogonal protection schemes, can produce volumes of compounds with relatively little effort. The scaffold used to illustrate a cystine-stabilized fold herein is derived from the sarafotoxin (SRT) peptide 6b (**Fig. 2**). Sarafotoxin is a member of the endothelin family of proteins, which contains 21 amino acids, incorporating cysteinyl residues at positions 1, 3, 11, and 15. These vasoactive proteins contain disulfide bonds between Cys^1–Cys^{15} and Cys^3–Cys^{11}, and an α-helical region between residues 9 and 15 *(52)*. This scaffold has been used extensively and has proven to be robust enough so that numerous modifications can be made to the primary structure without impeding the folding *(52,53)* (Lauer-Fields et al., unpublished results).

Matrix metalloproteinases (MMPs) are involved in the physiological remodeling of tissues as well as pathological destruction of extracellular matrix components *(54)*. Dysregulation of MMP activity has been correlated with pathologies such as arthritis, cardiovascular disease, tumor cell metastasis, and gingivitis *(55)*. The activity of 25 MMPs in humans can be regulated in numerous ways including gene expression, activation of the zymogen precursors, association with endogenous inhibitors such as the tissue inhibitors of metalloproteinases (TIMPs), or the physical removal of enzyme via proteolysis or cellular internalization. The approach described below focuses on inhibiting MMPs using synthetic mini-proteins that incorporate portions of the TIMP active site (**Fig. 2**) into a scaffold based on an endothelin family protein.

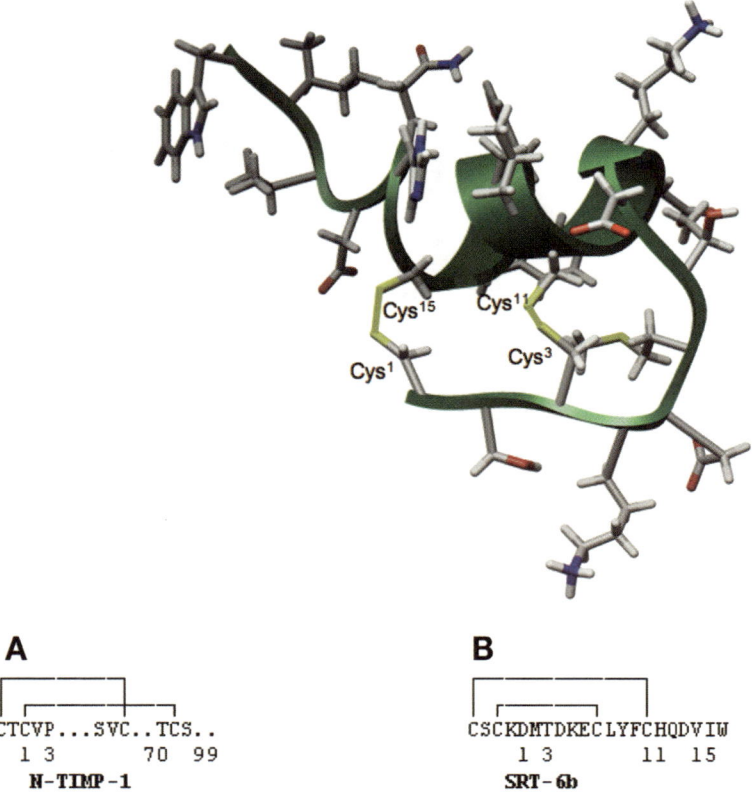

A

CTCVP...SVC..TCS..
1 3   70 99
N-TIMP-1

B

CSCKDMTDKEC LYFCHQDVIW
1 3   11 15
SRT-6b

Fig. 2. Structure of SRT 6b (top) and comparison of N-tissue inhibitor of metallo-proteinase (TIMP)-1 (bottom, **A**) and sarafotoxin 6b (bottom, **B**) sequences.

A comparison of the three-dimensional structures of SRT 6b (**Fig. 2**) and the reactive site of TIMPs (**Fig. 3**) suggest potentially similar interfaces with MMPs. Because of the apparent structural commonalities, SRT 6b was tested as a possible inhibitor of MMP activity. The K_i value was found to be in the micromolar range (K. Brew, unpublished results), less effective than the nanomolar inhibition by TIMPs. Nevertheless, it encouraged the development of SRT analogs as potential MMP inhibitors.

1.3. Peptide-Amphiphiles

The term "peptide-amphiphile" was first used in 1995 (*56*). This PA was utilized to mimic defined topological structures by incorporating an amino acid sequence with the propensity to form a triple-helix as the polar "head" group and

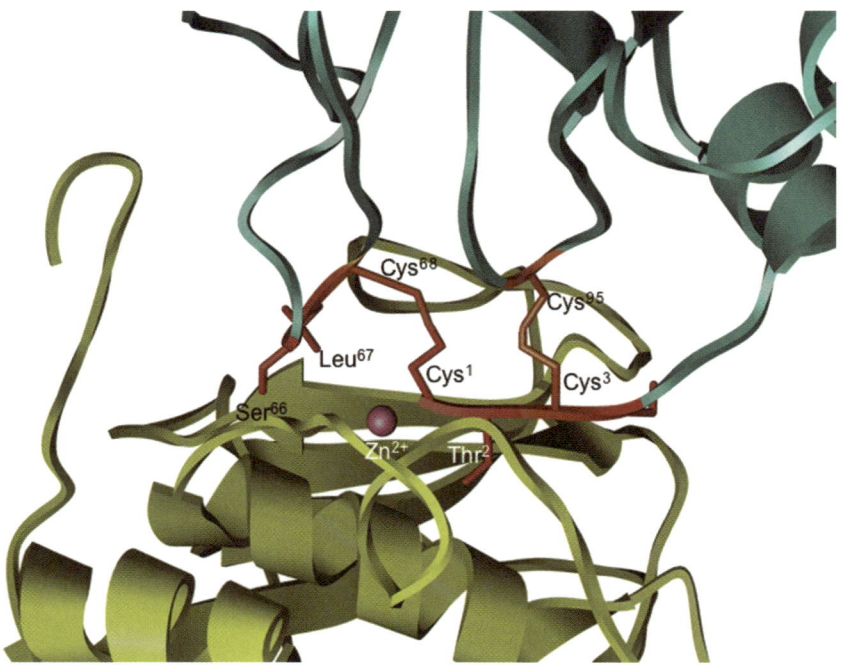

Fig. 3. Interface of tissue inhibitor of metalloproteinase (TIMP)-1 and matrix metalloproteinase (MMP)-3.

a dialkyl hydrocarbon chain as the nonpolar "tail." The concept of covalently linking a variety of peptide head groups to lipophilic tails has since broadened to include a vast array of structures, such as β-sheets based on β-amyloid, silk, or elastin sequences, as well as coiled-coils, and fibronectin-derived RGD turns *(57–60)*. The use of PAs to create defined topological structures has many advantages compared to template assisted or cystine-stabilized designs. The flexibility of this architecture is likely the greatest attribute. One can incorporate single binding sites, such as the collagen-derived CD44 chondroitin sulfate proteoglycan binding sequence *(61–65)*. Alternatively, a secondary binding site can be jointly presented to cells, such as the use of two fibronectin-derived sequences, RGD and PHSRN, to synergistically improve binding to the α5β1 integrin *(58,66,67)*. By using a variety of lipophilic tails, an array of structures can be created. These include, but are not limited to, sheets, spheres, rods, disks, and channels *(59,60,68)*. The field of biomaterials has benefited greatly from the use of PAs to create novel scaffolds *(66,69–71)*. Hartgerink et al. utilized a PA incorporating RGD and Ca²⁺ binding sites to initiate PA fiber assembly,

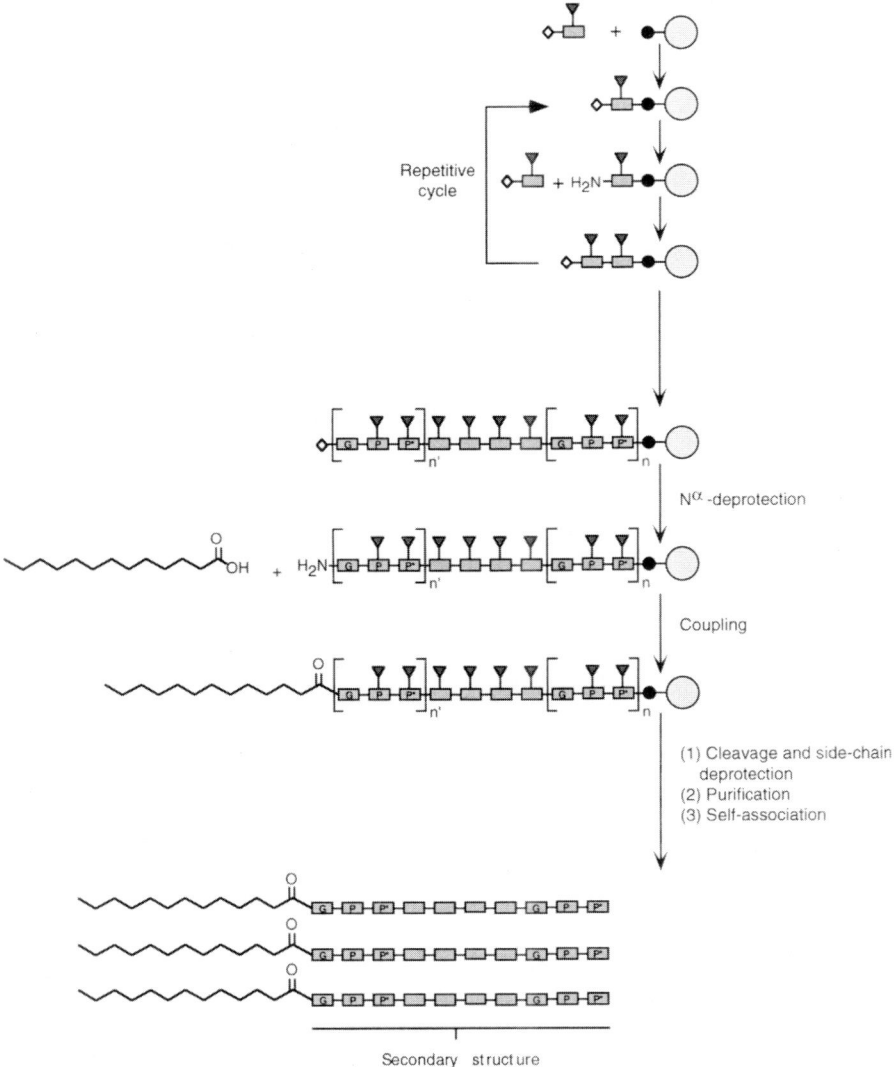

Fig. 4. Synthesis of peptide-amphiphiles (PAs). The desired peptide sequence is assembled by standard solid-phase stepwise synthesis methods. Once synthesis is complete, the N^{α}-amino protecting group is removed from the resin-bound peptide, and the lipophilic moiety is incorporated. In this particular example, an alkyl chain has been coupled to a collagen-like sequence to create a PA. The PA is then cleaved from the resin, which simultaneously removes the side-chain protecting groups, purified, and allowed to self-associate in an aqueous environment.

which in turn directed mineralization of hydroxyapatite in a three-dimensional structure that modeled bone tissue *(70)*. This molecule adds another dimension to the flexibility of PAs, namely the use of cystine-bridges to create reversible covalent cross-links to modulate structural stability. Coating surfaces of biomaterials with PAs also allows the use of additional lipophilic groups interspersed with the PA to modulate packing of head groups such that their presentation to enzymes, receptors, or cells can be optimized *(57,66,72)*. PAs have also been utilized in solution to create fluorogenic substrates incorporating triple-helices, polyPro(II) helices, and α-helices for studying enzyme–substrate interactions that more closely mimic enzyme interactions with native protein substrates *(73–77)*. Melting temperatures, and thus the head group stability, can be tightly regulated by modulating the length of the mono-alkyl hydrocarbon chain used for the tail, such that mechanistic studies illustrating an enzyme's ability to unwind a specific structure (i.e., triple-helix) can be performed *(73–77)*. Lastly, the relatively simple synthetic schemes used with PAs create an environment rich with versatility (**Fig. 4**). For these reasons, PAs have been used for an array of applications, including cell and receptor binding studies, mechanistic enzymology, antitumor therapy, targeted drug delivery, targeted oligonucleotide delivery, enhanced antimicrobial activity, and tissue engineering *(60,64,65,71–73,75–82)*. Two PA applications are described here: fluorogenic triple-helical substrates and triple-helical cell receptor ligands.

2. Materials

1. Dimethylformamide (DMF), *N*-methylpyrrolidone (NMP), *N*,*N*-diisopropylethylamine (DIEA), trifluoroacetic acid (TFA), methyl-*tert*-butyl ether (MTBE), 1,8-diaza-bicyclo[5.40.]undec-7-ene (DBU), dimethyl sulfoxide (DMSO), and acetonitrile, peptide synthesis- or high-performance liquid chromatography (HPLC)-grade, as well as cell culture media, bovine serum albumin (BSA), HEPES, glutaraldehyde, Coomassie blue R-250, formalin, Tris-HCl, NaCl, $CaCl_2$, and brij-35, are purchased from Fisher Scientific (Atlanta, GA).
2. Piperidine is purchased from AnaSpec (San Jose, CA).
3. Sodium acetate, Tris-HCl, ammonium acetate, NH_4HCO_3, isopropanol (Optima/ HPLC grade), acetonitrile (HPLC-grade), TFA (peptide synthesis-grade), β-mercaptoethanol (molecular biology grade), thioanisole, ethanedithiol, phenol, $CaCl_2$, brij-35, and NaCl are purchased from Fisher Scientific (Atlanta, GA).
4. Fmoc-amino acids, *N*-[(1H-benzotriazol-1-yl)(dimethylamino)methylene]-*N*-methylmethanaminium hexafluorophosphate *N*-oxide (HBTU), 1-H-benzotriazolium-1-[*bis*(dimethylamino)methylene]-5-chloro-hexafluorophosphate-(1-),

3-oxide (HCTU), 1-hydroxybenzotriazole (HOBt), Fmoc-Gly-Sasrin resin, Fmoc-Rink Amide 4-methylbenzhydrylamine (MBHA), or NovaSyn TGR PEG resin (*see* **Note 1**) are purchased from EMD Biosciences (San Diego, CA).

5. 5,5'-Dithiobis(2-nitrobenzoic acid), α-cyano-4-hydroxycinnamic acid (mass spectrometry [MS] grade), and 2-hydroxyethyl disulfide are purchased from Sigma Scientific (St. Louis, MO).

6. Alamar Blue is obtained from BioSource International (Camarillo, CA).

7. The monoalkyl chains hexanoic acid [CH_3-$(CH_2)_4$-CO_2H, designated C_6], octanoic acid [CH_3-$(CH_2)_6$-CO_2H, designated C_8], decanoic acid [CH_3-$(CH_2)_8$-CO_2H, designated C_{10}], dodecanoic acid [CH_3-$(CH_2)_{10}$-CO_2H, designated C_{12}], tetradecanoic acid [CH_3-$(CH_2)_{12}$-CO_2H, designated C_{14}], palmitic acid [CH_3-$(CH_2)_{14}$-CO_2H, designated C_{16}], and stearic acid [CH_3-$(CH_2)_{16}$-CO_2H, designated C_{18}] are purchased from Aldrich.

8. OGS lysis buffer: 50 mM Tris-HCl, 150 mM NaCl, pH 7.2, 0.5 mM CaCl$_2$, 0.5 mM MnCl$_2$, 1 μM phenylmethyl sulfonyfluoride (PMSF), 10 μg/mL aprotinin, 10 μg/mL, 10 μg/mL leupeptin, and 50 mM octyl-β-glucoside.

9. IP lysis buffer is prepared as 0.25% triton x-100, 75 mM NaCl, 25 mM Tris-HCl, 0.5 mM vanadate, and 2.5 mM ethylenediamine tetraacetic acid (EDTA).

10. Enzyme assay buffer (EAB)1: 50 mM tricine, 100 mM NaCl, 10 mM CaCl$_2$, and 0.05% brij-35.

11. EAB2: 50 mM Tris-HCl, pH 7.5, 50 mM NaCl, 10 mM CaCl$_2$, and 0.05% Brij-35.

12. Adhesion medium: 20 mM HEPES and 2 mg/mL BSA in basal media.

3. Methods

3.1. Template-Assembled Synthetic Protein Methods and Applications

3.1.1. Preparation of [N-tris(Fmoc-Ahx)-Lys-Lys]-Tyr-Gly Branched Template

1. Deprotect 1 g of Fmoc-Gly-Sasrin resin with 20% piperidine in DMF for 30 min followed by a second 5-min treatment.

2. Wash the resin three times with DMF. If the substitution level is higher than 0.5 mmol/g, add 0.2 mmol benzoic anhydride dissolved in 25 mL DMF and 0.4 mmol DIEA, mix for 1 h, and wash the resin three times with DMF.

3. Dissolve 3 eq of Fmoc-Tyr(*t*Bu) in approx 10 mL DMF. Add 2.7 eq of HBTU and 3 eq of HOBt, and then add the mixture to resin. Add 6 eq of DIEA, mix for 30–60 min, and wash the resin with DMF three times.

4. Remove Fmoc as indicated above and wash the resin three times with DMF.

5. Dissolve 3 eq of Fmoc-Lys(Dde) (*see* **Note 2**) in approx 10 mL DMF. Add 2.7 eq of HBTU and 3 eq of HOBt and then add the mixture to the resin. Add 6 eq of DIEA, shake for 30–60 min, and wash with DMF three times.

6. Repeat the Fmoc removal and Fmoc-Lys(Dde) coupling steps such that the resin contains two adjacent Lys(Dde) residues (*see* **Note 3**).

7. If the desired construct is homotrimeric, the Dde and Fmoc groups can be simultaneously removed by treatment with 2% hydrazine in DMF for 30 min followed by three rinses with DMF. An optional spacer of Fmoc-Ahx acid can be added to reduce the steric hindrance of the branch.

8. Keeping in mind that the chain is now trimeric, dissolve 3 eq of Fmoc-Ahx in approx 10 mL DMF, add 2.7 eq of HBTU and 3 eq of HOBt, and add the mixture to the resin. Add 6 eq of DIEA, mix for 60 min, and wash three times with DMF.

9. A small volume of template is cleaved from the resin using H_2O–TFA (5:95) for 1 h and analyzed by matrix-assisted laser desorption/ionization (MALDI)-time-of-flight (TOF) MS to confirm proper branch assembly. This scale of template is sufficient for three 0.25 mmol syntheses.

One can also utilize the branching strategy to construct heterotrimeric triple-helical peptides *(42)*. The synthesis of triple-helical peptides with one chain of different sequence from the other two requires a four-dimensional orthogonal scheme, where HMP is the linker and Dde or 1-(4,4-dimethyl-2,6-dioxocyclohex-1-ylidene)-3-methylbutyl (ivDde) *(83,84)* side-chain protection of Fmoc-Lys is used (**Fig. 1**). Synthesis by Fmoc strategy of one chain (i.e., the α2 collagen chain sequence) proceeds through the α-amino of Lys; thus, the Dde/ivDde and Fmoc groups are *not* removed simultaneously. The α2 chain sequence is assembled by Fmoc chemistry as described above. After incorporation of the α2 chain sequence, the peptide chain is "capped" by using allyloxycarbonyl (Aloc) chloride or Aloc-Gly, creating an *N*-terminal Aloc group. The Lys ε-amino Dde or ivDde groups are then removed with hydrazine (see above), which does not remove Aloc, and synthesis by Fmoc strategy proceeds for the other two chains (i.e., α1 collagen chain sequence). Once all three chains are equivalent in terms of number of residues incorporated, the Aloc group is removed by treatment with $(Ph_3P)_4Pd$ in $CHCl_3$–acetic acid–*N*-methyl morpholine (20:1:0.5) *(85)* and the Fmoc groups removed as described above. Finally, Fmoc-Gly, -Pro, and - Hyp(*t*Bu) are incorporated into all three chains. Cleavage and side-chain deprotection are by TFA as described below. This method was used previously to construct a heterotrimeric α1β1 integrin binding site from type IV collagen *(42)*.

3.1.2. Peptide Synthesis, Side-Chain Deprotection, and Cleavage

1. Incorporation of individual amino acids is by Fmoc solid-phase methodology on an Applied Biosystems 433A Peptide Synthesizer using cycles described previously *(9,10)*.

2. For Fmoc removal, a solution of DBU–piperidine–NMP (1:5:44) is used for 5–10 and 10–15 min.
3. All couplings are performed with HCTU/HOBt.
4. Peptide-resins are cleaved and side-chain deprotected by treatment with H_2O–TFA (5:95), H_2O–thioanisole–TFA (5:5:90), or ethanedithiol-H_2O–thioanisole-phenol–TFA (2.5:5:5:5:82.5) for ≥ 2 h depending upon the peptide sequence *(86,87)*.
5. Resins are filtered and rinsed with TFA, and the combined filtrate and wash reduced under vacuum at room temperature to approx 0.5 mL and precipitated with MTBE.
6. The precipitate is centrifuged and washed three times with MTBE, dissolved in acetonitrile-H_2O (1:9–1:4), and purified by preparative reversed-phase (RP)-HPLC.

3.1.3. Purification and Characterization of Peptide-Template

1. Preparative RP-HPLC is performed on a Rainin AutoPrep System with a Vydac 218TP152022 C18 column (15–20 μm particle size, 300 Å pore size, 250 × 22 mm) at a flow rate of 10 mL/min with 0.1% TFA in H_2O (A) and 0.1% TFA in acetonitrile (B), and detection at $\lambda = 220$ nm.
2. The eluent is lyophilized to a powder and repurified using a semipreparative Vydac 219TP54 diphenyl column (5 μm particle size, 300 Å pore size, 250 × 4.6 mm) at a flow rate of 2 mL/min with detection at $\lambda = 220$ nm.
3. Analytical HPLC is performed on a Hewlett-Packard 1100 Liquid Chromatograph equipped with an ODS Hypersil C-18 column (5 μm particle size, 100 × 2.1 mm).
4. Chain assembly is confirmed by Edman degradation and MALDI-TOF mass spectrometry using a specialized matrix mixture containing 2,5-dihydroxybenzoic acid/2-hydroxy-5-methoxybenzoic acid (9:1, v/v) *(88,89)*.

Triple-helicity is monitored by circular dichroism (CD) spectroscopy in the far ultraviolet (UV) wavelengths. Peptides are dissolved in the appropriate buffer at concentrations of 2–500 μM. Conditions should reflect final relevant assay conditions as much as possible based on limitations imposed by sample availability and/or instrumentation. For example, if the peptides are to be used in a cell-binding assay at 10 μM in phosphate-buffered saline (PBS), then the CD spectra should be acquired at that peptide concentration in a similar buffer. Because chloride ions interfere with the acquisition of data, a phosphate buffer can be substituted for PBS. A typical spectra for a triple-helix shows a positive molar ellipticity at λ approx 225 nm and a negative molar ellipticity at λ approx 205 nm. In addition to the far UV wavelength scan of λ approx 195–250 nm, one can determine the melting temperature by monitoring the change in molar ellipticity at $\lambda = 225$ nm with a constant change in temperature from 5–85°C. For samples exhibiting sigmoidal melting curves, the inflection point in the transition region (first derivative) is defined as T_m. Alternatively, T_m is evaluated from the midpoint of the transition.

3.1.4. Affinity Chromatography Utilizing a Branched Peptide

Identification of receptors binding to regions within type IV collagen has been achieved using template-assembled mini-collagens *(9)*.

1. Branched $\alpha1(IV)1263-1277$ triple-helical peptide (THP) is coupled to activated CH-Sepharose according to the instructions of the supplier (Pharmacia Biotech). In addition, a mock-coupled column is made without the peptide.
2. The mock column of CH-Sepharose is incubated with $1\,M$ Tris-HCl or $1\,M$ Gly to block all reactive sites.
3. Cells are extracted in OGS lysis buffer by shaking 30 min at 4°C.
4. The lysates are cleared by centrifugation at $36,500\,g$ for 60 min at 4°C.
5. For this experiment, buffer conditions are optimized for the potential recovery of cell surface proteoglyans. Alternate detergents and buffers can be utilized based on the relevant receptors for each experiment. Cell lysates were shaken with the mock beads for 4 h at 4°C to remove any materials that will bind nonspecifically to the Sepharose beads.
6. The unbound materials were collected and incubated with the peptide-Sepharose beads by rocking overnight at 4°C.
7. The beads are washed with 3 volumes of OGS lysis buffer, and the bound proteins are eluted with IP lysis buffer supplemented with $50\,mM$ EDTA and $1\,M$ NaCl.
8. The eluate is concentrated and the buffer exchanged with IP lysis buffer using Microsep Centrifugal Concentrators.
9. A portion of the sample can be electrophoresed by sodium dodecyl sulfate (SDS)-polyacrylamide gel electrophoresis (PAGE) and proteins visualized with a general protein stain. Once the content and molecular masses of bound proteins are determined, individual proteins are identified using immunoprecipitation or Western blotting.
10. Peptide-columns are washed extensively to remove all bound material and stored in a buffer containing an antimicrobial agent such as ethanol or sodium azide for further use.

Invasion of the basement membrane is believed to be a critical step in the metastatic process. Melanoma cells have been shown previously to bind distinct triple-helical regions within basement membrane (type IV) collagen. Additionally, tumor-cell binding sites within type IV collagen contain glycosylated hydroxylysine residues. Triple-helical models of the type IV collagen $\alpha1(IV)1263-1277$ sequence were utilized to (a) determine the melanoma cell receptor for this ligand and (b) analyze the results of single-site glycosylation on melanoma cell recognition. Branched $\alpha1(IV)1263-1277$ THP was immobilized to CH-Sepharose, and precleared human melanoma cell lysates were added to the beads. Following application of the cell lysates, the column was washed with 3 volumes of OGS lysis buffer, and then bound materials were eluted with IP lysis buffer. Eluants were incubated with monoclonal antibodies (mAbs) against either CD44 or the β1 integrin subunit, and precipitated proteins analyzed by SDS-PAGE. Under nonreducing conditions, a protein of approx 85–90 kDa was immunoprecipitated by the anti-CD44 mAb **(Fig. 5)**. This apparent molecular weight corresponds to melanoma CD44s core protein following chondroitinase treatment *(90,91)*. No corresponding proteins were observed using an anti-β1 integrin subunit mAb immunoprecipitation (data not shown). Immunoprecipitation analysis of whole cell lysates showed the presence of both CD44 **(Fig. 5)** and the β1 integrin subunit (data not shown), consistent with prior studies *(90)*.

To examine the role of chondroitin sulfate proteoglycan (CSPG) in the binding of melanoma cells to α1(IV)1263–1277, affinity chromatography was

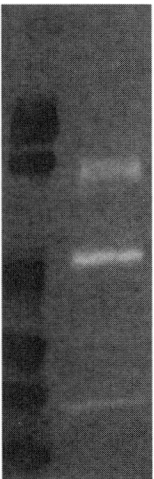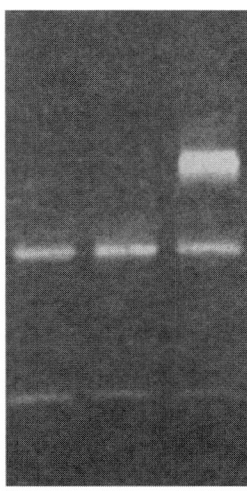

Fig. 5. Immunoprecipitation of immunoblot analysis of melanoma cell surface proteins eluted from the α1(IV)1263–1277 triple-helical peptide (THP) affinity column. Lane 1 contains the molecular size markers (108, 97, 48, 35, 28, and 20 kDa). Lane 2 contains proteins eluted by EDTA + NaCl from the THP column, immunoprecipitated with an anti-CD44 monoclonal antibody (mAb), and then treated with chondroitinase ACII. Lanes 3 and 4 contain lysis buffer immunoprecipitated with an anti-CD44 mAb, which serves as a negative control. Lane 5 contains melanoma whole cell lysate proteins immunoprecipitated with an anti-CD44 mAb, and then treated with chondroitinase ACII. In lanes 2 and 5, one protein of approx 85–90 kDa, corresponding to CD44, was immunoprecipitated. In lanes 2-5, the CD44 mAb appears at approx 50 and approx 25 kDa. No proteins were immunoprecipitated with the anti-β1 mAb (data not shown). (Reproduced from **ref. 65**, with permission of *J. Biol. Chem.*)

performed using branched α1(IV)1263–1277 THP and chondroitin-4-sulfate, chondroitin-6-sulfate, and dermatan sulfate. Both chondroitin-4-sulfate and chondroitin-6-sulfate were found to specifically bind to α1(IV)1263–1277 THP, whereas dermatan sulfate did not **(Fig. 6)**. The relative elution profiles of chondroitin-4-sulfate and chondroitin-6-sulfate made it appear that chondroitin-4-sulfate has a greater ability to bind branched α1(IV)1263–1277 THP, but a significant amount (> 4000 rfu) of chondroitin-6-sulfate remains bound to the THP and elutes only with successive washes with acetate buffer, pH 4.0, and Tris-HCl buffer, pH 8.0. Overall, α1(IV)1263–1277 is bound by CD44 in the CSPG modified form.

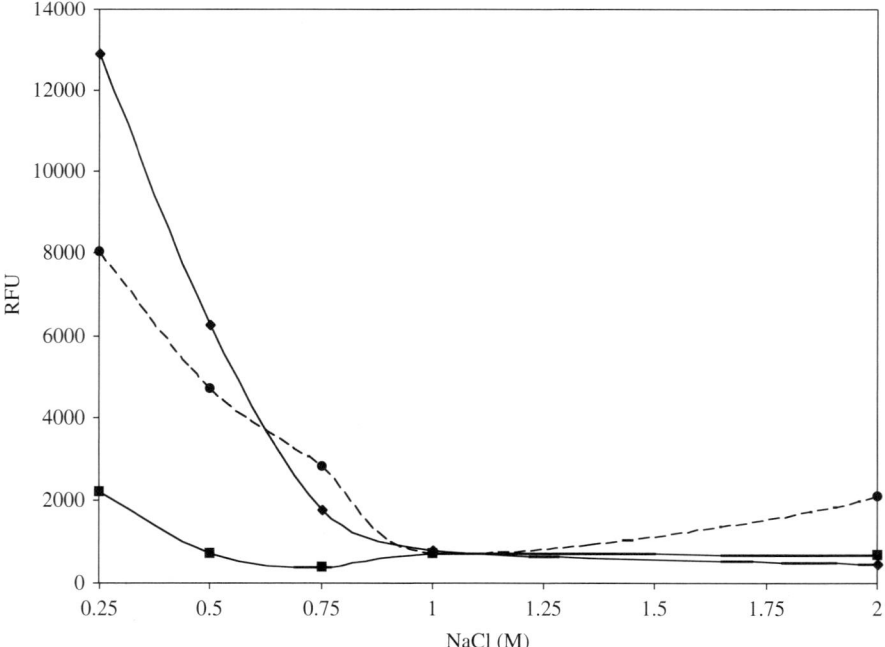

Fig. 6. Affinity chromatographic analysis of chondroitin-4-sulfate (closed diamonds, solid line), dermatan sulfate (closed squares, solid line), or chondroitin-6-sulfate (closed circles, dashed line) binding to the α1(IV)1263-1277 triple-helical peptide (THP) affinity column. The glycosaminoglycans were eluted by increasing NaCl concentrations, and detected by fluorometric analysis. Chondroitin-4-sulfate and chondroitin-6-sulfate bound specifically to the α1(IV)1263–1277 THP, while dermatan sulfate did not. (Reproduced from **ref. 65**, with permission of *J. Biol. Chem.*)

3.2. Disulfide-Stabilized Structures: Methods and Applications

3.2.1. Synthesis, Cleavage, and Side-Chain Deprotection of Sarafotoxin-Based Peptides

A comparison of sequences of endothelin-family members is shown in **Table 1**. A rough consensus sequences exists as follows: $C\delta CX\delta^5\phi XD\pm E^{10}C\phi\text{-}\phi\alpha C^{15}H\phi D\phi I^{20}W$, where δ = polar, ϕ = bulky/hydrophobic, α = aromatic, $\pm$ = charged, and X = any residue.

Previous work has shown that truncation at endothelin Asp[18] results in a peptide that loses all vasoconstrictive activity *(92)*. A family of analogs were created that retain the endothelin scaffold without unwanted vasoconstrictive activity, These peptides used residues 1–18 as a starting point, with substitutions at residues 2, 4, 5, 16, and 18, as well as extensions of nonendothelin sequences for residues 19–23. In all cases, the analogs folded in a proper manner (i.e., with disulfide bonds between Cys[1]–Cys[15] and Cys[3]–Cys[11]) with yields ranging from approx 60–100%. Data included herein are based on the following sequence, shown with protecting group strategy: Cys(Trt)-Ser(tBu)-Cys(Trt)-Val-Asp(OtBu)-Met-Thr(*t*Bu)-Asp(OtBu)-Lys(Boc)-Glu(OtBu)-Cys(Trt)-Leu-Tyr (*t*Bu)-Phe-Cys(Trt)-Val-Trp(Boc)-Ser(*t*Bu)-Glu(OtBu)-Met-Ala-OH. This sequence was designed using SRT 6b residues 1–15, with Lys[4] substituted by Val, and residues 16–21 derived from TIMP-1. The peptide-resin is cleaved and deprotected for 2 h with ethanedithiol–thioanisole–phenol–water–TFA

Table 1
Sequences of Endothelin (ET) and Sarafotoxin (SRT) Family Members

Species	Sequence
ET1	CSCSSLMDKECVYFCHLDIIW-OH
ET2	CSCSSWLDKECVYFCHLDIIW-OH
ET3	CTCFTYKDKECVYYCHLDIIW-OH
ET4/VIC	CSCNSWLDKECVYFCHLDIIW-OH
Bibrotoxin	CSCADMTDKECLYFCHQDVIW-NH$_2$
SRT 6a	CSCKDMTDKECLNFCHQDVIW-NH$_2$
SRT 6b	CSCKDMTDKECLYFCHQDVIW-OH
SRT 6c	CTCNDMTDEECLNFCHQDVIW-OH
SRT 6d	CTCKDMTDEECLYFCHQDIIW-NH$_2$
SRT 6e	CTCKDMTDEECLYFCHQGIIW-NH$_2$

Residues in bold are conserved.

(2.5:5:5:5:82.5) *(86,87)*, precipitated and washed repeatedly in cold MTBE, and dissolved in folding buffer as indicated below.

3.2.2. Peptide Oxidation and Folding

1. Crude peptides are either purified by RP-HPLC on a Rainin AutoPrep System with a Vydac 218TP152022 preparative C_{18} column (15–20 μm particle size, 300 Å pore size, 250 × 22 mm) at a flow rate of 10 mL/min with 0.1% TFA in H_2O (A) and 0.1% TFA in acetonitrile (B), and detection at λ = 220 nm, or folded without purification (*see* **Note 4**).
2. Peptides are dissolved at 100–600 μM in one of four buffers: (1) 0.1 M NH_4Ac, pH 8.5; (2) 0.1 M NH_4HCO_3, pH 8.5; (3) 0.1 M NH_4HCO_3, pH 8.5, 1 mM 2-hydroxyethyl disulfide, 5 mM β-mercaptoethanol; or (4) 0.1 M NH_4HCO_3, pH 8.5, 1 mM 2-hydroxyethyl disulfide, 5 mM β-mercaptoethanol supplemented with isopropanol to a final concentration of 30% (*see* **Notes 5** and **6**). All folding procedures are performed at 4°C with mixing.
3. Folding is monitored by analytical RP-HPLC on a Hewlett Packard 1100 Liquid Chromatograph equipped with a small-pore, narrow-bore C_{18} column (5 μm particle size, 120 Å pore size, 100 × 2.1 mm) with 0.1% TFA in H_2O (A) and 0.1% TFA in acetonitrile (B), and detection at λ = 220 and 280 nm.
4. Oxidized peptides are rotovapped and purified as discussed above for preparative RP-HPLC.

3.2.3. Peptide Characterization

MALDI-TOF mass spectrometry is performed using α-cyano-4-hydroxycinnamic acid on a Voyager-DE STR mass spectrometer (Applied Biosystems). Free sulfhydryl content is also quantified using a modified Ellman's test *(93)*:

1. 5,5′-Dithiobis(2-nitrobenzoic acid) (DTNB) is dissolved in 50 mM sodium acetate at a stock concentration of 2 mM.
2. The DTNB solution is diluted to a final concentration of 500 μM, in 100 mM Tris-HCl, pH 8.0.
3. 10 μL of buffer, sample, or standard is added to 190 μL of DTNB solution, mixed and absorbance is read at λ = 412 nm.
4. Free sulfhydryl content is calculated compared to a standard curve generated with D/L-Cys.
5. All reagents are made fresh immediately prior to use.

3.2.4. Sarafotoxin Scaffolds as Metalloprotease Inhibitors

1. Peptides are dissolved in EAB1 and their concentrations determined spectrophotometrically by A_{280}.

2. A relevant enzyme, for example, MMP-3, is diluted in EAB1 to a concentration that is twofold higher than the expected final concentration.
3. A serial dilution is performed with the peptide in EAB1 to create a range of concentrations.
4. One volume of enzyme is mixed with one volume of peptide and the samples are incubated for 1–2 h at 37°C. The actual incubation time is determined by the stability and final concentration of the enzyme. If the enzyme is unstable at 37°C, then incubation can be performed at 20–30°C. If the temperature is lowered, the incubation time must be extended. A positive control inhibitor, such as TIMP-1, can be used to determine appropriate assay conditions.
5. Upon completion of the enzyme-inhibitor incubation period, a small volume of fluorogenic substrate (10% of the final assay volume) is added to each well. The final concentration of substrate should be = 10% K_M.
6. The fluorescence is monitored at the appropriate wavelengths. A typical MMP substrate fluorophore/quencher pair is Mca/Dnp, which utilizes $\lambda_{excitation} = 324$ nm and $\lambda_{emission} = 393$ nm (73,76,94,95).
7. Initial velocities are obtained from plots of fluorescence vs time, using only data points corresponding to less than 20% full hydrolysis, which corresponds to a time frame in which the fluorescence change versus time is linear. The initial velocity is equal to the slope of that line. A plot of [inhibitor] versus initial velocity or percentage of enzyme activity will illustrate the extent of inhibition.

This particular sequence (CSCVDMTDKECLYFCVWSEMA) produced two fractions of completely oxidized peptide. Presumably fraction 1 retains the proper disulfide bonds between Cys^1–Cys^{15} and Cys^3–Cys^{11}, whereas fraction 2 contains the improperly folded species, Cys^1–Cys^{11} and Cys^3–Cys^{15}. CD spectroscopy (data not shown) and MMP-3 inhibition profiles (**Fig. 7**) support this assessment of folding. A full description of inhibition profiles and peptide characterization will be published elsewhere.

3.3. Peptide-Amphiphile Methods and Applications

3.3.1. Preparation of Peptide-Amphiphiles

1. Incorporation of individual amino acids is by Fmoc solid-phase methodology on an Applied Biosystems 433A Peptide Synthesizer using cycles described previously (9,10).
2. For Fmoc removal, a solution of DBU–piperidine–NMP (1:5:44) is used for 5–10 and 10–15 min.
3. All couplings are performed with HCTU/HOBt.
4. Peptide-resins are cleaved and side-chain deprotected by treatment with H_2O–TFA (5:95), H_2O–thioanisole–TFA (5:5:90), or ethanedithiol–H_2O–thioanisole–phenol–TFA (2.5:5:5:5:82.5) for ≥ 2 h depending upon the peptide sequence (86,87).

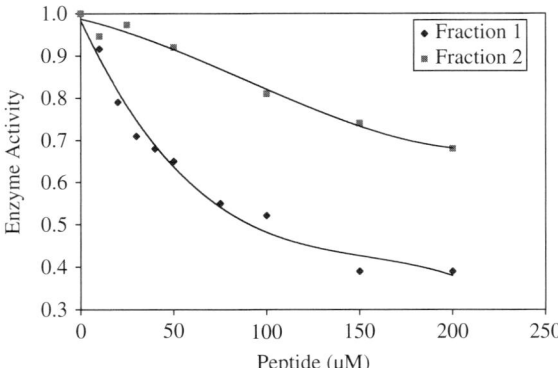

Fig. 7. Inhibition profiles for a selected sarafotoxin peptide with matrix metalloproteinase (MMP)-3. Both fractions are fully oxidized. Fraction 1 is presumably properly
folded, whereas fraction 2 is likely to be improperly folded. Full characterization and
activity profiles will be described elsewhere.

5. Resins are filtered and rinsed with TFA, and the combined filtrate and wash reduced
   under vacuum at room temperature to approx 0.5 mL and precipitated with MTBE.
6. The precipitate is centrifuged and washed three times with MTBE, dissolved in
   acetonitrile–H_2O (1:9–1:4), and purified by preparative RP-HPLC.

3.3.2. PA Synthesis

1. Acylation of the peptide-resin involves condensing a 4-fold molar excess of alkyl
   acid to the N^α-deprotected resin, with a 3.8-fold molar excess each of HBTU and
   HOBt in DMF for 2 h. The reaction is initiated by the addition of an 8-fold molar
   excess of DIEA and proceeds for 1 h.
2. Cleavage and side-chain deprotection of the lipidated peptide-resins is accomplished
   by treating the resin for 1–2 h with either H_2O–TFA (1:19) or ethanedithiol–
   thioanisole–phenol–H_2O–TFA (2.5:5:5:5:82.5) *(86,87)*.
3. PA cleavage solutions are extracted with methyl *tert*-butyl ether prior to purification.

3.3.3. Purification and Characterization of PA

Preparative RP-HPLC is performed on a Rainin AutoPrep System with a
Vydac 218TP152022 C_{18} or 214TP152022 C_4 column (15–20 μm particle size,
300 Å pore size, 250 × 22 mm) at a flow rate of 10 mL/min with 0.1% TFA in
H_2O (A) and 0.1% TFA in acetonitrile (B), and detection at $\lambda = 220$ nm. For
very hydrophobic samples, it is advantageous to use 0.1% TFA in isopropanol

instead of acetonitrile *(61,62)*. Analytical HPLC is performed on a Hewlett-Packard 1100 Liquid Chromatograph equipped with an ODS Hypersil C-18 column (5 μm particle size, 100 × 2.1 mm).

PA composition can be determined by amino acid analysis *(9)* and/or MALDI-TOF MS *(56,61,64,73,89,96,97)*. PA homogeneity can also be evaluated by diphenyl and nonporous C_{18} reversed-phase HPLC *(96)* and/or hydrophobic interaction HPLC *(10)*.

3.3.4. Structural Characterization of PAs

Triple-helicity is monitored by CD spectroscopy in the far UV wavelengths *(9,10,61,62,64,73,89,97,98)* as described in 3.1.3. Nuclear magnetic resonance (NMR) spectroscopy can also be used to study the relative thermal stability, alignment, and flexibility (backbone mobilities) of triple-helical PAs *(61–63)*.

3.3.5. Peptide-Amphiphiles as Substrates

A continuous assay method, such as one that utilizes an increase in fluorescence upon hydrolysis, allows for rapid and convenient kinetic evaluation of proteases. Quenched fluorescent triple-helical substrates that utilize fluorescence resonance energy transfer (FRET)/intramolecular fluorescence energy transfer (IFET) have been constructed by incorporation of a fluorophore [(7-methoxycoumarin-4-yl)acetyl; Mca] and a quencher (2,4-dinitrophenyl; Dnp) in the same peptide chain *(73,74,94,95,99–101)*. The Mca/Dnp fluorogenic triple-helical substrates have been utilized to determine individual kinetic parameters and activation energies for hydrolysis of triple-helices (see below) *(73–77)*. Additionally, both general MMP triple-helical substrates *(73,76)* and MMP-2/MMP-9 and MMP-14 selective triple-helical substrates *(75,77)* have been described. For the triple-helical constructs, Mca is incorporated by acylation onto the ε-amino group of Lys *(73)* or by utilizing the Fmoc-Lys(Mca) building block *(102)*.

1. Substrates are dissolved in EAB2. If solubility is poor, a concentrated stock solution can be prepared in DMSO and diluted in EAB2 such that the final concentration of DMSO does not exceed 0.3%.
2. Protease assays are carried out in EAB2 by incubating a range of substrate concentrations with 10 n*M* enzyme at 30–37°C. Substrates should be added to the 96 or 384-well plate and pre-incubated to the appropriate temperature for 20–30 min.
3. Add enzyme, mix the plate thoroughly and monitor fluorescence at the appropriate wavelengths (*see* **Subheading 3.2.4., step 6**).

4. Initial velocities are obtained from plots of fluorescence versus time, using only data points corresponding to less than 20% full hydrolysis, which corresponds to a time frame in which the fluorescence change vs time is linear. The slope from these plots is divided by the fluorescence change corresponding to complete hydrolysis for that substrate concentration and then multiplied by the substrate concentration to obtain initial velocity in units of μM/s.

A series of fluorogenic triple-helical MMP substrates were designed to incorporate known MMP cleavage sites in types I, II, III, and V collagen (**Table 2**) *(73–76,103)*. Fluorogenic THP (fTHP) substrates were initially based on a consensus sequence derived from the collagenolytic MMP cleavage sites in human types I–III collagen *(54)*. fTHP-3 and fTHP-4 *(73,74)* include a Gly~ Leu cleavage site and Gly-Pro-Hyp repeats to enhance triple-helical stability. To serve as the FRET fluorophore-quencher pair, Mca and Dnp are linked to Lys side-chains in the P_5 and P_5' subsites, respectively. Arg in the P_2' and P_8' subsites are favored by collagenolytic MMPs *(104)* while serving to enhance fTHP solubilities *(73)*.

MMP-1 hydrolysis of fTHP-3 at 30°C occurred exclusively at the Gly~ Leu bond. This is the analogous bond cleaved by MMP-1 in the native $\alpha 1$(II) collagen chain *(105)*. Thus, incorporation of the fluorophore/quenching pair of Mca and Dnp did not affect the ability of the enzyme to recognize or cleave the substrate. The relative order of k_{cat}/K_M values was MMP-13 > MMP-1 $\sim$ MMP-2 > MMP-3 (**Table 2**). MMP-13 had the highest k_{cat}/K_M value for hydrolysis of fTHP-3 primarily due to a lower K_M value. fTHP-4 was synthesized to create a more soluble version of fTHP-3, and was found to exhibit similar kinetics upon MMP treatment as fTHP-3 (**Table 2**). MMP-1 and MMP-2 triple-helicase activities were enhanced when the Mca fluorophore was replaced by 2-amino-3-(7-methoxy-4-coumaryl)propionic acid (Amp; fTHP-5 and fTHP-6) or 2-amino-3-(6,7-dimethoxy-4-coumaryl)propionic acid (Adp; fTHP-7 and fTHP-8) (**Table 2**), demonstrating that the shorter side-chain of Amp or Adp [compared with Lys(Mca)] in the substrate P_5 subsite was better tolerated by MMP-1 and MMP-2 *(76)*.

The consensus sequence was subsequently modified in the P_1' subsite, where Leu was replaced with Cys(Mob), to create fTHP-9. The Cys(Mob) residue in the P_1' subsite was reported to enhance MMP-14 activity and specificity *(106)*. It should be noted, however, the substrates from the prior MMP-14 study were neither triple-helical nor linear collagen models *(106)*. Both fTHP-4 and fTHP-9 were acylated with decanoic (C_{10}) and palmitic (C_{16}) acids to confer differential thermal stability and allow the correlation of substrate structural stability to enzyme activity. The effect of the single substitution of

Table 2

Kinetic Parameters for fTHP Hydrolysis by Matrix Metalloproteinases (MMPs) at 30°C

Enzyme	Substrate	k_{cat}/K_M (sec^{-1}M^{-1})	k_{cat} (sec^{-1})	K_M (μM)	E_a (kcal/mol)
MMP-1	fTHP-3[a]	1278	0.008	61.2	11.6
MMP-2	fTHP-3	1082	0.017	17.2	ND
MMP-13	fTHP-3	2273	0.045	20.5	ND
MMP-1	fTHP-4[b]	6400	0.070	11.0	ND
MMP-1[c]	fTHP-4	3900	0.060	15.2	20.0
MMP-1(ΔC)[c,d]	fTHP-4	1500	0.020	13.3	29.0
MMP-14(ΔC)[e]	fTHP-4	32000	0.48	15.1	ND
MMP-14(ΔC)[c,e]	fTHP-4	59000	1.37	23.1	8.8
MMP-14(ΔC)[c,e]	C$_{10}$-fTHP-4	4100	0.0033	0.80	ND
MMP-14(ΔC)[c,e]	C$_{16}$-fTHP-4	1200	0.0021	1.75	ND
MMP-1	fTHP-5[f]	4056	0.114	28.1	9.5
MMP-2	fTHP-5	24440	0.049	2.0	10.0
MMP-1	fTHP-6[g]	3678	0.146	39.8	9.3
MMP-2	fTHP-6	25490	0.048	1.9	7.0
MMP-1	fTHP-7[h]	2594	0.124	47.7	ND
MMP-2	fTHP-7	26430	0.053	2.0	ND
MMP-1	fTHP-8[i]	3065	0.103	33.5	ND
MMP-2	fTHP-8	27200	0.063	2.3	ND
MMP-1[c]	fTHP-9[j]	3000	0.060	20.3	ND
MMP-1[c]	C$_{10}$-fTHP-9	NC	NC	NC	NC
MMP-14(ΔC)[e]	fTHP-9	99000	1.15	11.6	ND
MMP-14(ΔC)[c,e]	fTHP-9	168000	1.34	8.0	11.4

(Continued)

Table 2
(Continued)

Enzyme	Substrate	k_{cat}/K_M (sec^{-1}M^{-1})	k_{cat} (sec^{-1})	K_M (μM)	E_a (kcal/mol)
MMP-14(ΔC)[c,e]	C$_{10}$-fTHP-9	29000	0.080	2.8	ND
MMP-14(ΔC)[c,e]	C$_{16}$-fTHP-9	30000	0.196	6.6	14.3
MMP-1	α1(V)436–447 fTHP[k]	NC	NC	NC	NC
MMP-2	α1(V)436–447 fTHP	14002	0.062	4.4	ND
MMP-3	α1(V)436–447 fTHP	NC	NC	NC	NC
MMP-9	α1(V)436–447 fTHP	5449	0.044	8.1	ND
MMP-13	α1(V)436–447 fTHP	NC	NC	NC	NC
MMP-14(ΔC)[d]	α1(V)436–447 fTHP	NC	NC	NC	NC

[a] fluorogenic triple-helical peptide (fTHP)-3 = C$_6$-(Gly-Pro-Hyp)$_5$-Gly-Pro-Gln-Gly~Leu-Arg-Gly-Gln-Lys(Dnp)-Gly-Val-Arg-(Gly-Pro-Hyp)$_5$-NH$_2$.

[b] fTHP-4 = (Gly-Pro-Hyp)$_5$-Gly-Pro-Lys(Mca)-Gly-Pro-Gln-Gly~Leu-Arg-Gly-Gln-Lys(Dnp)-Gly-Val-Arg-(Gly-Pro-Hyp)$_5$-NH$_2$.

[c] Assay includes 0.25% dimethylsulfoxide.

[d] MMP-1 with residues 243–450 deleted.

[e] MMP-14 with residues 279–523 deleted.

[f] fTHP-5 = C$_6$-(Gly-Pro-Hyp)$_5$-Gly-Pro-Lys(Mca)-Gly-Pro-Gln-Gly~Leu-Arg-Gly-Gln-Lys(Dnp)-Gly-Val-Arg-(Gly-Pro-Hyp)$_5$-NH$_2$.

[g] fTHP-6 = C$_6$-(Gly-Pro-Hyp)$_5$-Gly-Pro-L-Amp-Gly-Pro-Gln-Gly~Leu-Arg-Gly-Gln-Lys(Dnp)-Gly-Val-Arg-(Gly-Pro-Hyp)$_5$-NH$_2$.

[h] fTHP-7 = C$_6$-(Gly-Pro-Hyp)$_5$-Gly-Pro-D-Amp-Gly-Pro-Gln-Gly~Leu-Arg-Gly-Gln-Lys(Dnp)-Gly-Val-Arg-(Gly-Pro-Hyp)$_5$-NH$_2$.

[i] fTHP-8 = C$_6$-(Gly-Pro-Hyp)$_5$-Gly-Pro-L-Adp-Gly-Pro-Gln-Gly~Leu-Arg-Gly-Gln-Lys(Dnp)-Gly-Val-Arg-(Gly-Pro-Hyp)$_5$-NH$_2$.

[j] fTHP-9 = (Gly-Pro-Hyp)$_5$-Gly-Pro-Lys(Mca)-Gly-Pro-Gln-Gly~Cys(Mob)-Arg-Gly-Gln-Lys(Dnp)-Gly-Val-Arg-(Gly-Pro-Hyp)$_5$-NH$_2$.

[k] α1(V)436–447 fTHP = (Gly-Pro-Hyp)$_5$-Gly-Pro-Lys(Mca)-Gly-Pro-Pro-Gly~Val-Val-Gly-Glu-Lys(Dnp)-Gly-Glu-Gln-(Gly-Pro-Hyp)$_5$-NH$_2$. NC, not cleaved; ND, not determined.

Cys(Mob) for Leu in the P_1' subsite was substantially different for MMP-1 and MMP-14($\Delta_{279-523}$) (**Table 2**). MMP-1 preferred fTHP-4; the replacement of Leu by Cys(Mob) increased K_M, resulting in a lower k_{cat}/K_M value. In contrast, MMP-14($\Delta_{279-523}$) preferred fTHP-9, with Cys(Mob) causing decreased K_M values resulting in higher k_{cat}/K_M values. This indicates that the P_1' subsite *in a triple-helical context* provides selectivity within these collagenolytic MMPs.

The temperature-dependent hydrolysis of fTHP-4, fTHP-9, and C_{16}-fTHP-9 was evaluated for MMP-1, MMP-1($\Delta_{243-450}$), and MMP-14($\Delta_{279-523}$). Kinetic parameters were determined at 25, 30, 35, and 37°C, and the activation energies (E_a) were calculated (**Table 2**). The activation energy could not be determined for MMP-1 with fTHP-9 or C_{16}-fTHP-9 because there was very little or no hydrolysis even at higher temperatures. Comparison of MMP-1 to MMP-14($\Delta_{279-523}$) demonstrates a higher activation energy for triple-helix hydrolysis by MMP-1. This is consistent with the lower k_{cat}/K_M values observed for MMP-1 hydrolysis of fTHPs compared with MMP-14($\Delta_{279-523}$) and the greater sensitivity to triple-helical thermal stability exhibited by MMP-1. The C-terminal hemopexin-like domain contributes to the ease of hydrolysis, as MMP-1($\Delta_{243-450}$) has a higher activation energy than MMP-1. Finally, an increase in substrate thermal stability (C_{16}-fTHP-9 compared with fTHP-9) results in a higher activation energy for MMP-14($\Delta_{279-523}$) hydrolysis.

MMP-1 and MMP-14 have different collagen preferences, with MMP-1 favoring type III collagen over type I, and *vice versa* for MMP-14 (*54,107*). It has been previously observed that type III collagen is more susceptible to general proteolysis than type I collagen (*108,109*), and the susceptibility of interstitial collagens to cleavage by trypsin and trypsin/chymotrypsin mixtures appears to be dependent on the relative unwinding of the triple-helix (*109–113*). Thus, there may be a more prevalent local instability in the type III molecule compared to type I. Local unwinding or enhanced backbone mobilities have been proposed to facilitate collagenolysis, as the MMP cleavage site in interstitial collagens is distinguished by being "loosely" triple-helical (low imino acid content) following the cleavage site and "tightly" triple-helical (high imino acid content) prior to the cleavage site, as well as possessing a low content (<10%) of charged residues (*101,105*). Recent work utilizing THP models and NMR hydrogen exchange spectroscopy experiments (*114,115*) or molecular dynamics trajectories (*116*) concluded that such flexibility exists at the site of MMP hydrolysis. MMP-1 may prefer type III collagen as a result of the more mobile triple-helix at the cleavage site and MMP-1 having limited ability to hydrolyze more thermally stable local structures.

The Gly$_{439}$∼Val$_{440}$ bonds in type V collagen are cleaved by MMP-9, along with a similar cleavage site sequence in type XI collagen *(117)*. A fluorogenic substrate was designed based on this sequence and designated α1(V)436–447 fTHP [(Gly-Pro-Hyp)$_5$-Gly-Pro-Lys(Mca)-Gly-Pro-Pro-Gly∼ Val-Val-Gly-Glu-Lys(Dnp)-Gly-Glu-Gln-(Gly-Pro-Hyp)$_5$-NH$_2$] *(75)*. 1(V)436-447 fTHP was hydrolyzed efficiently by MMP-2 and MMP-9, but was not hydrolyzed by MMP-1, MMP-3, MMP-13, or MMP-14 (**Table 2**). Thus, overall the PA approach has been utilized to develop fluorogenic substrates that can conveniently monitor general triple-helical peptidase activity or selectively monitor MMP-2/MMP-9 or MMP-14 activity.

The a disintegrin and metalloproteinase with thrombospondin motifs (ADAMTS) family has been utilized to determine if a substrate can be designed based upon topology. Among the ADAMTS family members, several have been described to have activity against aggrecan, while others can cleave the *N*-propeptide region of native interstitial procollagens *(118)*. Interestingly, the processing of the *N*-propeptide region by ADAMTS-2 requires retention of substrate supersecondary structure *(119,120)*, and peptide models of aggrecan cleavage sites are not processed efficiently by ADAMTS-4 and ADAMTS-5 *(121)*. This suggests that members of the ADAMTS family require a distinct topology for efficient substrate hydrolysis. We have thus designed an ADAMTS-4 substrate that incorporates a known aggrecan cleavage site within a poly-Pro II structure [C$_6$-(Gly-Pro-Hyp-Pro-Hyp-Gly)$_2$-Gly-Pro-Hyp-Gly-Thr-Lys(Mca)-Gly-Glu-Leu-Glu-Gly-Arg-Gly-Thr-Lys(Dnp)-Gly-Ile-Ser-(Gly-Pro-Hyp-Pro-Hyp-Gly)$_2$-Gly-Pro-Hyp-NH$_2$; fSSPa] *(122)*. A second substrate was designed to incorporate an aggrecan cleavage site sequence within an α-helical structure [C$_{16}$-Lys-Lys(Mca)-Glu-Ile-Glu-Ala-Leu-Lys-Ala-Glu-Leu-Glu-Lys(Dnp)-Leu-Lys-Ala-Glu-Ala-Glu-Ala-Leu-Lys-Ala-NH$_2$; αHPa] *(122)*. fSSPa and αHPa were treated with a highly active ADAMTS-4 deletion mutant, ADAMTS-4-3, at 37°C. fSSPa was hydrolyzed, with K$_M$ = 308.0 μ*M* and k$_{cat}$ = 0.564 s^{-1}, while αHPa was not cleaved. The lack of activity towards the α-helical substrate indicates selectivity for the poly-Pro II topology.

3.3.6. Peptide-Amphiphiles as Cell-Binding Ligands

When layered on hydrophobic surfaces, peptide-amphiphiles appear to line up perpendicular to the surface, even if not tightly compressed *(72,123,124)*. This property allows peptide-amphiphiles to be used as cell ligands in multiwell plate assays. Cell adhesion to substrate-coated nontissue culture treated plates is performed as described previously *(64,65,79,80,125)*:

1. PAs dissolved in PBS are diluted in 70% ethanol, added to the 96-well plate, and allowed to adsorb overnight at room temperature. In some cases, cell binding can be enhanced by mixing varying amounts of fatty acids or other lipids with the PAs *(57,72)*. This produces a surface with ligands optimally spaced for cell binding. Ratio and content of fatty acid must be determined empirically.
2. Nonspecific binding sites are blocked with 2 mg/mL BSA in PBS for 2 h at 37°C.
3. Cells are released with 5 mM EDTA in PBS or trypsin-EDTA, and washed three times with adhesion medium (20 mM HEPES, 2 mg/mL BSA in basal media).
4. 5,000–10,000 cells diluted in adhesion medium are added to the plate. The plate is incubated 60 min at 37°C.
5. Nonadherent cells are removed by washing three times with 37°C adhesion medium.
6. Cells are quantified with adhesion medium containing 10% Alamar-Blue. Cells are incubated in this medium for 4–18 h and cell number determined either spectophotometrically using A_{570} and A_{600} or fluorescently with $\lambda_{excitation} = 530$–560 nm and $\lambda_{emission} = 590$ nm, according to the manufacturer's instructions.

The extent of cell spreading can be monitored as follows. Perform the assay exactly as indicated above with the exception of the last step, cell quantitation. After washing unbound cells, the remaining cells are fixed with 2.5% glutaraldehyde dissolved in formalin, and stained with R-250 Coomassie Blue. Digital photos of each well are taken and the area of the cells quantified with the assistance of Quantity One (BioRad) or similar imaging software.

Triple-helical models of the type IV collagen α1(IV)1263–1277 sequence have been used to analyze the results of single-site glycosylation on melanoma cell recognition. Human melanoma cell adhesion has been examined for C_{16}-(Gly-Pro-Hyp)$_4$-[α1(IV)1263–1277]-(Gly-Pro-Hyp)$_4$-NH$_2$ and C_{16}-[Hyl(Gal)^{1265}]-(Gly-Pro-Hyp)$_4$-[α1(IV)1263–1277]-(Gly-Pro-Hyp)$_4$-NH$_2$ at 37°C (**Fig. 8**). The C_{16}-(Gly-Pro-Hyp)$_4$-[α1(IV)1263–1277]-(Gly-Pro-Hyp)$_4$-NH$_2$ peptide-amphiphile promoted significant adhesion of melanoma cells, with an EC$_{50}$ value of approx 2.5 μM. The glycosylated peptide-amphiphile promoted very low levels of adhesion of melanoma cells at all concentrations tested. Neither the [α1(IV)1263–1277] peptide nor the C_{16} tail alone produced significant adhesion over the concentration range studied *(126)*. Prior studies had shown that the single-stranded [α1(IV)1263–1277] peptide promotes adhesion at concentrations greater than 50 μM (EC$_{50}$ ~ 170 μM) *(9,126)*.

The ability of C_{16}-(Gly-Pro-Hyp)$_4$-[α1(IV)1263–1277]-(Gly-Pro-Hyp)$_4$-NH$_2$ and C_{16}-[Hyl(Gal)^{1265}]-(Gly-Pro-Hyp)$_4$-[α1(IV)1263–1277]-(Gly-Pro-Hyp)$_4$-NH$_2$ to promote cell spreading was next studied. Spreading was quantitated over a ligand concentration range of 0.01–50 μM (**Fig. 9**). Melanoma cell

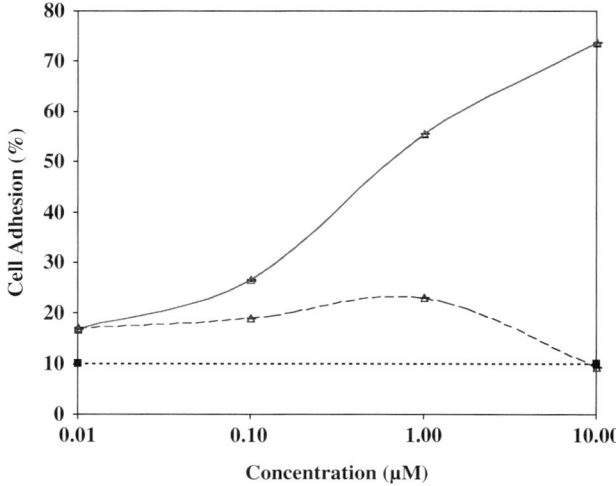

Fig. 8. Human melanoma cell adhesion to C_{16}-(Gly-Pro-Hyp)$_4$-[IV-H1]-(Gly-Pro-Hyp)$_4$-NH$_2$ (open triangles, solid line), C_{16}-[Hyl(Gal)^{1265}]-(Gly-Pro-Hyp)$_4$-[IV-H1]-(Gly-Pro-Hyp)$_4$-NH$_2$ (open triangles, dashed line), or bovine serum albumin (dashed line) at 37°C. Peptide-amphiphile concentrations were 0.01–10 μM. (Reproduced from **ref. 65**, with permission of *J. Biol. Chem.*)

spreading was more extensive on C_{16}-(Gly-Pro-Hyp)$_4$-[α1(IV)1263–1277]-(Gly-Pro-Hyp)$_4$-NH$_2$ compared with C_{16}-[Hyl(Gal)^{1265}]-(Gly-Pro-Hyp)$_4$-[α1(IV)1263–1277]-(Gly-Pro-Hyp)$_4$-NH$_2$. Representative microscopic images of melanoma cell spreading on 10 μM C_{16}-(Gly-Pro-Hyp)$_4$-[α1(IV) 1263–1277]-(Gly-Pro-Hyp)$_4$-NH$_2$ and 10 μM C_{16}-[Hyl(Gal)^{1265}]-(Gly-Pro-Hyp)$_4$-[α1(IV)1263–1277]-(Gly-Pro-Hyp)$_4$-NH$_2$ (**Fig. 10**) illustrate the modulation of cell activity based on glycosylation.

We have found that glycosylation inhibits CD44 interaction with the α1(IV)1263–1277 region derived from basement membrane collagen. This result is unexpected, as prior studies had shown that melanoma cell binding to α1(IV)1263–1277 is primarily via electrostatic interactions with Lys$_{1265}$ and Lys$_{1268}$ (**127**). Although it is possible that the glycosylation may mask the side-chain charge of residue 1265, such behavior seems unlikely given the small size of the carbohydrate. It is more likely that we have observed a specific, unfavorable carbohydrate-carbohydrate interaction between the CD44 CS and the α1(IV)1263–1277 galactose residue. This is the first demonstration of the prophylactic effects of ligand glycosylation on tumor cell interaction with the basement membrane, while a related prior report had shown that tumor cell

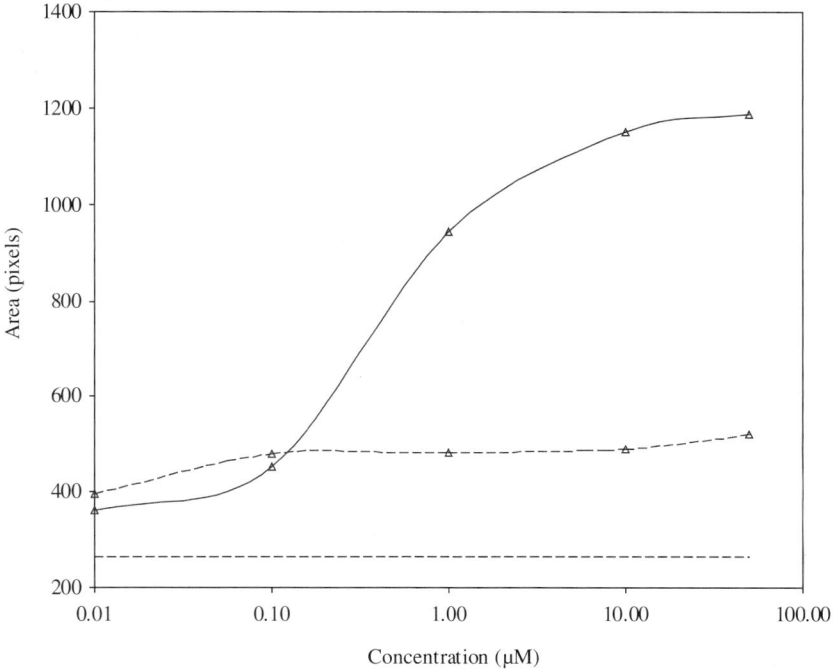

Fig. 9. Human melanoma cell spreading on C_{16}-(Gly-Pro-Hyp)$_4$-[IV-H1]-(Gly-Pro-Hyp)$_4$-NH$_2$ (open triangles, solid line), C_{16}-[Hyl(Gal)^{1265}]-(Gly-Pro-Hyp)$_4$-[IV-H1]-(Gly-Pro-Hyp)$_4$-NH$_2$ (open triangles, dashed line), or bovine serum albumin (dashed line) at 37°C. Peptide-amphiphile concentrations were 0.01–50 μM. (Reproduced from **ref. 65**, with permission of *J. Biol. Chem.*)

surface sialic acid reduced binding to type IV collagen *(128)*. Overall, little is known about how carbohydrates interact with cell surface receptors, particularly in the case of unfavorable associations *(129,130)*. More often, such interactions are favorable, as when carcinoma cell surface mucins associate with platelet P-selectin, creating a platelet "cloak" surrounding the tumor cells that aids in the metastatic process *(131)*. Although CD44 does bind certain carbohydrates (hyaluronic acid; HA), this interaction requires a minimum of six sugar residues (three repeating disaccharide units), with affinity increasing for longer HA molecules ($K_d \sim 0.3\,nM$) *(132)*. The present study suggests that glycosylation can be used for modulating tumor cell behaviors, based on carbohydrate structure and chain length.

The reduced binding of CD44/CSPG due to ligand glycosylation presents a possible "cryptic sites" mechanism by which tumor cells may invade the

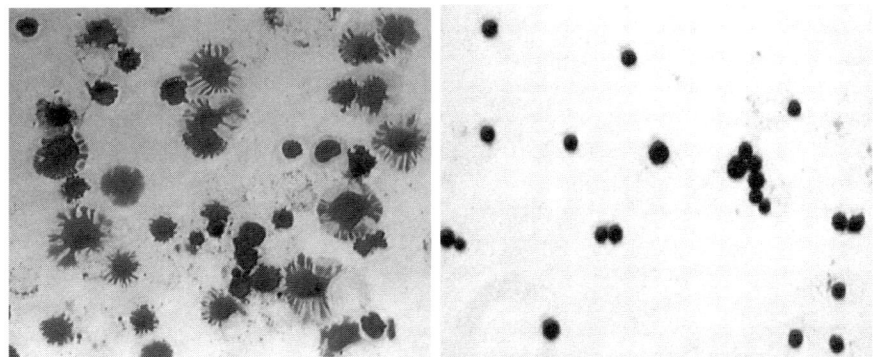

Fig. 10. Human melanoma cell spreading on (**A**) $10\,\mu M$ C_{16}-(Gly-Pro-Hyp)$_4$-[IV-H1]-(Gly-Pro-Hyp)$_4$-NH$_2$ or (**B**) $10\,\mu M$ C_{16}-[Hyl(Gal)^{1265}]-(Gly-Pro-Hyp)$_4$-[IV-H1]-(Gly-Pro-Hyp)$_4$-NH$_2$ at 37°C. (Reproduced from **ref. 65**, with permission of *J. Biol. Chem.*)

basement membrane. In the native, glycosylated state, regions within type IV collagen may have minimal interaction with receptors such as CD44/CSPG. After tumor cells bind to type IV collagen (presumably via integrins such as $\alpha 2\beta 1$), cell surface or secreted glycosidases could liberate the collagen-bound carbohydrates. This process would expose "cryptic sites" for interaction with CD44/CSPG and/or other cell surface receptors (such as the $\alpha 3\beta 1$ integrin, which also binds to a glycosylated region within type IV collagen [*15,133,134*]). Galactosylation has been shown previously to mask Lewis X antigens (*135*), and enzymes have been identified that "cycle" O-linked N-acetylglucosamine, potentially regulating nuclear and cytoplasmic signaling (*136*). In the latter case, glycosylation may be analogous to or compete with protein phosphorylation in signal transduction (*137,138*). For example, addition of O-linked β-N-acetylglucosamine to insulin receptor substrate-1 and -2 apparently decreases phosphorylation and affects insulin-mediated homeostatsis. Of direct relevance to collagen, specific enzymes have been characterized for (a) removal of glucose from disaccharide-modified Hyl (2-O-α-D-glucopyranosyl-O-β-D-galactopyranosyl-Hyl glucohydrolase [*139,140*]) and (b) transfer of galactose to Hyl (UDP-galactose:hydroxylysine-collagen (basement membrane) galactosyltransferase [*141*]). In addition, a cell surface galactosyltransferase that binds to type IV collagen and a cell-surface inactive β-galactosidase that binds elastin and laminin have been described (*142,143*). Although a deglycosylation/cryptic sites mechanism provides interesting speculation, it should also be noted that not all Lys residues in type IV collagen are fully hydroxylated and glycosylated *(133,144)*, and thus receptor interaction may just occur with the subpopulation of type IV collagen that does not contain carbohydrate.

Acknowledgments

We gratefully acknowledge the support of this work by the National Institutes of Health (CA 77402, EB 00289, and CA 98799 to G.B.F., AR 40994 to K.B.), the FAU Center of Excellence in Biomedical and Marine Biotechnology (contribution #P200505), and a Glenn/American Federation for Aging Research (AFAR) Scholarship (to J.L.L.-F.).

4. Notes

1. The use of lower substitution polyethylene glycol-based resins in combination with DBU/piperidine Fmoc deprotection and HCTU/HOBt couplings give improved yields for longer sequences or those incorporating difficult couplings.
2. By utilizing the array of readily available orthogonally protected Lys derivatives, including, but not limited to, Fmoc-Lys(Dde), Dde-Lys(Fmoc), Fmoc-Lys(Fmoc), Fmoc-Lys(Mmt), Fmoc-Lys(Mtt), or Fmoc-Lys(ivDde), an array of topological designs can be created. The dilute acid utilized for side-chain deprotection of the Mmt and Mtt groups will cleave the template from the Gly-SASRIN resin. Thus, a more acid-stable resin such as Rink Amide MBHA should be utilized in combination with Fmoc-Lys(Mmt) and Fmoc-Lys(Mtt).
3. One can create a variety of templates based on the simple Lys-branching method. The template can accommodate two to eight branches with ease and can also create a number of homo- or heteromultimeric sequences. However, as the number of branching points increases care must be taken during N-α-amino or N-ε-amino deprotection and subsequent couplings. The resin becomes increasingly "sticky" as the number of free amino groups increases, which can complicate subsequent deprotection and coupling reactions. Ensure that a proper solution volume or resin mixing action is taken to avoid deletions during peptide chain assembly.
4. Previously published reports suggest that purification prior to folding improves the success rate for disulfide formation *(145)*. Although this may be true for some specific sequences, peptides based on the sarafotoxin scaffold folded equally well without prior purification.
5. Isopropanol is included to increase α-helical content of the peptides *(51)*. The correctly folded sarafotoxin scaffold contains modest α-helical content *(52)*, thus increasing this content prior to oxidation is utilized to optimize the yields of correctly folded peptide. As discussed under **Subheading 1.**, the incorporation of an organic solvent has been shown to enhance the folding in other systems as well *(49)*.
6. Partly as a result of solubility considerations, NH_4HCO_3 proved to be slightly superior to NH_4Ac as a folding buffer. Buffers 2, 3, and 4 (*see* **Subheading 3.2.2.**)

show some differences in folding efficiency. In all cases peptides based on the sarafotoxin-scaffold are oxidized quantitatively using these conditions. Disulfide arrangement, however, varies slightly with buffer content. These variations appear to be sequence specific with a ratio of correctly to incorrectly folded peptide ranging from 3:1 to > 99:1. This project is ongoing, and thus clear-cut patterns are still emerging. For that reason, each peptide is tested under all three conditions prior to large-scale folding.

References

1. Sakakibara, S., Kishida, Y., Kikuchi, Y., Sakai, R., and Kakiuchi, K. (1968) Synthesis of poly-(L-prolyl-L-prolylglycyl) of defined molecular weights. *Bull. Chem. Soc. Jpn.* **41,** 1273.
2. DeGrado, W. F. (1980) Design of peptides and proteins. *Adv. Protein Chem.* **39,** 51–124.
3. Mayo, K. H. and Fields, G. B. (1997) Peptides as models for understanding protein folding, in *Protein Structural Biology in Bio-Medical Research* (Allewell, N., and Woodward, C., eds.) JAI Press, Inc., Greenwich, CT: pp. 567–612.
4. Lieberman, M. and Sasaki, T. (1991) Iron(II) organizes a synthetic peptide into three-helix bundles. *J. Am. Chem. Soc.* **113,** 1470–1471.
5. Mutter, M., Tuchscherer, G. G., Miller, C., et al. (1992) Template-assembled synthetic proteins with four-helix-bundle topology. *J. Am. Chem. Soc.* **114,** 1463–1470.
6. Ghadiri, M. R., Soares, C., and Choi, C. (1992) A convergent approach to protein design: metal ion-assisted spontaneous self-assembly of a polypeptide into a triple-helix bundle protein. *J. Am. Chem. Soc.* **114,** 825–831.
7. Ghadiri, M. R., Soares, C., and Choi, C. (1992) Design of an artificial four-helix bundle metalloprotein via a novel ruthenium(II)-assisted self-assembly process. *J. Am. Chem. Soc.* **114,** 4000–4002.
8. Dawson, P. E. and Kent, S. B. H. (1993) Convenient total synthesis of a 4-helix TASP molecule by chemoselective ligation. *J. Am. Chem. Soc.* **115,** 7263–7266.
9. Fields, C. G., Mickelson, D. J., Drake, S. L., McCarthy, J. B., and Fields, G. B. (1993) Melanoma cell adhesion and spreading activities of a synthetic 124-residue triple-helical "mini-collagen". *J. Biol. Chem.* **268,** 14,153–14,160.
10. Fields, C. G., Lovdahl, C. M., Miles, A. J., Matthias-Hagen, V. L., and Fields, G. B. (1993) Solid-phase synthesis and stability of triple-helical peptides incorporating native collagen sequences. *Biopolymers* 33, 1695–1707.
11. Vuilleumer, S. and Mutter, M. (1993) Synthetic peptide and template-assembled synthetic protein models of the hen egg white lysozyme 87–97 helix. *Biopolymers* **33,** 389–400.
12. Tuchscherer, G., Grell, D., Mathieu, M., and Mutter, M. (1999) Extending the concept of template-assembled synthetic proteins, *J. Peptide Res.* **54,** 185–194.

13. Dai, Q., Prorok, M., and Castellino, F. J. (2004) A new mechanism for metal ion-assisted interchain helix assembly in a naturally occurring peptide mediated by optimally spaced γ-carboxyglutamic acid residues. *J. Mol. Biol.* **336,** 731–744.

14. Becker, C. F. W., Oblatt-Montal, M., Kochendoerfer, G. G., and Montal, M. (2004) Chemical synthesis and single channel properties of tetrameric and pentameric TASPs (template-assembled synthetic proteins) derived from the transmembrane domain of HIV virus protein u (Vpu). *J. Biol. Chem.* **279,** 17,483–17,489.

15. Miles, A. J., Skubitz, A. P. N., Furcht, L. T., and Fields, G. B. (1994) Promotion of cell adhesion by single-stranded and triple-helical peptide models of basement membrane collagen α1(IV)531-543: evidence for conformationally dependent and corformationally independent type IV collagen cell adhesion sites. *J. Biol. Chem.* **269,** 30,939–30,945.

16. Barnes, M. J., Knight, C. G., and Farndale, R. W. (1996) The use of collagen-based model peptides to investigate platelet-reactive sequences in collagen. *Biopolymers (Peptide Sci.)* **40,** 383–397.

17. Knight, C. G., Morton, L. F., Onley, D. J., et al. (1998) Identification in collagen type I of an integrin α2β1-binding site containing an essential GER sequence. *J. Biol. Chem.* **273,** 33,287–33,294.

18. Emsley, J., Knight, C. G., Farndale, R. W., Barnes, M. J., and Liddington, R. C. (2000) Structural basis of collagen recognition by integrin α2β1. *Cell* **101,** 47–56.

19. Roth, W., and Heidemann, E. (1980) Triple helix-coil transition of covalently bridged collagen-like peptides. *Biopolymers* **19,** 1909–1917.

20. Mutter, M., Hersperger, R., Gubernator, K., and Muller, K. (1989) The construction of new proteins: a template-assembled synthetic protein (TASP) containing both a 4-helix bundle and β-barrel-like structure. *Proteins* **5,** 13–21.

21. Tuchscherer, G., Domer, B., Sila, U., Kamber, B., and Mutter, M. (1993) The TASP concept: mimetics of peptide ligands, protein surfaces and folding units. *Tetrahedron* **49,** 3559–3575.

22. Tuchscherer, G. (1993) Template assembled synthetic proteins: condensation of a multifunctional peptide to a topological template via chemoselective ligation. *Tetrahedron Lett.* **34,** 8419–8422.

23. Dumy, P., Eggleston, I. M., Cervigni, S., Sila, U., Sun, X., and Mutter, M. (1995) A convenient synthesis of cyclic peptides as regioselectively addressable functionalized templates (RAFT). *Tetrahedron Lett.* **36,** 1255–1258.

24. Peluso, S., Dumy, P., Eggleston, I. M., Garrouste, P., and Mutter, M. (1997) Protein mimetics (TASP) by sequential condensation of peptide loops to an immobilised topological template. *Tetrahedron* **53,** 7231–7236.

25. Goodman, M., Feng, Y., Melacini, G., and Taulane, J. P. (1996) A template-induced incipient collagen-like triple-helical structure. *J. Am. Chem. Soc.* **118,** 5156–5157.

26. Goodman, M., Melacini, G., and Feng, Y. (1996) Collagen-like triple helices incorporating peptoid residues. *J. Am. Chem. Soc.* **118,** 10,928–10,929.

27. Feng, Y., Melacini, G., Taulane, J. P., and Goodman, M. (1996) Acetyl-terminated and template-assembled collagen-based polypeptides composed of Gly-Pro-Hyp sequences. 2. Synthesis and conformational analysis by circular dichroism, ultraviolet absorbance, and optical rotation. *J. Am. Chem. Soc.* **118**, 10,351–10,358.

28. Feng, Y., Melacini, G., Taulane, J. P., and Goodman, M. (1996) Collagen-based structures containing the peptoid tesidue N-isobutylglycine (Nleu): synthesis and biophysical studies of Gly-Pro-Nleu sequences by circular dichroism, ultraviolet absorbance, and optical rotation. *Biopolymers* **39**, 859–872.

29. Melacini, G., Feng, Y., and Goodman, M. (1996) Collagen-based structures containing the peptoid residue N-isobutylglycine (Nleu). 6. Conformational analysis of Gly-Pro-Nleu sequences by 1H NMR, CD, and molecular modeling. *J. Am. Chem. Soc.* **118**, 10,725–10,732.

30. Melacini, G., Feng, Y., and Goodman, M. (1997) Collagen-based structures containing the peptoid residue N-isobutylene (Nleu): conformational analysis of Gly-Nleu-Pro sequences by 1H-NMR and molecular modeling. *Biochemistry* **36**, 8725–8732.

31. Feng, Y., Melacini, G., and Goodman, M. (1997) Collagen-based structures containing the peptoid residue N-isobutylglycine (Nleu): synthesis and biophysical studies of Gly-Nleu-Pro sequences by circular dichroism and optical rotation, *Biochemistry* **36**, 8716–8724.

32. Kwak, J., De Capua, A., Locardi, E., and Goodman, M. (2002) TREN (Tris(2-aminoethyl)amine): an effective scaffold for the assembly of triple helical collagen mimetic structures. *J. Am. Chem. Soc.* **124**, 14,085–14,091.

33. Oh, J., Takahashi, R., Kondo, S., et al. (2001) The membrane-anchored MMP inhibitor RECK is a key regulator of extracellular matrix integrity and angiogenesis. *Cell* **107**, 789–800.

34. Kinberger, G. A., Cai, W., and Goodman, M. (2002) Collagen mimetic dendrimers. *J. Am. Chem. Soc.* **124**, 15,162–15,163.

35. Rump, E. T., Rijkers, D. T. S., Hilbers, H. W., de Groot, P. G., and Liskamp, R. M. J. (2002) Cyclotriveratrylene (CTV) as a new chiral triacid scaffold capable of inducing triple helix formation of collagen peptides containing either a native sequence or Pro-Hyp-Gly repeats. *Chem. Eur. J.* **8**, 4613–4621.

36. Brask, J. and Jensen, K. J. (2001) Carboproteins: a 4-α-helix bundle protein model assembled on a D-galactopyranoside template. *Bioorg. Med. Chem. Lett.* **11**, 697–700.

37. Thulstrup, P. W., Brask, J., Jensen, K. J., and Larsen, E. (2005) Synchroton radiation circular dichroism spectroscopy applied to metmyoglobin and a 4-α-helix bundle carboprotein. *Biopolymers* **78**, 46–52.

38. Diekmann, G. R., McRorie, D. K., Lear, J. D., Sharp, K. A., DeGrado, W. F., and Pecoraro, V. L. (1998) The role of protonation and metal chelation preferences in defining the properties of mercury-binding coiled coils. *J. Mol. Biol.* **280**, 897–912.

39. Kohn, W. D., Kay, C. M., Sykes, B. D., and Hodges, R. S. (1998) Metal ion induced folding of a de novo designed coiled-coil peptide. *J. Am. Chem. Soc.* **120,** 1124–1132.
40. Li, X., Suzuki, K., Kanaori, K., Tajima, K., Kashiwada, A., and Hiroaki, H. (2000) Soft metal ions, Cd(II) and Hg(II), induce triple-stranded alpha-helical assembly and folding of a de novo designed peptide in their trigonal geometries. *Protein Sci.* **9,** 1327–1333.
41. Cai, W., Kwok, S. W., Taulane, J. P., and Goodman, M. (2004) Metal-assisted assembly and stabilization of collagen-like triple helices. *J. Am. Chem. Soc.* **126,** 15,030–15,031.
42. Fields, C. G., Grab, B., Lauer, J. L., Miles, A. J., Yu, Y.-C., and Fields, G. B. (1996) Solid-phase synthesis of triple-helical collagen-model peptides. *Lett. Peptide Sci.* **3,** 3–16.
43. Barthe, P., Rochette, S., Vita, C., and Roumestand, C. (2000) Synthesis and NMR solution structure of an α-helical hairpin stapled with two disulfide bridges. *Protein Sci.* **9,** 942–955.
44. Blandl, T., Cochran, A. G., and Skelton, N. J. (2003) Turn stability in β-hairpin peptides: investigation of peptides containing 3:5 type I G1 bulge turns. *Protein Sci.* **12,** 237–247.
45. Weston, C. J., Cureton, C. H., Calvert, M. J., Smart, O. S., and Allemann, R. K. (2004) A stabile miniature protein with oxaloacetate decarboxylase activity. *ChemBioChem* **5,** 1075–1080.
46. Heitz, A., Le-Nguyen, D., and Chiche, L. (1999) Min-21 and Min-23, the smallest peptides that fold like a cystine-stabilized β-sheet motif: design, solution structure, and thermal stability. *Biochemistry* **38,** 10,615–10,625.
47. Craik, D. J., Daly, N. L., and Waine, C. (2001) The cystine knot motif in toxins and implications for drug design. *Toxicon* **39,** 43–60.
48. Aumelas, A., Chiche, L., Kubo, S., Chino, N., Watanabe, T. X., and Kobayashi, Y. (1999) The chimeric peptide [Lys(-2)-Arg(-1)]-sarafotoxin-S6b, composed of the endothelin pro-sequence and sarafotoxin, retains the salt-bridge staple between Arg(-1) and Asp8 previously observed in [Lys(-2)-Arg(-1)]-endothelin. Implications of this salt-bridge in the contractile activity and the oxidative folding reaction. *Eur. J. Biochem.* **266,** 977–985.
49. Nielsen, J. S., Buczek, P., and Bulaj, G. (2004) Cosolvent-assisted oxidative folding of a bicyclic α-conotoxin ImI. *J. Peptide Sci.* **10,** 249–256.
50. Tam, J. P., Dong, X., and Wu, C.-R. (1993) Solvent chaperone in protein folding: selective enhancement of disulfide isomers of endothelin, in *Peptide Chemistry 1992: Proceedings of the 2nd Japan Symposium on Peptide Chemistry* (Yanaihara, N., ed.). Escom, Leiden, The Netherlands: pp. 24–26.
51. Kumaran, S. and Roy, R. P. (1999) Helix-enhancing propensity of fluoro and alkyl alcohols: influence of pH, temperature and cosolvent concentration on the helical conformation of peptides. *J. Peptide Res.* **53,** 284–293.

52. Mills, R. G., Atkins, A. R., Harvey, T., Junius, F. K., Smith, R., and King, G. F. (1991) Conformation of sarafotoxin-6b in aqueous solution determined by NMR spectroscopy and distance geometry. *FEBS Lett.* **282,** 247–252.

53. Kubo, S., Chino, N., Nakajima, K., et al. (1997) Improvement in the oxidative folding of endothelin-1 by a Lys-Arg extension at the amino terminus: implication of a salt bridge between Arg-1 and Asp8. *Lett. Peptide Sci.* **4,** 185–192.

54. Woessner, J. F. and Nagase, H. (2000) *Matrix Metalloproteinases and TIMPs.* Oxford University Press, Oxford.

55. Birkedal-Hansen, H., Moore, W. G. I., Bodden, M. K., et al. (1993) Matrix metalloproteinases: a review. *Crit. Rev. Oral Biol. Med.* **4,** 197–250.

56. Berndt, P., Fields, G. B., and Tirrell, M. (1995) Synthetic lipidation of peptides and amino acids: monolayer structure and properties. *J. Am. Chem. Soc.* **117,** 9515–9522.

57. Pakalns, T., Haverstick, K. L., Fields, G. B., McCarthy, J. B., Mooradian, D. L., and Tirrell, M. (1999) Cellular recognition of synthetic peptide amphiphiles in self-assembled monolayer films. *Biomaterials* **20,** 2265–2279.

58. Dillow, A. K., Ochsenhirt, S. E., McCarthy, J. B., Fields, G. B., and Tirrell, M. (2001) Adhesion of $\alpha5\beta1$ receptors to biomimetic substrates constructed from peptide amphiphiles. *Biomaterials* **22,** 1493–1505.

59. Rosler, A., Klok, H. A., Hamlye, I. W., Castelletto, V., and Mykhaylyk, O. O. (2003) Nanoscale structure of poly(ethylene glycol(hybrid block copolymers containing amphiphilic β-strand peptide sequences. *Biomacromolecules* **4,** 859–863.

60. Vandermeulen, G. W. M., and Klok, H. A. (2004) Peptide/protein hybrid materials: enhanced control of structure and improved performance through conjugation of biological and synthetic polymers. *Macromol. Biosci.* **4,** 383–398.

61. Yu, Y.-C., Berndt, P., Tirrell, M., and Fields, G. B. (1996) Self-assembling amphiphiles for construction of protein molecular architecture. *J. Am. Chem. Soc.* **118,** 12,515–12,520.

62. Yu, Y.-C., Tirrell, M., and Fields, G. B. (1998) Minimal lipidation stabilizes protein-like molecular architecture. *J. Am. Chem. Soc.* **120,** 9979–9987.

63. Yu, Y.-C., Roontga, V., Daragan, V. A., Mayo, K. H., Tirrell, M., and Fields, G. B. (1999) Structure and dynamics of peptide-amphiphiles incorporating triple-helical proteinlike molecular architecture. *Biochemistry* **38,** 1659–1668.

64. Malkar, N. B., Lauer-Fields, J. L., Borgia, J. A., and Fields, G. B. (2002) Modulation of triple-helical stability and subsequent melanoma cellular responses by single-site substitution of fluoroproline derivatives. *Biochemistry* **41,** 6054–6064.

65. Lauer-Fields, J. L., Malkar, N. B., Richet, G., Drauz, K., and Fields, G. B. (2003) Melanoma cell CD44 interaction with the $\alpha1(IV)1263$-1277 region from basement membrane collagen is modulated by ligand glycoslyation. *J. Biol. Chem.* **278,** 14,321–14,330.

66. Mardilovich, A. and Kokkoli, E. (2004) Biomimetic peptide-amphiphiles for functional biomaterials: the role of GRGDSP and PHSRN. *Biomacromolecules* **5,** 950–957.
67. Kokkoli, E., Ochsenhirt, S. E., and Tirrell, M. (2004) Collective and single-molecule interactions of α5β1 integrins. *Langmuir* **20,** 2397–2404.
68. Silva, G. A., Czeisler, C., Niece, K. L., et al. (2004) Selective differentiation of neural progenitor cells by high-epitope density nanofibers. *Science* **303,** 1352–1355.
69. Forns, P., Lauer-Fields, J. L., Gao, S., and Fields, G. B. (2000) Induction of protein-like molecular architecture by monoalkyl hydrocarbon chains. *Biopolymers* **54,** 531–546.
70. Hartgerink, J. D., Beniash, E., and Stupp, S. I. (2001) Self-assembly and mineralization of peptide-amphiphile nanofibers. *Science* **297,** 1684–1688.
71. Malkar, N. B., Lauer-Fields, J. L., Juska, D., and Fields, G. B. (2003) Characterization of peptide-amphiphiles possessing cellular activation sequences. *Biomacromolecules* **4,** 518–528.
72. Yu, Y.-C., Pakalns, T., Dori, Y., McCarthy, J. B., Tirrell, M., and Fields, G. B. (1997) Construction of biologically active protein molecular architecture using self-assembling peptide-amphiphiles. *Meth. Enzymol.* **289,** 571–587.
73. Lauer-Fields, J. L., Broder, T., Sritharan, T., Nagase, H., and Fields, G. B. (2001) Kinetic analysis of matrix metalloproteinase triple-helicase activity using fluorogenic substrates. *Biochemistry* **40,** 5795–5803.
74. Lauer-Fields, J. L. and Fields, G. B. (2002) Triple-helical peptide analysis of collagenolytic protease activity. *Biol. Chem.* **383,** 1095–1105.
75. Lauer-Fields, J. L., Sritharan, T., Stack, M. S., Nagase, H., and Fields, G. B. (2003) Selective hydrolysis of triple-helical substrates by matrix metalloproteinase-2 and -9. *J. Biol. Chem.* **278,** 18,140–18,145.
76. Lauer-Fields, J. L., Kele, P., Sui, G., Nagase, H., Leblanc, R. M., and Fields, G. B. (2003) Analysis of matrix metalloproteinase activity using triple-helical substrates incorporating fluorogenic L- or D-amino acids. *Anal. Biochem.* **321,** 105–115.
77. Minond, D., Lauer-Fields, J. L., Nagase, H., and Fields, G. B. (2004) Matrix metalloproteinase triple-helical peptidase activities are differentially regulated by substrate stability. *Biochemistry* **43,** 11,474–11,481.
78. Tu, R., Mohanty, K., and Tirrell, M. (2004) Liposomal targeting through peptide-amphiphile functionalization. *Am. Pharm. Rev.* **7(2),** 36–41.
79. Baronas-Lowell, D., Lauer-Fields, J. L., and Fields, G. B. (2004) Induction of endothelial cell activation by a triple-helical α2β1 integrin ligand derived from type I collagen α1(I)496-507. *J. Biol. Chem.* **279,** 952–962.
80. Baronas-Lowell, D., Lauer-Fields, J. L., Borgia, J. A., et al. (2004) Differential modulation of human melanoma cell metalloproteinase expression by α2β1 integrin and CD44 triple-helical ligands derived from type IV collagen. *J. Biol. Chem.* **279,** 43,503–43,513.

81. Lockwood, N. A., Haseman, J. R., Tirrell, M. V., and Mayo, K. H. (2004) Acylation of SC dodecapeptide increases bactericidal potency against Gram-positive bacteria, including drug-resistant strains. *Biochem. J.* **378,** 93–103.
82. Chu-Kung, A. F., Bozzelli, K. N., Lockwood, N. A., Haseman, J. R., Mayo, K. H., and Tirrell, M. V. (2004) Promotion of peptide antimicrobial activity by fatty acid conjugation. *Bioconjugate Chem.* **15,** 530–535.
83. Chhabra, S. R., Hothi, B., Evans, D. J., White, P. D., Bycroft, B. W., and Chan, W. C. (1998) An appraisal of new variants of Dde amine protecting group for solid phase peptide synthesis. *Tetrahedron Lett.* **39,** 1603–1606.
84. Rohwedder, B., Mutti, Y., Dumy, P., and Mutter, M. (1998) Hydrazinolysis of Dde: complete orthogonality with Aloc protecting groups. *Tetrahedron Lett.* **39,** 1175–1178.
85. Kates, S. A., Daniels, S. B., and Albericio, F. (1993) Automated allyl cleavage for continuous-flow synthesis of cyclic and branched peptides. *Anal. Biochem.* **212,** 303–310.
86. King, D. S., Fields, C. G., and Fields, G. B. (1990) A cleavage method which minimizes side reactions following Fmoc solid phase peptide synthesis. *Int. J. Peptide Protein Res.* **36,** 255–266.
87. Fields, C. G. and Fields, G. B. (1993) Minimization of tryptophan alkylation following 9-fluorenylmethoxycarbonyl solid-phase peptide synthesis. *Tetrahedron Lett.* **34,** 6661–6664.
88. Henkel, W., Vogl, T., Echner, H., et al. (1999) Synthesis and folding of native collagen III model peptides. *Biochemistry* **38,** 13,610–13,622.
89. Lauer-Fields, J. L., Nagase, H., and Fields, G. B. (2000) Use of Edman degradation sequence analysis and matrix-assisted laser desorption/ionization mass spectrometry in designing substrates for matrix metalloproteinases. *J. Chromatogr. A.* **890,** 117–125.
90. Knutson, J. R., Iida, J., Fields, G. B., and McCarthy, J. B. (1996) CD44/chondroitin sulfate proteoglycan and α2β1 integrin mediate human melanoma cell migration on type IV collagen and invasion of basement membranes. *Mol. Biol. Cell* **7,** 383–396.
91. Takahashi, K., Eto, H., and Tanabe, K. K. (1999) Involvement of CD44 in matrix metalloproteinase-2 regulation in human melanoma cells. *Int. J. Cancer* **80,** 387–395.
92. Kimura, S., Kasuya, Y., Sawamura, T., et al. (1988) Structure-activity relationships of endothelin: importance of the C-terminal moiety. *Biochem. Biophys. Res. Commun.* **156,** 1182–1186.
93. Ellman, G. L. (1959) Tissue sulfhydryl groups. *Arch. Biochem. Biophys.* **82,** 70–77.
94. Knight, C. G., Willenbrock, F., and Murphy, G. (1992) A novel coumarin-labelled peptide for sensitive continuous assays of the matrix metalloproteinases. *FEBS Lett.* **296,** 263–266.

95. Nagase, H., Fields, C. G., and Fields, G. B. (1994) Design and characterization of a fluorogenic substrate selectively hydrolyzed by stromelysin 1 (matrix metalloproteinase-3). *J. Biol. Chem.* **269**, 20,952–20,957.
96. Fields, C. G., Grab, B., Lauer, J. L., and Fields, G. B. (1995) Purification and analysis of synthetic, triple-helical "minicollagens" by reversed-phase high-performance liquid chromatography. *Anal. Biochem.* **231**, 57–64.
97. Grab, B., Miles, A. J., Furcht, L. T., and Fields, G. B. (1996) Promotion of fibroblast adhesion by triple-helical peptide models of type I collagen-derived sequences. *J. Biol. Chem.* **271**, 12,234–12,240.
98. Lauer-Fields, J. L., Tuzinski, K. A., Shimokawa, K., Nagase, H., and Fields, G. B. (2000) Hydrolysis of triple-helical collagen peptide models by matrix metalloproteinases. *J. Biol. Chem.* **275**, 13,282–13,290.
99. Knight, C. G. (1991) A quenched fluorescent substrate for thimet peptidase containing a new fluorescent amino acid, DL-2-amino-3-(7-methoxy-4-coumaryl)propionic acid. *Biochem. J.* **274**, 45–48.
100. Anastasi, A., Knight, C. G., and Barrett, A. J. (1993) Characterization of the bacterial metalloendopeptidase pitrilysin by use of a continuous fluorescence assay. *Biochem. J.* **290**, 601–607.
101. Lauer-Fields, J. L., Juska, D., and Fields, G. B. (2002) Matrix metalloproteinases and collagen catabolism. *Biopolymers (Peptide Sci.)* **66**, 19–32.
102. Malkar, N. B. and Fields, G. B. (2001) Synthesis of $N\alpha$-(fluoren-9-ylmethoxycarbonyl)-$N\varepsilon$-[(7-methoxycoumarin-4-yl)acetyl]-L-lysine for use in solid-phase synthesis of fluorogenic substrates. *Lett. Peptide Sci.* **7**, 263–267.
103. Hurst, D. R., Schwartz, M. A., Ghaffari, M. A., et al. (2004) Catalytic- and ecto-domains of membrane type 1-matrix metalloproteinase have similar inhibition profiles but distinct endopeptidase activities. *Biochem. J.* **377**, 775–779.
104. Nagase, H. and Fields, G. B. (1996) Human matrix metalloproteinase specificity studies using collagen sequence-based synthetic peptides. *Biopolymers* **40**, 399–416.
105. Fields, G. B. (1991) A model for interstitial collagen catabolism by mammalian collagenases. *J. Theor. Biol.* **153**, 585–602.
106. Mucha, A., Cuniasse, P., Kannan, R., et al. (1998) Membrane type-1 matrix metalloproteinase and stromelysin-3 cleave more efficiently synthetic substrates containing unusual amino acids in their P'_1 positions. *J. Biol. Chem.* **273**, 2763–2768.
107. Ohuchi, E., Imai, K., Fujii, Y., Sato, H., Seiki, M., and Okada, Y. (1997) Membrane type I matrix metalloproteinase digests intersitial collagens and other extracellular matrix macromolecules. *J. Biol. Chem.* **272**, 2446–2451.
108. Miller, E. J., Finch, J. E., Jr., Chung, E., Butler, W. T., and Robertson, P. B. (1976) Specific cleavage of the native type III collagen molecule with trypsin. *Arch. Biochem. Biophys.* **173**, 631–637.

109. Birkedal-Hansen, H., Taylor, R. E., Bhown, A. S., Katz, J., Lin, H.-Y., and Wells, B. R. (1985) Cleavage of bovine skin type III collagen by proteolytic enzymes. *J. Biol. Chem.* **260**, 16,411–16,417.

110. Bächinger, H. P., Bruckner, P., Timpl, R., Prockop, D. J., and Engel, J. (1980) Folding mechanism of the triple helix in type-III collagen and type-III pN-collagen. *Eur. J. Biochem.* **106**, 619–632.

111. Bruckner, P. and Prockop, D. J. (1981) Proteolytic enzymes as probes for the triple-helical conformation of procollagen. *Anal. Biochem.* **110**, 360–368.

112. Ryhänen, L., Zaragoza, E. J., and Uitto, J. (1983) Conformational stability of type I collagen triple helix: evidence for temporary and local relaxation of the protein conformation using a proteolytic probe. *Arch. Biochem. Biophys.* **223**, 562–571.

113. Sieron, A. L., Fertala, A., Ala-Kokko, L., and Prockop, D. J. (1993) Deletion of a large domain in recombinant human procollagen II does not alter the thermal stability of the triple helix. *J. Biol. Chem.* **268**, 21,232–21,237.

114. Fan, P., Li, M. H., Brodsky, B., and Baum, J. (1993) Backbone dynamics of (Pro-Hyp-Gly)$_{10}$ and a designed collagen-like triple-helical peptide by ^{15}N NMR relaxation and hydrogen-exchange measurements. *Biochemistry* **32**, 13,299–13,309.

115. Fiori, S., Saccá, B., and Moroder, L. (2002) Structural properties of a collagenous heterotrimer that mimics the collagenase cleavage site of collagen type I. *J. Mol. Biol.* **319**, 1235–1242.

116. Stultz, C. M. (2002) Localized unfolding of collagen explains collagenase cleavage near imino-poor sites. *J. Mol. Biol.* **319**, 997–1003.

117. Niyibizi, C., Chan, R., Wu, J.-J., and Eyre, D. (1994) A 92 kDa gelatinase (MMP-9) cleavage site in native type V collagen. *Biochem. Biophys. Res. Commun.* **202**, 328–333.

118. Nagase, H. and Kashiwagi, M. (2003) Aggrecanases and cartilage matrix degradation. *Arthritis Res. Ther.* **5**, 94–103.

119. Tanzawa, K., Berger, J., and Prockop, D. J. (1985) Type I procollagen N-proteinase from whole chick embryos: cleavage of a homotrimer of pro-α1(I) chains and the requirement for procollagen with a triple-helical conformation. *J. Biol. Chem.* **260**, 1120–1126.

120. Arnold, W. V., Fertala, A., Sieron, A. L., et al. (1998) Recombinant procollagen II: deletion of D period segments identifies sequences that are required for helix stabilization and generates a temperature-sensitive N-proteinase cleavage site. *J. Biol. Chem.* **273**, 31,822–31,828.

121. Miller, J. A., Liu, R.-Q., Davis, G. L., Pratta, M. A., Trzaskos, J. M., and Copeland, R. A. (2003) A microplate assay specific for the enzyme aggrecanase. *Anal. Biochem.* **314**, 260–265.

122. Lauer-Fields, J. L., Sritharan, T., Kashiwagi, M., Nagase, H., and Fields, G. B. (2007) Substrate conformation modulates aggrecanase (ADAMTS-4) affinity and sequence specificity: suggestion of a common topological specificity of functionally diverse proteases. *J. Biol. Chem.* **282**, in press.

123. Dori, Y., Bianco-Peled, H., Satija, S. K., Fields, G. B., McCarthy, J. B., and Tirrell, M. (2000) Ligand accessibility as a means to control cell response to bioactive bilayer membranes. *J. Biomed. Mater. Res.* **50**, 75–81.
124. Bianco-Peled, H., Dori, Y., Schneider, J., Sung, L.-P., Satija, S., and Tirrell, M. (2001) Structural study of langmuir monolayers containing lipidated poly(ethylene glycol) and peptides. *Langmuir* **17**, 6931–6937.
125. Lauer, J. L., Gendron, C. M., and Fields, G. B. (1998) Effect of ligand conformation on melanoma cell α3β1 integrin-mediated signal transduction events: implications for a collagen structural modulation mechanism of tumor cell invasion. *Biochemistry* **37**, 5279–5287.
126. Fields, G. B., Lauer, J. L., Dori, Y., Forns, P., Yu, Y.-C., and Tirrell, M. (1998) Proteinlike molecular architecture: biomaterial applications for inducing cellular receptor binding and signal transduction. *Biopolymers* **47**, 143–151.
127. McCarthy, J. B., Mickelson, D. J., Fields, C. G., and Fields, G. B. (1993) The use of collagen-model peptides to correlate collagen primary and secondary structural effects with the mechanisms of tumor cell adhesion, motility and invasion, in *Peptides 1992* (Schneider, C. H., and Eberle, A. N., eds.). Escom Science Publishers, Leiden, The Netherlands: pp. 109–110.
128. Dennis, J., Waller, C., Timpl, R., and Schirrmacher, V. (1982) Surface sialic acid reduces attachment of metastatic tumour cells to collagen type IV and fibronectin. *Nature* **300**, 274–276.
129. Hakomori, S. (1990) Bifunctional role of glycosphingolipids: modulators for transmembrane signaling and mediators for cellular interactions. *J. Biol. Chem.* **265**, 18,713–18,716.
130. Spillmann, D. and Burger, M. M. (1996) Carbohydrate-carbohydrate interactions in adhesion. *J. Cell. Biochem.* **61**, 562–568.
131. Borsig, L., Wong, R., Feramisco, J., Nadeau, D. R., Varki, N. M., and Varki, A. (2001) Heparin and cancer revisited: mechanistic connections involving platelets, P-selectin, carcinoma mucins, and tumor metastasis. *Proc. Natl. Acad. Sci. USA* **98**, 3352–3357.
132. Naor, D., Slonov, R. V., and Ish-Shalom, D. (1997) CD44: structure, function, and association with the malignant process, in *Advances in Cancer Research* (Vande Woude, G. F. and Klein, G., eds.) Academic, Orlando, FL: pp. 241–319.
133. Babel, W. and Glanville, R. W. (1984) Structure of human-basement-membrane (type IV) collagen: complete amino-acid sequence fo a 914-residue-long pepsin fragment from the α1(IV) chain. *Eur. J. Biochem.* **143**, 545–556.
134. Miles, A. J., Knutson, J. R., Skubitz, A. P. N., Furcht, L. T., McCarthy, J. B., and Fields, G. B. (1995) A peptide model of basement membrane collagen α1(IV) 531-543 binds the α3β1 integrin. *J. Biol. Chem.* **270**, 29,047–29,050.
135. Cho, S. K., Yeh, J.-C., Cho, M., and Cummings, R. D. (1996) Transcriptional regulation of α1,3-galactosyltransferase in embryonal carcinoma cells by retinoic acid. *J. Biol. Chem.* **271**, 3238–3246.

136. Iyer, S. P. and Hart, G. W. (2003) Dynamic nuclear and cytoplasmic glycosylation: enzymes of O-GlcNAc cycling. *Biochemistry* **42**, 2493–2499.
137. Haltiwanger, R. S., Kelly, W. G., Roquemore, E. P., et al. (1992) Glycosylation of nuclear and cytoplasmic proteins is ubiquitous and dynamic. *Biochem. Soc. Trans.* **20**, 264–269.
138. Wells, L., Vosseller, K., and Hart, G. W. (2001) Glycosylation of nucleoplasmic proteins: signal transduction and O-GlcNAc. *Science* **291**, 2376–2378.
139. Hamazaki, H. and Hotta, K. (1980) Enzymatic hydrolysis of disaccharide unit of collagen: isolation of 2-O-alpha-D-glucopyranosyl-O-beta-D-galactopyranosyl-hydroxylysine glucohydrolase from rat spleens. *Eur. J. Biochem.* **111**, 587–591.
140. Ishii, I., Iwase, H., Hamazaki, H., and Hotta, K. (1987) Comparative study of specific alpha-1,2-glucosidase activity toward glucosyl galactosyl hydroxylysine in various animal species. *Comp. Biochem. Physiol. B* **88**, 313–316.
141. Spiro, M. J. and Spiro, R. G. (1971) Studies on the biosynthesis of the hydroxylysine-linked disaccharide unit of basement membranes and collagens II: kidney galactosyltransferase. *J. Biol. Chem.* **246**, 4910–4918.
142. Babiarz, B. and Cullen, E. (1992) 3T3 Cell surface galactosyltransferase is a calcium-dependent adhesion molecule for collagen type IV. *Exp. Cell Res.* **203**, 276–279.
143. Privitera, S., Prody, C. A., Callahan, J. W., and Hinek, A. (1998) The 67-kDa enzymatically inactive alternatively spliced variant of β-galactosidase is identical to the elastin/laminin-binding protein. *J. Biol. Chem.* **273**, 6319–6326.
144. Nayak, B. R. and Spiro, R. G. (1991) Localization and structure of the asparagine-linked oligosaccharides of type IV collagen from glomerular basement membrane and lens capsule. *J. Biol. Chem.* **266**, 13,978–13,987.
145. Andreu, D., Albericio, F., Sole, N. A., Munson, M. C., Ferrer, M., and Barany, G. (1994) Formation of disulfide bonds in synthetic peptides and proteins, in *Peptide Synthesis Protocols: Methods in Molecular Biology, Vol. 35* (Pennington, M. W. and Dunn, B. M., eds.). Humana, Totowa, NJ: pp. 91–169.

6

Proteolytic Profiling of the Extracellular Matrix Degradome

Diane Baronas-Lowell, Janelle L. Lauer-Fields, Mohammad Al-Ghoul, and Gregg B. Fields

Abstract

The profiling of protein function is one of the most challenging scientific tasks in the postgenomic age. Traditional protein expression methodologies have focused only on the quantification of proteins under varying conditions or pathologies. Determining the functional differences between protein populations allows for a more accurate view of the outcomes in normal vs diseased proteomes. Because the presence or absence of a protein's function can affect its complex surroundings (consisting of multiple other proteins and substrates), the study of proteome functionality yields information on protein-protein interactions, amplification cascades, signaling pathways, and posttranslational modifications. Of significant interest are proteinases, as proteolysis is responsible for tight regulation of various cellular and tissue processes. Proteinase activities, or lack there of, alter the proteome makeup by regulating other proteins or by generating cleavage products. This chapter describes current proteolytic profiling technologies using activity or target-based formats. In particular, the analysis of collagenolytic matrix metalloproteinase activity using fluorogenic triple-helical substrates is discussed.

Key Words: Proteolytic profiling; proteomics; fluorogenic substrates; matrix metalloproteinases; collagen; protein expression.

1. Introduction

This is the era of analyzing global cellular processes. Efforts have generated information on whole genome sequences and their expression profiles, including their transcriptomes and proteomes. However, most medicinal questions, including those in drug discovery, ultimately focus on altered or altering protein activity. Protein activity is often not directly related to protein expression. In

From: *Methods in Molecular Biology, vol. 386: Peptide Characterization and Application Protocols*
Edited by: G. Fields © Humana Press Inc., Totowa, NJ

order for a protein to be functional, it is subjected to regulation by posttranslational modifications, associations with other proteins resulting in collective activities/functions, and/or interactions with mechanism-based inhibitors.

Traditional abundance-based proteomic techniques are riddled with technical limitations. For instance, two-dimensional gel electrophoresis (2DE) has limited separation capabilities and is technically challenging for analyzing membrane proteins and proteins of low abundance. Coupling 2DE with mass spectrometry (MS) and/or liquid chromatography (LC) has advanced proteomic analyses, but even so, these methods fail to measure posttranslational modifications and, ultimately, protein activity *(1)*.

Because proteinases operate in a complex milieu of other proteins, substrates, and cleavage products, and since these components of the proteinase's environment often regulate or are involved in amplification cascades of the proteinase, there have been substantial efforts made towards understanding proteinase in vivo activities. Proteinase substrates cannot be simply thought of as "substrates," but rather as cleavage products with functional roles. For example, some proteinases (matrix metalloproteinase [MMP]-14) cleave proteinase inhibitors, resulting in activation of the proteinases inhibited by those inhibitors and subsequent cleavage of other substrates by those proteinases, etc. *(2)*.

Different methodologies are currently being used to understand the expression profile of proteinases at a specific time or under a particular cellular treatment, in addition to the range of substrates on which a given proteinase will act. Determining a proteinase's substrates will contribute to the understanding of that proteinase's normal function, as well as its function under various pathological conditions. This knowledge also contributes to the mechanism of action behind proteinase inhibitors/activators, required for drug development. Perhaps, the greatest level of complexity arises from understanding an individual proteinase's role in the entire proteolytic system as well as the role of its resulting cleavage products in the proteinase degradome.

Due to the diversity of the proteinase population, no solitary technique can be used to understand the roles of all proteinases at one time. A single (or a class of) proteinase(s) must be excavated out of the proteome, retaining its (their) function, and then its (their) function must be quantified.

1.1. Proteolytic Profiling Technologies

Current proteolytic profiling can be organized into two categories: (1) activity-based protein profiling (ABPP) (also known as mechanism-based profiling) using an enzyme active site-directed chemical probe to assess the proteolytic activity of a class of enzymes in a complex sample; and (2) target-based profiling using targets based either on (a) a given proteinase or (b) a given substrate in a complex sample.

ABPP probes have three characteristic elements: a binding group that binds the active site of enzymes within a given enzyme class, a reactive group for labeling ABPP-bound enzymes, and a reporter group that allows observation and isolation of the ABPP-bound enzyme *(1,3)* (**Fig. 1**). ABPP probes have been used for identification of upregulated metalloproteinases (MPs) in invasive cancer cells *(4)* (**Fig. 1A**). Using probes with built-in hydroxamate (which selectively binds MPs and chelates zinc), benzophenone (which modifies MP active sites by photo-crosslinking), and the reporter groups rhodamine (for visualization) and biotin (for affinity purification), the Cravatt laboratory affinity-purified neprilysin (an integral membrane metalloproteinase), leucine aminopeptidase (LAP), and dipeptidylpeptidase III (DPPIII) as enzymes that are upregulated in invasive melanoma. Furthermore, they demonstrated that neprilysin, LAP, and DPPIII were inhibited by GM6001, an MMP inhibitor, suggesting that nonspecific inhibition of MPs (not limited to the MMP family) may explain the toxicity observed with MMP inhibitors in clinical trials.

Previously, Cravatt and co-workers utilized an active site directed probe to analyze functional serine hydrolases in crude rat cell and tissue extracts *(5)*. Their probe was a biotinylated long-chain fluorophosphonate (an irreversible serine hydrolase inhibitor) that specifically labeled catalytically active serine hydrolases and thus recorded changes in functional status and expression levels (**Fig. 1B**).

The Cravatt laboratory described a method that merges ABPP probes with a microarray format *(6)*. First, they treat the proteome with fluorescent ABPPs and then trap and visualize the labeled enzymes on glass slides using anti-enzyme antibodies. Two breast cancer cell lines were studied, MDA-MB-231 and 231 mfp. The 231 mfp line is the more tumorigenic of the cell lines. Microarrays correctly showed that tissue plasminogen activator (tPA) and urokinase plasminogen activator (uPA) are more active in the 231 mfp line. When specific inhibitors were added to the proteomes, the corresponding enzyme activities were suppressed. The sensitivity of ABPP microarrays was compared to gel-based methods, and ABPP microarrays were found to be approx 30-fold more sensitive, have better resolving power, and use 10 times less proteomic sample.

Using the structure of a general cysteine proteinase inhibitor, E-64, Greenbaum et al. synthesized analogous probes with added biotin or radiolabeled-iodine affinity tags to track cysteine proteinases *(7)* (**Fig. 1C**). Cysteine

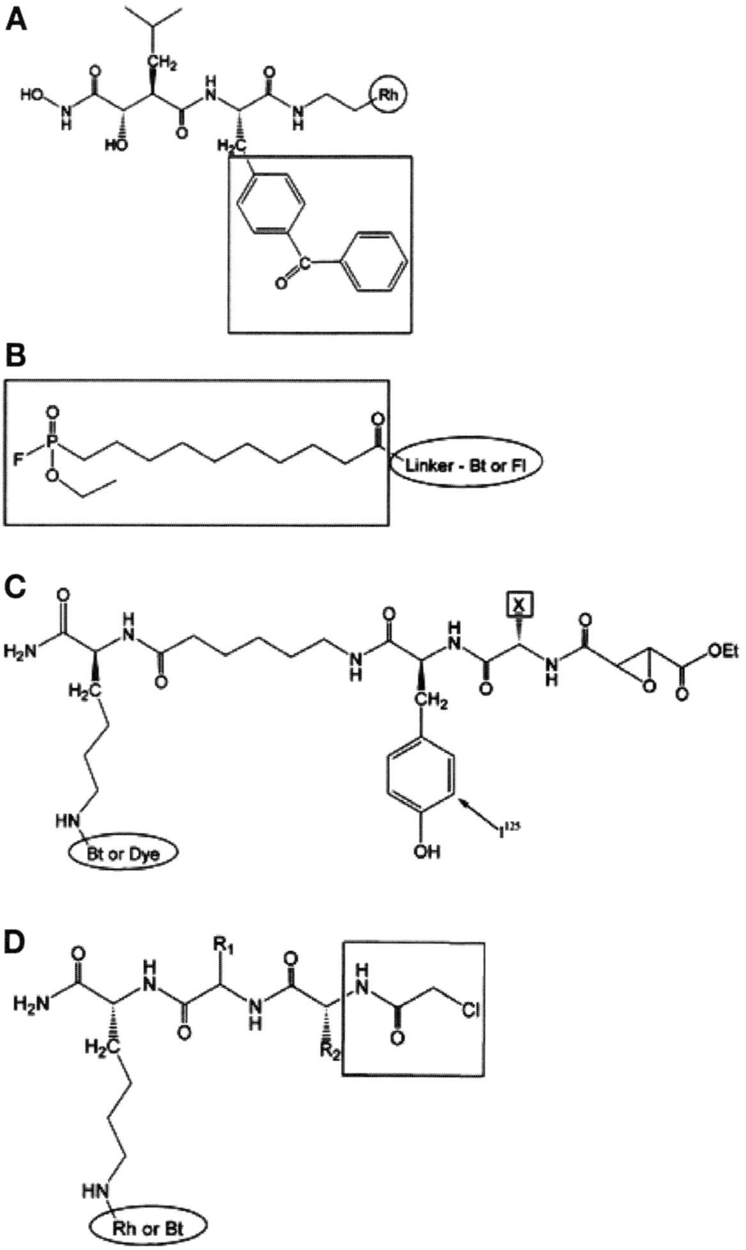

Fig. 1. Structures of activity-based protein profiling (ABPP) probes. ABPP probes have been used to identify active metalloproteinases (**A**), serine hydrolases (**B**) and cysteine proteases (**C**). A combinatorial set of ABPP probes was used to identify various disease-associated enzymes (**D**).

	Target	Binding Group	Reactive Group (Boxed)	Reporter Group (Circled)
A	Metalloproteinases	Hydroxamate	Benzophenone	Rhodamine (Rh)
B	Serine Hydrolases	Fluorosphonate	Fluorosphonate	Fp-Biotin or FP-Fluorescein
C	Cysteine Proteases	E-64	Amino acid (X)	Radioactive Iodine
D	Non-Directed	variable Dipeptides	α-Chlorocetamide	Rhodamine or Biotin

Fig. 1. (*Continued*)

proteinases were profiled in crude extracts from cells and tissues, in addition to cell lines from a multistage carcinogenesis model. By changing amino acids in the peptide portion of the inhibitor, an inhibitor library was constructed to look at both known and novel cysteine proteinase activities. Furthermore, it was demonstrated that these functional probes could be used in affinity purification of cysteine proteinases. The same laboratory used one of the analogs to construct four fluorescently-labeled probes (that can be multiplexed) for cysteine proteinases *(8)* (**Fig. 1C**). These probes were cell-permeable, allowing proteinase activity *and proteinase localization* in intact cells and tissue sections to be examined. A series of (small molecule) inhibitor libraries were designed by altering amino acid sequences in the core peptide backbone. By treating rat liver extracts with a fluorescent probe, then an inhibitor library, then another fluorescent probe, and finally two-dimensional electrophoresis, proteinase spots could be identified that were actively binding the probe and inhibited by different inhibitors. As done previously, biotin-tagged probes were used to subsequently affinity purify the various proteinases.

Increasing the versatility of this technology, the Cravatt laboratory developed *nondirected* ABPP probes consisting of a dipeptide probe library with an α-chloroacetamide reactive group to characterize proteomic differences between slender and obese *(ob/ob)* mice *(1)* (**Fig. 1D**). The ABPP library was initially narrowed down into a smaller subset of probes (termed the "optimal probe set") that was capable of identifying most of the library's targets using affinity isolation followed by LC-tandem mass spectrometry (MS/MS) *(1)*. By applying this optimal probe set to analyze liver tissues from wild-type and *ob/ob* mice, differential expression of multiple enzymes was observed, including the upregulation in *ob/ob* livers of hydroxypyruvate reductase (an enzyme involved in glucose synthesis) *(1)*.

ABPP probes are capable of identifying active enzymes. They can discriminate between active enzymes and their inactive counterparts (e.g., zymogens, inhibitor-bound forms, or non-posttranslationally modified forms). One of the drawbacks of ABPP technology is that a large proteomic sample is still required to ultimately purify the target enzyme.

Proteinase substrates can be identified from a complex, cell-free protein mixture using proteomic techniques. The substrates for MMP-14 in plasma were identified by comparing plasma proteins treated with or without the MMP-14 catalytic domain *(2)*. 2DE protein spots that differed between the two conditions were in-gel trypsin digested and analyzed by matrix-assisted laser desorption/ionization (MALDI)-time-of-flight (TOF) MS. This proteomic approach identified cleavage products from both previously known (including vitronectin, α1-antitrypsin, and α2-macroglobulin) and novel (including apolipoprotein [apo]A-I, apoE, and gelsolin) MMP-14 substrates *(2)*. Like other proteolytic enzyme substrates, some of the substrates of MMP-14 proteolysis are proteinase inhibitors that are inactivated (α1-antitrypsin) or activated (α2-macroglobulin) upon cleavage *(2)*.

Another technique for identifying proteinase substrates uses proteinase-transfected cells and isotope-coded affinity tag (ICAT) labeling with tandem MS sequencing *(9)* (**Fig. 2**). Conditioned media from MMP-14 transfected breast carcinoma cells was treated with isotopically heavy $[^{13}C]_9$ ICAT, whereas conditioned media from catalytically-inactive mutant MMP-14 transfected cells was treated with light $[^{13}C]_0$ ICAT. Following trypsin digestion, multidimensional LC and MS/MS of the combined media were used to identify potential MMP-14 substrates. Surprisingly, most of the identified MMP-14 substrates were not components of the extracellular matrix, and included interleukin (IL)-8, secretory leukocyte proteinase inhibitor, pro-tumor necrosis factor (TNF)α, death receptor 6, and connective tissue growth factor, suggesting that MMP-14 may be an important cell signaling proteinase.

ICAT techniques measure substrate availability, spatially and temporally, in vivo, which may prove to be more physiologically relevant than ABPP techniques. Also, ICAT techniques allow biological circumstances to dictate the relevant proteinase and substrate conformations, incorporating effects such as posttranslational modifications and proteinase exosites (substrate binding sites distant from the enzyme's active site) on proteinase activities. ICAT techniques allow the identification of novel substrates (which, in turn, reshape our opinions of what proteases do) and they allow the quantitative analysis of proteolysis in defined cell cultures.

Drawbacks for ICAT techniques are that some cleaved products do not change ICAT levels. For instance, low abundance cleavage products are not detected by MS as a result of sample preparation and/or chromatography prior to MS. Also, the level of the recombinant protein expression (or catalytic

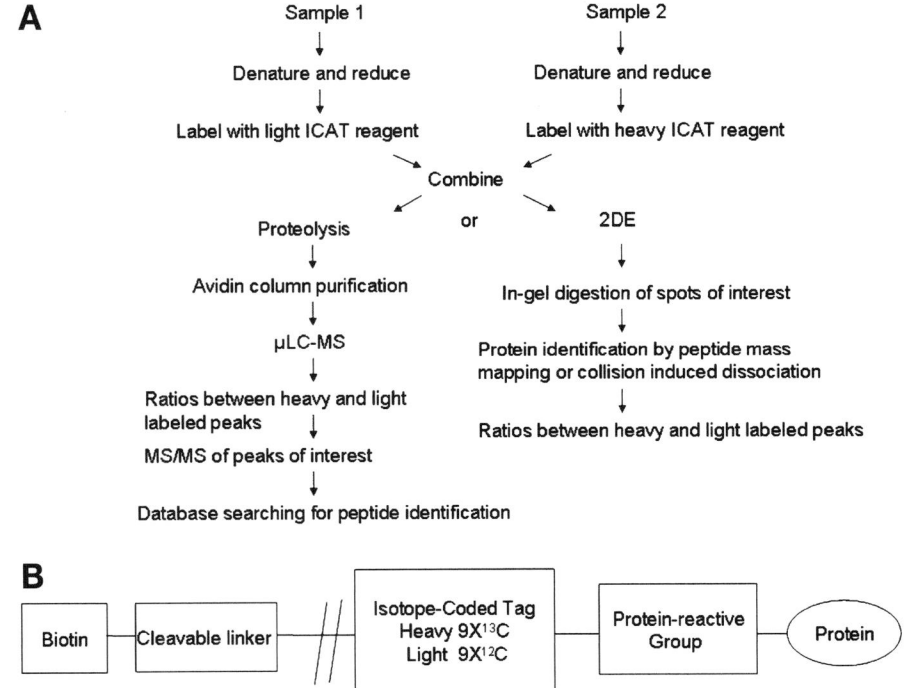

Fig. 2. Isotope-coded affinity tag (ICAT) methodology. (**A**) Outline showing treatment of two samples with different ICAT reagents resulting in proteolytic substrate identification, as well as protein quantification. (**B**) Basic ICAT reagent configurations.

domain addition) may be higher than physiologically relevant, resulting in more cleaved substrates than normal. Finally, the rate of substrate turnover (kinetics) is left unknown by this technique.

There are more than 500 genes that encode proteinases and proteinase-like proteins in the human genome; metalloproteinases represent approximately one-third of these genes *(10)*. The characterized proteolytic roles of these enzymes are growing; they regulate many bioactive molecules that in turn control basic cellular processes, including cell signaling, proliferation, differentiation, migration, morphogenesis, and apoptosis. These proteolytic activities offer accurate levels of control in the processes that they regulate. By understanding which proteinases are expressed and active at a given time, as well as their available substrates (with their respective activities), our insight into proteinase function in both normal and pathological conditions will become more organized. The Fields laboratory has developed methods to differentiate between proteinase presence and collagenolytic activity, thus allowing the key proteinase players to be identified in melanoma metastasis.

1.2. Collagen in Cancer Progression

Collagen assumes a vital role in stabilization of connective tissue and maintaining the structural integrity of the human body (reviewed in **ref. 11**). The natural breakdown of collagen is critical to physiological processes such as embryogenesis and bone remodeling. On the other hand, the destruction of collagen's triple-helical structure can also give rise to a variety of pathologies, including tumor metastasis, arthritis, glomerulonephritis, periodontal disease, and tissue ulcerations *(12–17)*.

The interactions of collagen and other extracellular matrix (ECM) proteins with various melanoma cell surface receptors trigger a series of intracellular events. These signaling pathways result in the release of proteinases, cell surface receptor shedding, and growth factor and cytokine activation, and ultimately promote tumor cell progression. The hydrolysis of collagen (collagenolysis) is one of the committed steps in basement membrane turnover *(18)*, and it has long been demonstrated that tumor extracts can possess collagenolytic activity *(17)*. The triple-helical structure of collagen bestows resistance to degradation by many mammalian proteinases. However, a proteinase family with members that can cleave collagen is the MMP family.

1.3. MMPs and Melanoma Metastasis

The roles that MMPs play in the metastatic process are diverse and include involvement in primary and metastatic tumor growth, angiogenesis, and degradation of basement membrane barriers (such as collagen) during tumor cell invasion *(17)*. Multiple studies have correlated MMP production and melanoma metastasis. Melanoma cells have been found to express MMPs with collagenolytic activity, including MMP-1, MMP-2, MMP-9, MMP-13, and MMP-14 *(19)*. Elevated expression of MMP-1, MMP-2, and MMP-9 correlates to an invasive melanoma phenotype *(19)*. In addition, MMP-14, when localized to invadopodia, activates MMP-2 on the cell surface *(19)*. Furthermore, MMP-14 has been demonstrated to catalyze CD44 shedding and promote cell migration *(20)* by co-localization at the lamellipodia *(21)*. Several membrane-type (MT)-MMPs, including MMP-14, MMP-15, and MMP-16, are all more intensely expressed in metastatic melanoma *(22)*. MMP-3 and MMP-10 (both stromelysins) are also strongly expressed in metastatic melanoma, particularly in the ECM adjacent to blood vessels *(23)*. Metastatic melanoma is distinct from early stage melanoma by either the constitutive or inducible expression of

MMP-9, which cleaves types V and XI collagens but does not cleave interstitial collagens *(24)*. Melanoma invasion of the basement membrane is dependent on MMP-1 expression *(25)*.

1.4. Profiling MMP Activities Using Fluorogenic Triple-Helical Substrates

The Fields laboratory has developed triple-helical peptide (THP) substrates in order to profile and quantify collagenolytic activity (**Table 1**). Initially, fluorogenic triple-helical substrates incorporating the type II collagen sequence 769–783 (Gly-Pro-Pro-Gly-Pro-Gln-Gly~Leu-Ala-Gly-Gln-Arg-Gly-Ile-Val) were constructed *(26)*, where the Gly~ Leu bond is cleaved by MMP-1, -8, and -13 in native type II collagen *(27,28)*. A continuous assay method, such as one that utilizes an increase in fluorescence upon hydrolysis, allows for rapid and convenient kinetic evaluation of collagenolytic proteases. Quenched fluorescent triple-helical substrates that utilize fluorescence resonance energy transfer (FRET)/intramolecular fluorescence energy transfer (IFET) are constructed by incorporation of a fluorophore and a quencher on neighboring peptide chains or in the same peptide chain. The quenching group, Lys(Dnp) (where Dnp is 2,4-dinitrophenyl), replaced the Arg in the P_5' position, and the fluorophore, Lys(Mca) [where Mca is (7-methoxycoumarin-4-yl)acetyl], replaced the Pro in the P_5 position, allowing FRET to occur when the substrate is cleaved *(26,29–31)*. To improve solubility, the Ala in the P_2' position and the Val in the P_8' position were each replaced with Arg. The second MMP-1 cleavage site was removed by replacing Ile in the P_7' position with Val. The resulting peptide sequence was prepared either with a branching method *(32,33)* or with a "peptide-amphiphile" method *(34,35)*. Circular dichroism (CD) spectroscopy showed that the sequence prepared as a peptide-amphiphile (herein referred to as fluorogenic [f]THP-3) had a melting point suitable for MMP kinetic analysis and a sharp thermal transition. Edman degradation sequencing and MALDI-MS analyses demonstrated that MMP-1 hydrolysis of fTHP-3 occurred exclusively at the Gly~Leu bond, indicating that the fluorophore/quenching pair of Mca and Dnp did not affect enzyme sequence specificity. fTHP-3 hydrolysis was observed for MMP-2 and MMP-13 at two loci, Gly~Leu and Gly~Gln. Furthermore, studies of temperature effects on fTHP-3 and its single-stranded analog revealed that activation energies for MMP-1 hydrolysis differed by 3.4-fold, similar to activation energy differences between MMP-1 hydrolysis of type I collagen and gelatin.

Many studies have focused on the roles of the two gelatinases, MMP-2 and MMP-9, in cancer. MMP-2 has been implicated in cell migration, growth,

Table 1
Fluorogenic Triple-Helical Substrates.

Enzyme target	Designation	Sequence
MMP-2, MMP-9	α1(V)436-447 fTHP	(GPP*)$_5$-GPK(Mca)GPPG~VVGEK(Dnp)GEQ-(GPP*)$_5$-NH$_2$
MMP-1, MMP-2, MMP-8, MMP-13, MMP-14	fTHP-3	C$_6$-(GPP*)$_5$-GPK(Mca)GPQG~LRGQK(Dnp)GVR-(GPP*)$_5$-NH$_2$
MMP-1, MMP-2, MMP-8, MMP-13, MMP-14	fTHP-4	(GPP*)$_5$-GPK(Mca)GPQG~LRGQK(Dnp)GVR-(GPP*)$_5$-NH$_2$
MMP-1, MMP-2, MMP-8, MMP-13, MMP-14	fTHP-5	(GPP*)$_5$-GPK(Amp)GPQG~LRGQK(Dnp)GVR-(GPP*)$_5$-NH$_2$
MMP-1, MMP-2, MMP-8, MMP-13, MMP-14	fTHP-7	(GPP*)$_5$-GPK(Adp)GPQG~LRGQK(Dnp)GVR-(GPP*)$_5$-NH$_2$
MMP-14	fTHP-9	(GPP*)$_5$-GPK(Mca)GPQG~C(Mob)RGQK(Dnp)GVR-(GPP*)$_5$-NH$_2$

MMP, matrix metalloproteinase; fTHP, fluorogenic triple-helical peptide. P*, 4-hydroxyproline; Amp, L-2-amino-3-(7-methoxy-4-coumaryl)propionic acid; Adp, L-2-amino-3-(6,7-dimethoxy-4-coumaryl)propionic acid.

angiogenesis, and enhanced ECM remodeling by tumors *(36–40)*. MMP-9 induces angiogenesis, tumor cell invasion, shedding of cellular adhesion molecules and becomes more abundant in advanced stage melanoma cells *(24,41–43)*. The Fields laboratory has constructed THP models containing the MMP-9 cleavage sites from types V and XI collagen. Each triple helical substrate again contains a fluorophore (Mca) and quencher (Dnp). Using the collagen α1(V)436–447 sequence as the substrate template and analyzing the cleavage products by Edman degradation sequencing and MALDI-MS, MMP-2 and MMP-9 selectively cleaved the Gly~Val bond that is cleaved by MMP-9 in native type V collagen *(29)*. No substrate hydrolysis was observed for MMP-1, -3, -13, or -14. It was determined, using single-stranded substrates, that primary structure was not the only determinant for MMP-2/-9 selectivity, as triple-helical conformation of the substrate enhanced hydrolysis rates. This would be expected of true collagenase behavior.

By using triple-helical peptide-amphiphiles, in which the thermal stability of the triple-helix is modulated by pseudo-lipids attached to the *N*-terminus of the peptide *(35)* and comparing substrates that differ by one amino acid residue in the P_1' position, the triple-helical peptidase activities of MMP-1 and MMP-14 were examined *(31)*. A consensus sequence derived from the collagenolytic MMP cleavage sites in human types I-III collagen (see earlier discussion) was used as a template. Comparison of different substrate hydrolysis by MMP-1 and MMP-14 indicated that (1) the P_1' substitution of Cys(Mob) for Leu is disfavored by MMP-1 and greatly favored by MMP-14; and (2) MMP-14 has much greater triple-helical peptidase activity than MMP-1. As previously shown, MMP-14 has a deeper S_1' pocket than MMP-1 *(44)*, and thus Cys(Mob) probably introduces additional favorable interactions in the S_1' environment of MMP-14 while creating steric clashes in the MMP-1 S_1' subsite. The favorable interactions of the Cys(Mob) side-chain may well facilitate MMP-14 unwinding the triple-helix. This feature, along with other unique sequence preferences *(45–47)*, can be explored farther in designing MMP-14 specific substrates or inhibitors. Additionally, a more thermally stable triple-helix translated into a decreased ability of either MMP-1 or MMP-14 to cleave the triple-helix *(31)*. MMP-1 showed much less activity towards all substrates compared with MMP-14, and also had higher triple-helical peptidase activation energies than MMP-14 *(31)*. MMPs clearly differ in their abilities to efficiently hydrolyze a triple-helix; differential thermal stability, as well as triple-helical sequence specificity, can be used to create MMP selective substrates. Such studies need to be extended to other collagenolytic MMPs (MMP-8, MMP-13).

The Fields laboratory has also developed a solid-phase activity assay for MMPs that is specific and can be used to quantify active enzyme concentration **(Fig. 3)** *(48)*. The assay has two principal components: a capture antibody that immobilizes the MMP without perturbing the enzyme active site; and a fluorescent substrate for monitoring proteolysis at low enzyme concentrations. The assay was standardized for MMP-1, MMP-3, MMP-13, and MMP-14. The efficiency of the assay was found to be critically dependent upon the antibody quality, highly specific substrates, and fresh enzyme samples.

1.5. MMPs and Cell Signaling Pathways in Melanoma

Melanoma cell interactions with the microenvironment (which includes extracellular matrix proteins such as collagen) are mediated by various receptors, including integrins and proteoglycans (*see* Chapter 10, Fig. 3). The $\alpha1\beta1$, $\alpha2\beta1$, and $\alpha3\beta1$ integrins are present on melanoma cells and bind to collagen *(49,50)*. The $\alpha1\beta1$ and $\alpha2\beta1$ integrins are upregulated in metastatic

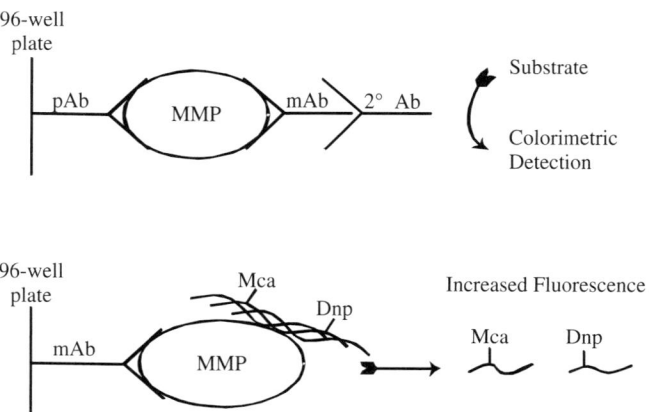

Fig. 3. Schematic diagrams for (top) conventional indirect sandwich enzyme-linked immunosorbent assay (ELISA) and (bottom) solid-phase matrix metalloproteinase (MMP) activity assay. For both assays, MMPs are captured by an antibody on the solid-phase. The indirect sandwich ELISA then requires a detection antibody and an enzyme-labeled antibody. The sandwich ELISA can also be performed directly, in which case the detection antibody is also enzyme-labeled. In the solid-phase MMP activity assay, the fluorogenic substrate serves as the detection system. Dnp is 2,4-dinitrophenyl, Mca is (7-methoxycoumarin-4-yl)acetyl. (Reproduced from **ref. 48**, with permission of *J. Biomol. Tech.*)

melanoma, whereas the α3β1 integrin is upregulated in both primary and metastatic melanoma *(51–54)*.

Integrins are well characterized cell surface adhesion molecules. They are heterodimeric proteins (composed of one α and one β subunit) that possess distinct functions. The α2β1 integrin is the primary melanoma cell adhesion molecule for type IV collagen *(55,56)* and is important in metastatic melanoma migration on type IV collagen, whereas the α3β1 integrin functions in both primary and metastatic melanoma migration *(52,55,57)*. Melanoma cells use the α2β1 integrin to contract (remodel) type I collagen *(51,58)*. Upon binding to their specific ligands, integrins trigger intracellular pathways *(59,60)*, which can promote tumor cell progression *(61)*. Direct associations between integrins and proteinase production have been reported. For example, the α2β1 integrin, when bound to type I collagen gels, is a positive regulator for MMP-1 gene expression *(62)*. Antibodies against the α3β1 integrin stimulate the expression of MMP-2 *(63–65)*, MMP-9 *(66)*, and MMP-14 *(65)*. Integrins have also been shown to physically interact with proteinases: MMP-1 binds to the α2β1 integrin via its α2 subunit *(67)*, while the αvβ3 integrin binds to MMP-2 *(68)*. It is clear that integrin attachment to the ECM produces signals that coordinate with growth factor pathways and regulate cell proliferation and survival (reviewed in **ref.** *69*).

Melanoma cells have been shown to possess at least two distinct cell surface chondroitin sulfate proteoglycans *(70,71)*. Several lines of evidence indicate that one of these, CD44/CSPG, plays an important role in melanoma progression. CD44 is a family of isoforms generated by alternative splicing and differences in posttranslational modifications. The standard form of CD44 (known as CD44s) is upregulated at both mRNA production and cell surface protein expression in highly metastatic melanoma as compared with lowly metastatic melanoma or nontransformed melanocytes *(72–75)*. Enhanced expression of CD44s is also found in the tumor vasculature (endothelial cells) compared to endothelial cells from normal tissue *(76)*. When nonmetastatic cells are transfected with CD44 isolated from metastatic cells, metastatic behavior is induced *(77)*. Additionally, tumor cells expressing CD44s display accelerated tumor growth and metastatic spread in immunodeficient mice compared with parental cells *(77)*. Inhibition of CD44 binding to its ligand with, for instance, hyaluronic acid (HA), anti-CD44 mAb, or a CD44-receptor globulin, reduces tumor formation in the lung of animal models established from CD44-expressing tumor cell lines *(78)*. Finally, proteolytic removal of CD44 inhibits the growth of primary tumors and curtails metastasis in a mouse B16 melanoma model *(79)*.

CD44 binds directly to type IV collagen in a CS-dependent manner *(56)*. In general, basement membrane invasion by melanoma cells is blocked by cell surface CS removal *(56)*. The type IV collagen α1(IV)1263–1277 sequence promotes melanoma cell adhesion, spreading, and signaling *(32,80–85)*. Loss of triple-helical structure dramatically reduces melanoma cell adhesion, spreading, and signaling modulated by a ligand that contains this CD44 binding site *(32,83,84)*.

Melanoma cell surface CD44 can associate with active MMP-7, MMP-9, and MMP-14, potentially enhancing invasion of the basement membrane *(20,21,86–88)*. Although CD44 binds to types I, IV, VI, and XIV collagen, it is not a primary receptor for cell adhesion to collagen *(56,89–91)*. Thus, one possible role for CD44 may be the triggering of signal transduction. One result of CD44 signal transduction is upregulation and activation of integrins *(92)* and MMPs *(93,94)*. It is possible that CD44 works coordinately with the α2β1 integrin to efficiently bind to type IV collagen and subsequently trigger cell signaling pathways.

In an attempt to dissect the roles of the α2β1 integrin and CD44 cell surface receptors in melanoma progression, the Fields laboratory has constructed type IV collagen-derived THP ligands for these receptors. The Gly-Phe-Hyp-Gly-Glu-Arg motif, in triple-helical conformation, has been shown to bind to the α2β1 integrin *(95–97)*. This motif is found within type IV collagen at α1(IV)382–393 *(29,96)*. The type IV collagen α1(IV)1263–1277 sequence Gly-Val-Lys-Gly-Asp-Lys-Gly-Asn-Pro-Gly-Trp-Pro-Gly-Ala-Pro is bound by melanoma cell CD44 receptors, in the chondroitin sulfate proteoglycan (CSPG) form *(85,98,99)*. Loss of triple-helical structure dramatically reduces melanoma cell adhesion, spreading, and signaling modulated by this ligand *(32,83,84)*. These THP ligands were employed individually, along with their linear counterparts, to study temporal changes in gene expression, at the levels of both mRNA and protein. Reverse-transcription (RT)-PCR was used to measure the levels of MMP-1, MMP-2, MMP-3, MMP-8, MMP-9, MMP-13, and MMP-14 gene expression over 24 h. Protein expression of targets modulated by either ligand was examined by enzyme-linked immunosorbent assay (ELISA) analysis. Finally, fluorogenic THP substrates were used to quantify active proteinase production.

The analyses revealed trends in MMP mRNA, protein expression and the production of active enzymes (**Table 2**) (**Figs. 4** and **5**). MMPs initially exist as pro-forms (zymogens) and may be rendered inactive by binding to TIMPs or other more general proteinase inhibitors. For MMP-1, active enzyme was

Table 2
Relative Induction of Target Genes by the α2β1 Integrin and
CD44/Chondroitin Sulfate Proteoglycan (CSPG) Triple-Helical Ligands

Target	Receptor engaged	mRNA	Protein	Active enzyme
MMP-1	α2β1 integrin	++	+	++
"	CD44/CSPG	+	+	+
MMP-2	α2β1 integrin	$- \rightarrow ++$	ND	++
"	CD44/CSPG	$- \rightarrow +$	ND	++
MMP-3	α2β1 integrin	+	$++ \rightarrow -$	ND
"	CD44/CSPG	NM	+	ND
MMP-8	α2β1 integrin	NM	−	ND
"	CD44/CSPG	$+++ \rightarrow -$	++	ND
MMP-13	α2β1 integrin	+++	+++	ND
"	CD44/CSPG	+	++	ND
MMP-14	α2β1 integrin	$+ \rightarrow -$	ND	$+ \rightarrow -$
"	CD44/CSPG	$+ \rightarrow -$	ND	$+ \rightarrow -$

MMP, matrix metalloproteinase; NM, negligible modulation; ND, not determined. "+" Indicates relative upregulation; "−" indicates relative down regulation; "$+ \rightarrow -$" indicates upregulation followed by down regulation; "$- \rightarrow +$" indicates down regulation followed by upregulation.

quantified by solid-phase mAb immobilization of the enzyme followed by reaction with the fluorogenic substrate fTHP-4 (**Table 1**) *(48)*. The solid-phase assay showed more activity induced by the α2β1 integrin than by CD44, and treatment of samples with an activator of proMMPs (4-aminophenylmercuric acetate [APMA]) resulted in a further increase in MMP-1 activity (**Fig. 4**) *(48)*, indicative of the production of both MMP-1 and proMMP-1 by engagement of either the α2β1 integrin or CD44.

General triple-helical peptidase activity was evaluated using fTHP-4 in solution. Engagement of either the α2β1 integrin or CD44 resulted in significant triple-helical peptidase activity detected in melanoma cell conditioned media (**Fig. 5A**), with greater activity found in response to CD44. This activity was completely inhibited by ethylenediamine tetraacetic acid (EDTA), suggesting metalloproteinase activity *(29)*. The MMP-14 selective substrate C_{10}-fTHP-9 (**Table 1**) was used for comparison to general triple-helical peptidase activity. Soluble MMP-14 activity, which can be generated by non-autocatalytic shedding of MMP-14 *(100,101)*, was significant at early time points and then decreased over 8 h in response to the α2β1 integrin and CD44 ligands (**Fig. 5B**).

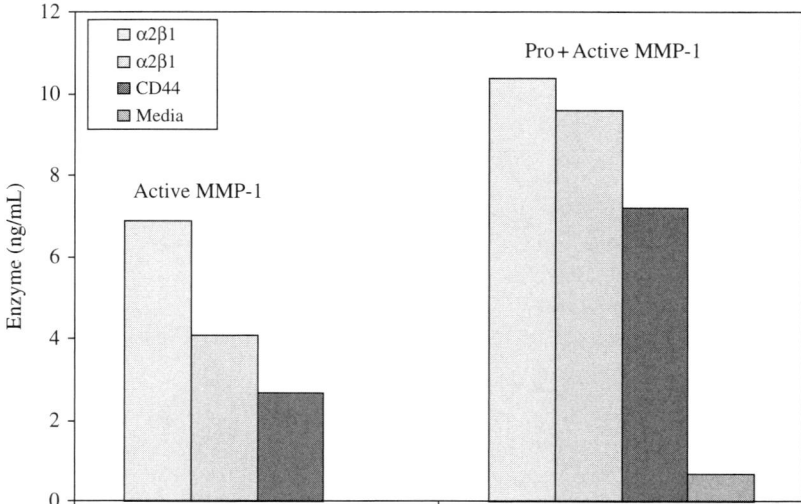

Fig. 4. Matrix metalloproteinase (MMP)-1 activity levels in conditioned media following melanoma cell adhesion to the α2β1 integrin specific C_{10}-[α1(IV)382-393]-NH_2 triple-helical peptide (THP) (light yellow bar) and C_{16}-[α1(IV)382-393]-NH_2 THP (yellow bar) and the CD44/ chondroitin sulfate proteoglycan specific C_{16}-[α1(IV)1263-1277]-NH_2 THP (pink bar). Cell supernatants were added to 96 well plates containing 2.5 μg/mL MMP-1 mAb. Samples were either untreated (left panels) or activated with 2 mM p-aminophenylmercuric acetate (right panels). Three washes of enzyme assay buffer were used to remove unbound proteins. The fluorogenic THP-4 substrate was added, and hydrolysis was monitored at 18 h. The light grey bar at the far right was media alone. Results are the mean of triplicate assays. (Reproduced from **ref. 48**, with permission of *J. Biomol. Tech.*)

The MMP-14 activity profiles correlate well with the mRNA expression profiles, in that MMP-14 is induced at early time points. The subsequent decrease in MMP-14 activity may be due to degradation of MMP-14. Other MMP activity profiles do not decrease at later time points, and exogenous MMP activity is not significantly affected over a 24-h period by melanoma-conditioned media. These results suggest that degradation of MMP-14 is specific, which is not surprising considering the multitude of MMP-14 shedding processes *(101)*.

Gelatinase activity was initially evaluated using the MMP-2/MMP-9 selective substrate α1(V)436–447 fTHP **(Table 1)** *(29)*. Gelatinase activity showed virtually identical increases in response to the $\alpha2\beta1$ integrin and CD44 ligands (**Fig. 5C**). Because the MMP-2/MMP-9 selective substrate does not differentiate between the two gelatinases, additional activity assays were performed using gelatin zymography. Zymography indicated that the $\alpha2\beta1$ integrin and CD44 ligands produced predominantly MMP-2 *(94)*. Treatment with 1,10-phenanthroline resulted in complete loss of gelatinolysis *(94)*, indicative of metalloproteinase activity. Ligands to the $\alpha2\beta1$ integrin or CD44 were found to induce different proteolytic profiles, suggesting that the extracellular matrix can modulate melanoma invasion. These initial studies warrant further investigation of the interactive role of CD44, MMP-13, and MMP-14 in melanoma progression and comparison between surface-bound and soluble active MMP-14 levels. Analysis of surface-bound MMP-14 activity is currently under investigation in our laboratory.

1.6. Melanoma Interactions with Neighboring Cell Types During Metastasis

The microenvironment's influence has been suggested to be more important in initiating and affecting the development of cancer than the genetic lesions found in the cancer cells *(102–104)*. Tumor cells have altered expression patterns of cell surface receptors and intracellular signaling factors that respond differently to their microenvironment than their normal counterparts. Cancer cells are well known for their adaptability, yet their ability to proliferate, mobilize, and invade relies on signaling pathways, some of which are initiated by neighboring stomal cells. In fact, the stromal part of the tumor supplies some of the most damaging traits to cancer cells (reviewed in **refs.** *18,105*). The stromal cells contribute most of the proteinases, allowing the cancer cells to destroy basement membranes, invade tissue boundaries, and metastasize *(18)*. These secreted proteinases degrade stromal components and introduce drastically changed interfaces between tumor and host, including the generation of mobile growth factors, angiogenic factors, and bioactive ECM fragments. The cancer cells, in turn, produce growth factors, chemokines, and cytokines that regulate the stromal cells.

Cell types that are drafted to the tumor mass include endothelial cells, fibroblasts, inflammatory cells, and smooth muscle cells (reviewed in **ref.** *61*). Cell communication between these various cell types leads to dynamic signaling cross talk networks and multiple levels of regulation. For instance, elevated levels of MMP-1 and ICAM-1 (but not TIMP-1), mRNA and protein, were found in fibroblasts that were located closer to melanoma tumors, as opposed to

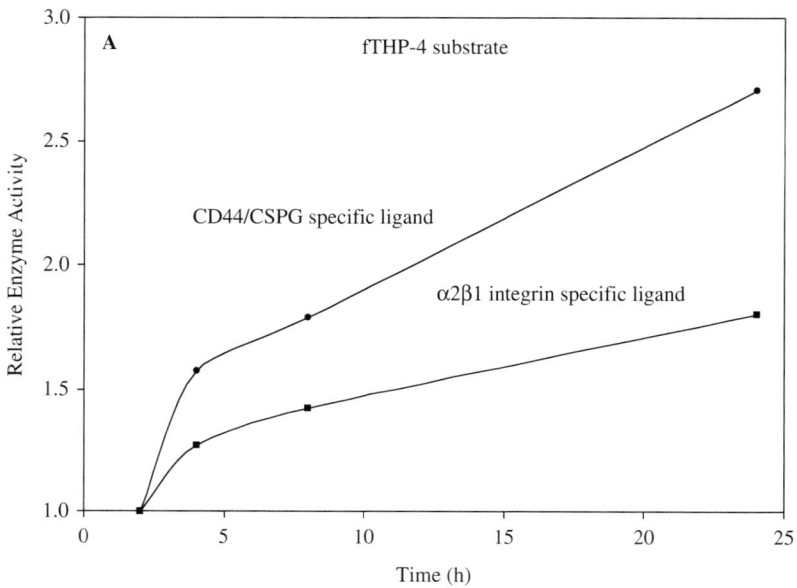

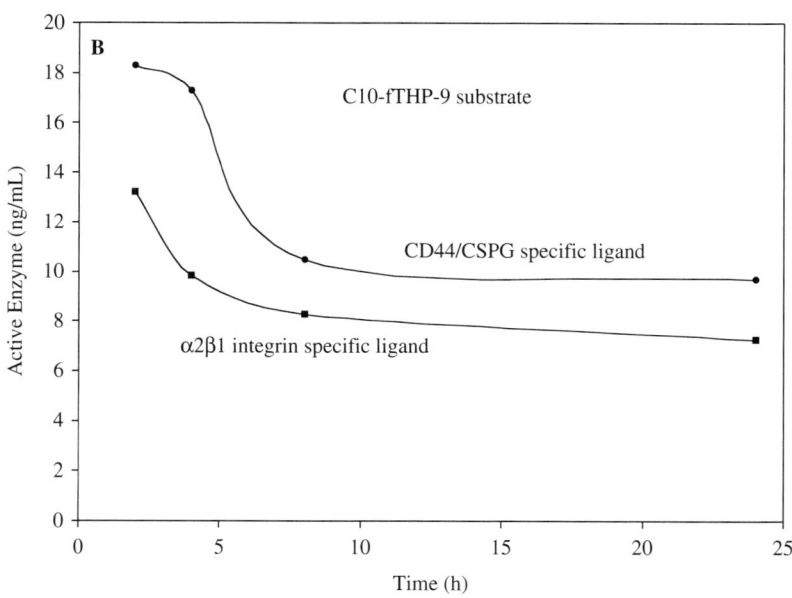

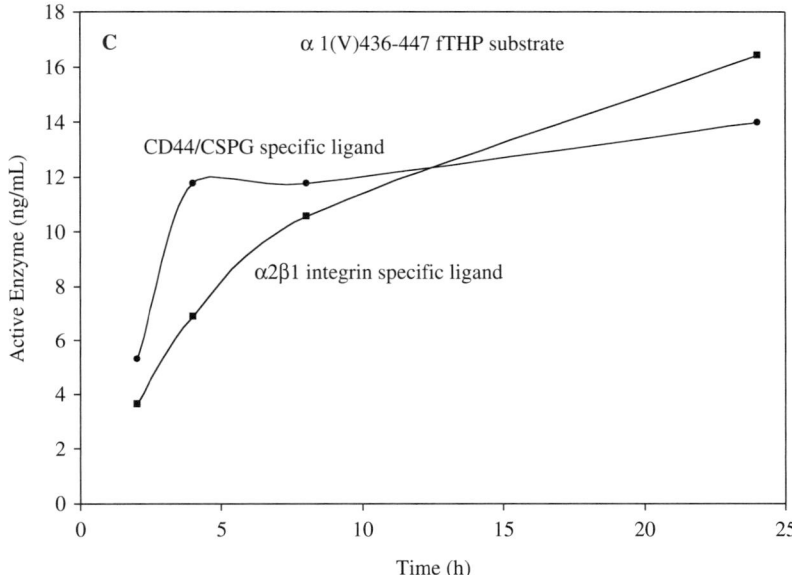

farther away *(106)*. Similar expression profiles are induced in fibroblasts treated with 24-h serum-starved melanoma cell supernatants *(106)*. Co-cultivating fibroblasts and breast carcinoma cells leads to an increase in MMP-9 protein production by the cancer cells, presumably induced by fibroblast production of thrombospondin-1 *(107)*.

Melanoma cells constitutively produce autocrine growth factors (hepatocyte growth factor [HGF], platelet-derived growth factor [PDGF]-A, basic fibroblast

Fig. 5. General triple-helical peptidase (**A**), matrix metalloproteinase (MMP)-14 (**B**), or MMP-2/MMP-9 (**C**) activity levels in conditioned media following melanoma cell adhesion to the α2β1 integrin specific C_{16}-[α1(IV)382-393]-NH$_2$ triple-helical peptide (THP) (squares) and the CD44/ chondroitin sulfate proteoglycan specific C_{16}-[α1(IV)1263-1277]-NH$_2$ THP (circles). M14 melanoma cells were allowed to adhere to 10 μ*M* THP ligands for 60 min at 37°C. Three washes of adhesion media were used to remove nonadherent cells and cells were grown for a total of 24 h. Cell supernatants were added to 96 well plates containing the fluorogenic (f)THP-4 (**A**), C_{10}-fTHP-9 (**B**), or α1(V)436-447 fTHP (**C**) substrate. Hydrolysis was monitored at 18 h using $\lambda_{excitation} = 325$ nm and $\lambda_{emission} = 393$ nm. Results are the mean of duplicate assays. (Reproduced from **ref. 94**, with permission of *J. Biol. Chem.*)

growth factor [bFGF], and IL-8), fibroblast-activating factors (transforming growth factor [TGF]-β and PDGF-A/B), and endothelial cell activating factors (vascular endothelial growth factor [VEGF] and bFGF) (reviewed in **ref. *104***). PDGF-A/B-activated fibroblasts produce feedback factors such as bFGF, insulin-like growth factor (IGF)-1, HGF, and endothelin (ET)-3 that stimulate melanoma cell growth and survival (reviewed in **ref. *104***). Melanoma-stimulated fibroblasts also produce seprase (a.k.a. fibroblast-activating protein); this enzyme is selectively expressed in fibroblasts involved in tissue remodeling, wound healing, and some cancers *(105)*, and has been implicated in collagen degradation *(108)*. Melanoma-stimulated endothelial cells produce ephrins and Eph receptors that may bring on angiogenic or invasive properties *(105)*.

In context-dependent ways, different signaling molecules, like MMPs, elicit different events, such as adhesion or movement, of tumor cells. In certain carcinomas, stromal cells, but not tumor cells, localize MMP-1, MMP-2, MMP-3, MMP-9, MMP-11, MMP-12, MMP-13, and MMP-14 *(88,109)*. These types of changes in the microenvironment surrounding cancer cells are instrumental in initiating and affecting cancer cell development *(102,103)*. There remains much to learn about how cancer cells affect the stroma and how the stroma affects cancer cells. The Fields laboratory has a unique system to study these complex interactions in a highly systematic manner, by singling out and/or combining receptor-ligand interactions and studying the resulting gene and protein expression profiles, as well as proteinase activity profiles. The methods for doing such are described below. Fluorogenic substrates have several advantages over other profiling methods, in that they (a) can be monitored continuously and at reasonably low concentration ranges, (b) can be used to determine individual kinetic parameters, and (c) are readily accommodated by high-throughput analytical techniques.

2. Materials

2.1. Equipment

1. SpectraMAX plus UV/VIS spectrophotometer (Molecular Devices Corp., Sunnyvale, CA).
2. Mastercycler gradient thermal cycler (Eppendorf, Westbury, NY).
3. Large horizontal electrophoresis system (Fisher Scientific, Pittsburgh, PA).
4. Fluor-S™ MutiImager (Bio-Rad, Hercules, CA).
5. Quantity One® v.4.2.2 software (Bio-Rad).
6. LightCycler (Roche Applied Sciences, Indianapolis, IN).

diluted in collagenase buffer), or 0.1 m*M* 1,10-phenanthroline (Sigma) (dissolved in 50% ethanol, diluted in collagenase buffer).

3. Methods

3.1. Preparation of Peptide-Amphiphiles: Synthesis, Purification, Analytical and Structural Characterization.

Refer to the methods described in Chapter 5 of this book.

3.2. Induction of Melanoma Cells

1. Pro-Bind™ 96-well assay plates are conditioned at room temperature overnight prior to initiation of the induction experiment with the appropriate concentration (10–100 μ*M*) of peptide-amphiphile *(48,94)*.
2. The plates are blocked by adding 2 mg/mL BSA in PBS and incubating overnight at room temperature.
3. Cell cultures used for induction experiments are typically 60–80% confluent before release from growth flasks with trypsin-EDTA (pH 7.4).
4. Cells are washed with adhesion media and seeded at approx 7500 cells/well. This cell density has been previously shown to produce efficient cell adhesion with minimization of cell–cell contacts *(85,94)*.
5. Aliquots of the melanoma-conditioned media are harvested at regular intervals over a 24-h period for later determination of metalloproteinase levels. Total RNA ($\sim$ 1–2 μg) is isolated at regular intervals over a 24 h period with the S.N.A.P. eukaryotic total RNA isolation kit and treated with DNase I (to remove genomic DNA) as directed by the manufacturer. To evaluate changes in gene expression for cells grown on peptide-amphiphiles, RNA is isolated from 90,000 cells per time point up to 24 h. RNA yield is determined by measuring the OD_{260} of an aliquot of the final preparation.
6. Both sets of samples are stored at $-20°C$ until analyses are performed.

3.3. Semi-Quantitative RT-PCR

Initially, semi-quantitative RT-PCR was used to measure changes in MMP RNA expression. However, as the technology advanced, real-time quantification of MMP RNA expression was implemented. Both methods are described here and the later is more preferred.

1. DNase I-treated, total RNA is reverse transcribed in a total of 20 μL containing 125 ng of RNA following the manufacturer's recommendations for SUPER-SCRIPT II.
2. PCR is performed on 10% of the resulting cDNA in a total of 50 μL following the manufacturer's recommendations for Platinum Taq DNA Polymerase. Platinum Taq DNA Polymerase is activated for 2 min at 94°C prior to the first amplification cycle.

A typical PCR amplification cycle consists of a 30 s melting at 94°C, annealing at an optimized temperature (approximately equal to T_m [primers −10°C]) for 30–45 s, and extension at 72°C for 60 s. Annealing temperatures, total amplification cycles performed, and amplicon sizes are listed in **Table 3** for the primers used in this study (*see also* **refs. *115,116***). "Primer dropping" PCR is accomplished by providing an internal GAPDH control to co-amplify with the target gene for the last 20 or 21 cycles of the total number of amplification reactions.

3. PCR products are electrophoresed in 3% agarose gels at 90 V for 45 min, and stained with SYBR green I.

4. Images of stained gels are captured on a Fluor-S MutiImager and quantified using the Quantity One v4.3.0 software package. Briefly, analysis consists of first determining the target to GAPDH ratio followed by normalization to the initial time point.

5. PCR amplicons are gel-purified using the S.N.A.P. gel purification kit. The isolated DNA is sent to the Biotechnology Program at the University of Florida for sequencing.

3.4. Quantitative RT-PCR

1. Quantitative real-time PCR (qPCR) is performed on a LightCycler.

2. Custom primers are designed using Invitrogen's LUX Designer. Primer sequences and melting temperatures are located in **Table 3**.

3. Amplifications utilize Platinum Quantitative PCR SuperMix-UDG following the manufacturers recommendations. The cycling program is UDG inactivation at 50°C for 2 min, 95°C for 2 min, followed by 45 cycles of (a) 5 s denaturation at 94°C, (b) annealing at 60°C for 10 s, and (c) extension at 72°C for 15 s. Melting curve analysis is performed using continuous acquisition and a slope of 0.1°C/s. The human GAPDH-certified LUX primer set is used as a housekeeping control for all amplifications.

3.5. Cell-Specific Protein Expression

1. Conditioned media is isolated by withdrawal of the media from growing cells (on tissue culture treated plates or from surfaces coated with peptide-amphiphiles) and centrifuging at 1000 *g* (to remove any floating cells).

2. ELISA is performed as described (*117*) using appropriate antibodies purchased from Chemicon International or ELISA kits obtained from R&D Systems (as directed by the manufacturer).

3.6. Cellular MMP Assays

Two assays are utilized to evaluate active MMP production by melanoma cells or fibroblasts: solid-phase MMP fluorogenic substrate assays and solution MMP fluorogenic substrate assays.

Table 3
Primers Designed for α2β1 Integrin and CD44/Chondroitin Sulfate Proteoglycan Studies

Target	Sequence	Amplicon size (bp)	Cycles	Annealing/ melting Temper- ature (°C)
GAPDH (F)	GGTGAAGGTCGGAGTCAACG	496	21	62
GAPDH (R)	CAAAGTTGTCATGGATGACC			
GAPDH-1 (F)	CGA AGT CAA CGG ATT TGG TCG TAT	306	20	62
GAPDH-1 (R)	AGC CTT CTC CAT GGT GGT GAA GAC			
MMP-1 (F)	ATG CGC ACA AAT CCC TTC TAC C	246	27	60
MMP-1 (R)	TTT CCT CAG AAA GAG CAG CAT C			
MMP-2 (F)	CAC CTG TCT CTG GGT CCA GAT CAG G[FAM]G	73	N/A[a]	60
MMP-2 (R)	CCC AAG TGG GAC AAG AAC CA			
MMP-3 (F)	CAC ATC TCC TAC AGG ATT GTG A	355	35	60
MMP-3 (R)	ACC AGG AAT AGG TTG GTA CCT G			
MMP-8 (F)	GAT GCT ATC ACC ACA CTC CGT	283	34	60
MMP-8 (R)	GCT GCG TCA ATT GCT TGG A			
MMP-8 (F)	GAC CAA GTA ATT TGC GGA GGT GTT GG[FAM] C	99	N/A[a]	60
MMP-8 (R)	CCT TGC TCA TGC CTT TCA GC			
MMP-9 (F)	CAG GCT CCG GTG GAC GAT GCC [FAM]G	112	N/A[a]	60
MMP-9 (R)	GCC CTC AGA GAA TCG CCA GTA			
MMP-13 (F)	GGT CTG TCA ATG AGA GCA TAA	514	38	60
MMP-13 (R)	GCC ATG GCC TTA GAC TAT TTT			
MMP-14 (F)	CCC TAT GCC TAC ATC CGT GA	569	35	58
MMP-14 (R)	TCC ATC CAT CAC TTG GTT AT			

[a] Real-time PCR was performed and threshold cycle number (C_T) was determined per sample. MMP, matrix metalloproteinase.

3.6.1. Solid-Phase MMP Fluorogenic Substrate Assays

1. These assays are performed as previously described *(48,94)*. The 96-well plate is incubated with the 100 μL of 5 μg/mL MMP capture mAb in PBS for at least 18 h at 4°C with mixing.
2. Nonspecific binding sites are blocked by incubating with 100 μL MMP blocking buffer for at least 4 h at 4°C with mixing, and the plate washed 3 times with enzyme assay buffer.
3. One hundred fifty microliters of either MMP standards or unknown samples are added to each well, followed by 50 μL 4X MMP blocking buffer (yielding final concentrations of 0.05% Tween 20 and 2 mg/mL BSA).
4. The plate is mixed for at least 18 h at 4°C.
5. All liquid is removed and 200 μL enzyme assay buffer or 2 mM APMA (where applicable) is added to each well.
6. The plate is incubated for 2 h at 37°C.
7. The wells are washed three times with enzyme assay buffer and 200 μL of the appropriate fluorogenic substrate (5–10 μM of fTHP-4, fTHP-5, or fTHP-7) is added to each well.
8. The plate is incubated at 37°C in a humidified atmosphere for 0–18 h.
9. Fluorescence readings ($\lambda_{excitation} = 324$ nm and $\lambda_{emission} = 393$ nm for fTHP-4 and fTHP-5; $\lambda_{excitation} = 348$ nm and $\lambda_{emission} = 436$ nm for fTHP-7) are taken at appropriate intervals.
10. A standard curve is generated by plotting the increase in fluorescence vs concentration of active enzyme. If enzyme activity is anticipated to be high, data collection is best optimized by monitoring activity over 1–4 h. Substrate cleavage is found only for the specific mAb and the corresponding MMP; crossreactivity between MMPs of similar function (i.e., MMP-1 and MMP-13) is not observed.

3.6.2. Solution MMP Fluorogenic Substrate Assays

1. Fluorogenic substrates are hydrolyzed in solution by conditioned media *(29,118)*. One volume of the appropriate fluorogenic substrate is added to each well in a 384-well plate. Where applicable, EDTA is added to each well.
2. The plate is incubated at 30°C in a humidified atmosphere for 30 min
3. One volume of conditioned media, adhesion media (as a negative control), or MMP (as a positive control) is added to each well.
4. The plate is incubated at 30°C in a humidified atmosphere for at least 18 h.
5. Fluorescence readings ($\lambda_{excitation} = 325$ nm and $\lambda_{emission} = 393$ nm) are taken.
6. A standard curve is created by plotting the increase in fluorescence vs concentration of MMP standard. This standard curve is used to calculate the active enzyme concentration in the conditioned media.

3.7. Protein Electrophoresis

1. Proteins in the cell-conditioned medium are resolved by SDS-PAGE with 4-15% Tris-HCl gels.
2. Gels are silver stained (following company recommended procedures) and gel images are captured with a Bio-Rad Fluor-S MutiImager.
3. Results are analyzed using the Quantity One v4.3.0 software package.

3.8. Gelatin Zymography

1. Conditioned medium samples are first concentrated 15 times using spin columns at 3000 rpm for 90 min.
2. Fifteen microliters of sample is mixed with 5 μL of 2X gel sample buffer.
3. Fifteen microliters of each sample is electrophoresed for 1 h at 180 V on SDS-polyacrylamide gels containing 1 mg/mL gelatin.
4. Gels are washed once with deionized water, three times with 2.5% (w/v) triton X-100 to remove SDS, and three times with collagenase buffer.
5. Gels are incubated in collagenase buffer at 37°C for 16 h.
6. Staining of the gels is with Coomassie brilliant blue (0.1% Coomassie, 40% methanol, 10% acetic acid) for 6–8 h.
7. Destaining is performed in 20% methanol and 10% acetic acid until the lysis zones are clear against a blue background.
8. Inhibition studies to determine proteinase class are performed by incubating gels in collagenase buffer containing 1 mM PMSF, 1 mM N-ethylmaleimide, or 0.1 mM 1,10-phenanthroline.

Acknowledgments

We gratefully acknowledge the support of the National Institutes of Health CA77402 and CA98799 (to G.B.F.) and the FAU Center of Excellence in Biomedical and Marine Biotechnology (contribution #P200506).

References

1. Barglow, K. T. and Cravatt, B. F. (2004) Discovering disease-associated enzymes by proteome reactivity profiling. *Chem. Biol.* **11**, 1523–1531.
2. Hwang, I. K., Park, S. M., Kim, S. Y., and Lee, S.-T. (2004) A proteomic approach to identify substrates of matrix metalloproteinase-14 in human plasma. *Biochim. Biophys. Acta* **1702**, 79–87.
3. Chan, E. W. S., Chattopadhaya, S., Panicker, R. C., Huang, X., and Yao, S. Q. (2004) Developing photoactive affinity probes for proteomic profiling: hydroxamate-based probes for metalloproteases. *J. Am. Chem. Soc.* **126**, 14, 435–14,446.

4. Saghatelian, A., Jessani, N., Joseph, A., Humphrey, M., and Cravatt, B. F. (2004) Activity-based probes for the proteomic profiling of metalloproteases. *Proc. Natl. Acad. Sci. USA* **101**, 10,000–10,005.

5. Liu, Y., Patricelli, M. P., and Cravatt, B. F. (1999) Activity-based protein profiling: the serine hydrolases. *Proc. Natl. Acad. Sci. USA* **96**, 14,694–14,699.

6. Sieber, S. A., Mondala, T. S., Head, S. R., and Cravatt, B. F. (2004) Microarray platform for profiling enzyme activities in complex proteomes. *J. Am. Chem. Soc.* **126**, 15,640–15,641.

7. Greenbaum, D., Medzihradszky, K. F., Burlingame, A., and Bogyo, M. (2000) Epoxide electrophiles as activity-dependent cysteine protease profiling and discovery tools. *Chem. Biol. 7*, 569–581.

8. Greenbaum, D., Baruch, A., Hayrapetoan, L., Darula, Z., Burlingame, A., Medzihradszky, K., and Bogyo, M. (2002) Chemical approaches for functionally probing the proteome. *Mol. Cell. Proteomics* **1**, 60–68.

9. Tam, E. M., Morrison, C. J., Wu, Y. I., Stack, M. S., and Overall, C. M. (2004) Membrane protease proteomics: isotope-coded affinity tag MS identification of undescribed MT1-matrix metalloproteinase substrates. *Proc. Natl. Acad. Sci. USA* **101**, 6917–6922.

10. Lopez-Otin, C., and Overall, C. M. (2002) Protease degradomics: a new challenge for proteomics. *Nat. Rev. Mol. Cell. Biol.* **3**, 509–519.

11. Baronas-Lowell, D., Lauer-Fields, J. L., and Fields, G. B. (2003) Defining the roles of collagen and collagen-like proteins within the proteome. *J. Liq. Chromatogr. Rel. Technol.* **26**, 2225–2254.

12. Liotta, L. A. (1992) Cancer cell invasion and metastasis. *Scientific American* **266(2)**, 54–63.

13. Birkedal-Hansen, H., Moore, W. G. I., Bodden, et al. (1993) Matrix metalloproteinases: a review, *Crit. Rev. Oral Biol. Med.* **4**, 197–250.

14. Nagase, H. (1996) Matrix metalloproteinases, in *Zinc Metalloproteases In Health and Disease* (Hooper, N. M., ed.). Taylor & Francis, London: pp. 153–204.

15. Chambers, A. F. and Matrisian, L. M. (1997) Changing views of the role of matrix metalloproteinases in metastasis. *J. Nat. Cancer Inst.* **89**, 1260–1270.

16. Kleiner, D. E. and Stetler-Stevenson, W. G. (1999) Matrix metalloproteinases and metastasis, *Cancer Chemother. Pharmacol.* **43(Suppl.)**, S42–S51.

17. Nelson, A. R., Fingleton, B., Rothenberg, M. L., and Matrisian, L. M. (2000) Matrix metalloproteinases: biologic activity and clinical implications. *J. Clin. Oncol.* **18**, 1135–1149.

18. Chang, C. and Werb, Z. (2001) The many faces of metalloproteases: cell growth, invasion, angiogenesis and metastasis. *Trends Cell Biol.* **11**, S37–S43.

19. Hofmann, U. B., Westphal, J. R., van Muijen, G. N. P., and Ruiter, D. J. (2000) Matrix metalloproteinases in human melanoma. *J. Invest. Dermatol.* **115**, 337–344.

20. Kajita, M., Itoh, Y., Chiba, T., et al. (2001) Membrane-type 1 matrix metallproteinase cleaves CD44 and promotes cell migration. *J. Cell Biol.* **153**, 893–904.

21. Mori, H., Tomari, T., Koshifumi, I., et al. (2002) CD44 directs membrane-type I matrix metalloproteinase to lamellipodia by associating with its hemopexin-like domain. *EMBO J.* **21**, 3949–3959.
22. Ohnishi, Y., Tajima, S., and Ishibashi, A. (2001) Coordinate expression of membrane type-matrix metalloproteinases-2 and 3 (MT2-MMP and MT3-MMP) and matrix metalloproteinase-2 (MMP-2) in primary and metastatic melanoma cells. *Eur. J. Dermatol.* **11**, 420–423.
23. Bodey, B., Bodey, J., B., Siegel, S. E., and Kaiser, H. F. (2001) Matrix metalloproteinase expression in malignant melanomas: tumor-extracellular matrix interactions in invasion and metastasis. *In Vivo* **15**, 57–64.
24. MacDougall, J. R., Bani, M. R., Lin, Y., Rak, J., and Kerbel, R. S. (1995) The 92-kDa gelatinase B is expressed by advanced stage melanoma cells: suppression by somatic cell hybridization with early stage melanoma cells. *Cancer Res.* **55**, 4174–4181.
25. Durko, M., Navab, R., Shibata, H. R., and Brodt, P. (1997) Suppression of basement membrane type IV collagen degradation and cell invasion in human melanoma cells expressing an antisense RNA for MMP-1. *Biochim. Biophys. Acta* **1356**, 271–280.
26. Lauer-Fields, J. L., Broder, T., Sritharan, T., Nagase, H., and Fields, G. B. (2001) Kinetic analysis of matrix metalloproteinase triple-helicase activity using fluorogenic substrates. *Biochemistry* **40**, 5795–5803.
27. Fields, G. B. (1991) A model for interstitial collagen catabolism by mammalian collagenases. *J. Theor. Biol.* **153**, 585–602.
28. Billinghurst, R. C., Dahlberg, L., Ionescu, M., et al. (1997) Enhanced cleavage of type II collagen by collagenases in osteoarthritic articular cartilage. *J. Clin. Invest.* **99**, 1534–1545.
29. Lauer-Fields, J. L., Sritharan, T., Stack, M. S., Nagase, H., and Fields, G. B. (2003) Selective hydrolysis of triple-helical substrates by matrix metalloproteinase-2 and -9. *J. Biol. Chem.* **278**, 18,140–18,145.
30. Lauer-Fields, J. L., Kele, P., Sui, G., Nagase, H., Leblanc, R. M., and Fields, G. B. (2003) Analysis of matrix metalloproteinase activity using triple-helical substrates incorporating fluorogenic L- or D-amino acids. *Anal. Biochem.* **321**, 105–115.
31. Minond, D., Lauer-Fields, J. L., Nagase, H., and Fields, G. B. (2004) Matrix metalloproteinase triple-helical peptidase activities are differentially regulated by substrate stability. *Biochemistry* **43**, 11,474–11,481.
32. Fields, C. G., Mickelson, D. J., Drake, S. L., McCarthy, J. B., and Fields, G. B. (1993) Melanoma cell adhesion and spreading activities of a synthetic 124-residue triple-helical "mini-collagen". *J. Biol. Chem.* **268**, 14,153–14,160.
33. Fields, C. G., Lovdahl, C. M., Miles, A. J., Matthias-Hagen, V. L., and Fields, G. B. (1993) Solid-phase synthesis and stability of triple-helical peptides incorporating native collagen sequences. *Biopolymers* **33**, 1695–1707.

34. Yu, Y.-C., Berndt, P., Tirrell, M., and Fields, G. B. (1996) Self-assembling amphiphiles for construction of protein molecular architecture. *J. Am. Chem. Soc.* **118**, 12,515–12,520.

35. Yu, Y.-C., Tirrell, M., and Fields, G. B. (1998) Minimal lipidation stabilizes protein-like molecular architecture. *J. Am. Chem. Soc.* **120**, 9979–9987.

36. Corcoran, M. L., Hewitt, R. E., Kleiner, D. E., and Stetler-Stevenson, W. G. (1996) MMP-2: expression, activation and inhibition. *Enzyme Protein* **49**, 7–19.

37. Giannelli, G., Falk-Marzillier, J., Schiraldi, O., Stetler-Stevenson, W. G., and Quaranta, V. (1997) Induction of cell migration by matrix metalloprotease-2 cleabage of laminin-5. *Science* **277**, 225–228.

38. Xu, J., Rodriguez, D., Petitclerc, E., et al. (2001) Proteolytic exposure of a cryptic site within collagen type IV is required for angiogenesis and tumor growth in vivo. *J. Cell Biol.* **154**, 1069–1080.

39. Young, T. N., Pizzo, S. V., and Stack, M. S. (1995) A plasma membrane-associated component of ovarian adenocarcinoma cells enhances the catalytic efficiency of matrix metalloproteinase-2. *J. Biol. Chem.* **270**, 999–1002.

40. Deryugina, E. I., Bourdon, M. A., Reisfeld, R. A., and Strongin, A. (1998) Remodeling of collagen matrix by human tumor cells requires activation and cell surface association of matrix metalloproteinase-2. *Cancer Res.* **58**, 3743–3750.

41. Bergers, G., Brekken, R. A., McMahon, G., et al. (2000) Matrix metalloproteinase-9 triggers the angiogenic switch during carcinogenesis. *Nature Cell Biol.* **2**, 737–744.

42. Ramos-DeSimone, N., Hahn-Dantona, E., Sipley, J., Nagase, H., French, D. L., and Quigley, J. P. (1999) Activation of matrix metalloproteinase-9 (MMP-9) via a converging plasmin/stromelysin-1 cascade enhances tumor cell invasion. *J. Biol. Chem.* **274**, 13,066–13,076.

43. Fiore, E., Fusco, C., Romero, P., and Stamenkovic, I. (2002) Matrix metalloproteinase 9 (MMP-9/gelatinase B) proteolytically cleaves ICAM-1 and participates in tumor cell resistance to natural killer cell-mediated cytotoxicity. *Oncogene* **21**, 5213–5223.

44. Terp, G. E., Cruciani, G., Christensen, I. T., and Jorgensen, F. S. (2002) Structural differences of matrix metalloproteinases with potential implications for inhibitor selectivity examined by the GRID/CPCA approach. *J. Med. Chem.* **45**, 2675–2684.

45. Mucha, A., Cuniasse, P., Kannan, R., et al. (1998) Membrane type-1 matrix metalloproteinase and stromelysin-3 cleave more efficiently synthetic substrates containing unusual amino acids in their P_1' positions. *J. Biol. Chem.* **273**, 2763–2768.

46. Ohkubo, S., Miyadera, K., Sugimoto, Y., Matsuo, K.-I., Wierzba, K., and Yamada, Y. (1999) Identification of substrate sequences for membrane type-1 matrix metalloproteinase using bacteriophage peptide display library. *Biochem. Biophys. Res. Commun.* **266**, 308–313.

47. Kridel, S. J., Sawai, H., Ratnikov, B. I., et al. (2002) A unique substrate binding mode discriminates membrane type 1-matrix metalloproteinase (MT1-MMP) from other matrix metalloproteinases. *J. Biol. Chem.* **277**, 23,788–23,793.

48. Lauer-Fields, J. L., Nagase, H., and Fields, G. B. (2004) Development of a solid-phase assay for analysis of matrix metalloproteinase activity. *J. Biomolecular Techniques* **15**, 305–316.
49. Kramer, R. H. and Marks, N. (1989) Identification of intracellular collagen receptor on human melanoma cells. *J. Biol. Chem.* **264**, 4684–4688.
50. Miles, A. J., Knutson, J. R., Skubitz, A. P. N., Furcht, L. T., McCarthy, J. B., and Fields, G. B. (1995) A peptide model of basement membrane collagen α1(IV) 531–543 binds the α3β1 integrin. *J. Biol. Chem.* **270**, 29,047–29,050.
51. Klein, C. E., Dressel, D., Steinmayer, T., et al. (1991) Integrin α2β1 is upregulated in fibroblasts and highly aggressive melanoma cell in three-dimensional collagen lattices and mediates the reorganization of type I collagen fibrils. *J. Cell Biol.* **115**, 1427–1436.
52. Yoshinaga, I. G., Vink, J., Dekker, S. K., Mihm, M.C., Jr., and Byers, H. R. (1993) Role of α3β1 and α2β1 integrins in melanoma cell migration, *Melanoma Res.* **3**, 435–441.
53. Heino, J. (1996) Biology of tumor cell invasion: interplay of cell adhesion and matrix degradation. *Int. J. Cancer* **65**, 717–722.
54. Mizejewski, G. J. (1999) Role of integrins in cancer: survey of expression patterns. *Proc. Soc. Exp. Biol. Med.* **222**, 124–138.
55. Etoh, T., Thomas, L., Pastel-Levy, C., Colvin, R. B., Mihm, M. C., Jr., and Byers, H. R. (1993) Role of integrin α2β1 (VLA-2) in the migration of human melanoma cells on laminin and type IV collagen. *J. Invest. Dermatol.* **100**, 640–647.
56. Knutson, J. R., Iida, J., Fields, G. B., and McCarthy, J. B. (1996) CD44/chondroitin sulfate proteoglycan and α2β1 integrin mediate human melanoma cell migration on type IV collagen and invasion of basement membranes. *Mol. Biol. Cell* **7**, 383–396.
57. Melchiori, A., Mortarini, R., Carlone, S., et al. (1995) The α3β1 integrin is involved in melanoma cell migration and invasion. *Exp. Cell Res.* **219**, 233–242.
58. Schön, M., Schön, M. P., Kuhröber, A., Schirmbeck, R., Kaufmann, R., and Klein, C. E. (1996) Expression of the human α2 integrin subunit in mouse melanoma cell confers the ability to undergo collagen-directed adhesion, migration, and matrix reorganization. *J. Invest. Dermatol.* **106**, 1175–1181.
59. Schwartz, M. A. (2001) Integrin signaling revisited. *Trends Cell Biol.* **11**, 466–470.
60. Hood, J. D. and Cheresh, D. A. (2002) Role of integrins in cell invasion and migration. *Nature Rev. Cancer* **2**, 91–100.
61. Alessandro, R. and Kohn, E. C. (2002) Signal transduction targets in invasion. *Clin. Exp. Metastasis* **19**, 265–273.
62. Riikonen, T., Westermarck, J., Koivisto, L., Broberg, A., Kähäri, V.-M., and Heino, J. (1995) Integrin α2β1 is a positive regulator of collagenase (MMP-1) and collagen α1(I) gene expression. *J. Biol. Chem.* **270**, 13,548–13,552.

63. Chintala, S. K., Sawaya, R., Gokaslan, Z. L., and Rao, J. S. (1996) Modulation of matrix metalloprotease-2 and invasion in human glioma cells by α3β1 integrin. *Cancer Lett.* **103**, 201–208.

64. Kubota, S., Ito, H., Ishibashi, Y., and Seyama, Y. (1997) Anti-α3 integrin antibody induces the activated form of matrix metalloprotease-2 (MMP-2) with concomitant stimulation of invasion through matrigel by human rhabdomyosarcoma cells. *Int. J. Cancer* **70**, 106–111.

65. Ellerbroek, S. M., Wu, Y. I., Overall, C. M., and Stack, M. S. (2001) Functional interplay between type I collagen and cell surface matrix metalloproteinase activity. *J. Biol. Chem.* **276**, 24,833–24,842.

66. Larjava, H., Lyons, J. G., Salo, T., et al. (1993) Anti-integrin antibodies induce type IV collagenase expression in keratinocytes. *J. Cell. Physiol.* **157**, 190–200.

67. Stricker, T. P., Dumin, J. A., Dickeson, S. K., et al. (2001) Structural analysis of the α2 integrin I domain/procollagenase-1 (matrix metalloproteinase-1) interaction. *J. Biol. Chem.* **276**, 29,375–29,381.

68. Brooks, P. C., Strömblad, S., Sanders, L. C., et al. (1996) Localization of matrix metalloproteinase MMP-2 to the surface of invasive cells by interaction with integrin αvβ3. *Cell* **85**, 683–693.

69. Miranti, C. K. and Brugge, J. S. (2002) Sensing the environment: a historical perspective on integrin signal transduction. *Nat. Cell Biol.* **4**, E83–E90.

70. Faassen, A. E., Drake, S. L., Iida, J., Knutson, J. R., and McCarthy, J. B. (1992) Mechanisms of normal cell adhesion to the extracellular matrix and alterations associated with tumor invasion and metastasis. *Adv. Pathol. Lab. Med* **5**, 229–259.

71. Iida, J., Meijne, A. M. L., Knutson, J. R., Furcht, L. T., and McCarthy, J. B. (1996) Cell surface chondroitin sulfate proteoglycans in tumor cell adhesion, motility and invasion. *Semin. Cancer Biol.* **7**, 155–162.

72. Leigh, C. J., Palechek, P. L., Knutson, J. R., McCarthy, J. B., Cohen, M. B., and Argenyi, Z. B. (1996) CD44 expression in benign and malignant nevomelanocytic lesions. *Hum. Pathol.* **27**, 1288–1294.

73. Lesley, J., Hyman, R., English, N., Catterall, J. B., and Turner, G. A. (1997) CD44 in inflammation and metastasis, *Glycoconjugate J.* **14**, 611–622.

74. Ahrens, T., Assmann, V., Fieber, C., et al. (2001) CD44 is the principal mediator of hyaluronic-acid-induced melanoma cell proliferation, *J. Invest. Dermatol.* **116**, 93–101.

75. Ranuncolo, S. M., Ladeda, V., Gorostidy, S., et al. (2002) Expression of CD44s and CD44 splice variants in human melanoma. *Oncology Reports* **9**, 51–56.

76. Griffioen, A. W., Coenen, M. J. H., Damen, C. A., et al. (1997) CD44 is involved in tumor angiogenesis; an activation antigen on human endothelial cells. *Blood* **90**, 1150–1159.

77. Naor, D., Slonov, R. V., and Ish-Shalom, D. (1997) CD44: structure, function, and association with the malignant process, in *Advances in Cancer Research* (Vande Woude, G. F., and Klein, G., eds.) Academic, Orlando, FL: pp. 241–319.
78. Eliaz, R. E. and Szoka, F.C., Jr. (2001) Liposome-encapsulated doxorubicin targeted to CD44: a strategy to kill CD44-overexpressing tumor cells, *Cancer Res.* **61**, 2592–2601.
79. Wald, M., Olejár, T., Sebková, V., Zadinová, M., Boubelík, M., and Poucková, P. (2001) Mixture of trypsin, chymotrypsin and papain reduces formation of metastases and extends survival time of C57Bl6 mice with syngeneic melanoma B16. *Cancer Chemother. Pharmacol.* **47**, S16–S22.
80. Chelberg, M. K., McCarthy, J. B., Skubitz, A. P. N., Furcht, L. T., and Tsilibary, E. C. (1990) Characterization of a synthetic peptide from type IV collagen that promotes melanoma cell adhesion, spreading, and motility. *J. Cell Biol.* **111**, 261–270.
81. Mayo, K. H., Parra-Diaz, D., McCarthy, J. B., and Chelberg, M. (1991) Cell adhesion promoting peptide GVKGDKGNPGWPGAP from the collagen type IV triple helix. *Biochemistry* **30**, 8251–8267.
82. Yu, Y.-C., Pakalns, T., Dori, Y., McCarthy, J. B., Tirrell, M., and Fields, G. B. (1997) Construction of biologically active protein molecular architecture using self-assembling peptide-amphiphiles. *Meth. Enzymol.* **289**, 571–587.
83. Fields, G. B., Lauer, J. L., Dori, Y., Forns, P., Yu, Y.-C., and Tirrell, M. (1998) Proteinlike molecular architecture: biomaterial applications for inducing cellular receptor binding and signal transduction. *Biopolymers* **47**, 143–151.
84. Malkar, N. B., Lauer-Fields, J. L., Borgia, J. A., and Fields, G. B. (2002) Modulation of triple-helical stability and subsequent melanoma cellular responses by single-site substitution of fluoroproline derivatives. *Biochemistry* **41**, 6054–6064.
85. Lauer-Fields, J. L., Malkar, N. B., Richet, G., Drauz, K., and Fields, G. B. (2003) Melanoma cell CD44 interaction with the α1(IV)1263-1277 region from basement membrane collagen is modulated by ligand glycosylation. *J. Biol. Chem.* **278**, 14,321–14,330.
86. Bourguignon, L. Y. W., Gunja-Smith, Z., Iida, N., et al. (1998) CD44v3,8-10 is involved in cytoskeleton-mediated tumor cell migration and matrix metalloproteinase (MMP-9) association in metastatic breast cancer cells. *J. Cell. Physiol.* **176**, 206–215.
87. Yu, Q. and Stamenkovic, I. (1999) Localization of matrix metalloproteinase 9 to the cell surface provides a mechanism for CD44-mediated tumor invasion. *Genes Develop.* **13**, 35–48.
88. Lynch, C. C. and Matrisian, L. M. (2002) Matrix metalloproteinases in tumor-host cell communication. *Differentiation* **70**, 561–573.
89. Carter, W. G. and Wayner, E. A. (1988) Characterization of the class III collagen receptor, a phosphorylated, transmembrane glycoprotein expressed in nucleated human cells. *J. Biol. Chem.* **263**, 4193–4201.

90. Ehnis, T., Dieterich, W., Bauer, M., von Lampe, B., and Shuppan, D. (1996) A chondroitin/dermatan sulfate form of CD44 is a receptor for collagen XIV (undulin). *Exp. Cell Res.* **229**, 388–397.

91. Faassen, A. E., Schrager, J. A., Klein, D. J., Oegema, T. R., Couchman, J. R., and McCarthy, J. B. (1992) A cell surface chondroitin sulfate proteoglycan, immuno-logically related to CD44, is involved in type I collagen-mediated melanoma cell motility and invasion. *J. Cell Biol.* **116**, 521–531.

92. Fujisaki, T., Tanaka, Y., Fujii, K., et al. (1999) CD44 stimulation induces integrin-mediated adhesion of colon cancer cell lines to endothelial cells by up-regulation of integrins and c-Met activation of integrins. *Cancer Res.* **59**, 4427–4434.

93. Takahashi, K., Eto, H., and Tanabe, K. K. (1999) Involvement of CD44 in matrix metalloproteinase-2 regulation in human melanoma cells. *Int. J. Cancer* **80**, 387–395.

94. Baronas-Lowell, D., Lauer-Fields, J. L., Borgia, J. A., et al. (2004) Differ-ential modulation of human melanoma cell metalloproteinase expression by α2β1 integrin and CD44 triple-helical ligands derived from type IV collagen. *J. Biol. Chem.* **279**, 43,503–43,513.

95. Knight, C. G., Morton, L. F., Onley, D. J., et al. (1998) Identification in collagen type I of an integrin α2β1-binding site containing an essential GER sequence. *J. Biol. Chem.* **273**, 33,287–33,294.

96. Knight, C. G., Morton, L. F., Peachey, A. R., Tuckwell, D. S., Farndale, R. W., and Barnes, M. J. (2000) The collagen-binding A-domains of integrin α1β1 and α2β1 recognize the same specific amino acid sequence, GFOGER, in native (triple-helical) collagens. *J. Biol. Chem.* **275**, 35–40.

97. Emsley, J., Knight, C. G., Farndale, R. W., Barnes, M. J., and Liddington, R. C. (2000) Structural basis of collagen recognition by integrin α2β1. *Cell* **101**, 47–56.

98. Mickelson, D. J., Faassen, A. E., and McCarthy, J. B. (1991) A cell surface chondroitin sulfate proteoglycan mediates melanoma cell motility and adhesion to a helical domain of type IV collagen. *J. Cell Biol.* **115**, 287a.

99. Knutson, J. R., Fields, G. B., Iida, J., Miles, A. J., and McCarthy, J. B. (1995) A type IV collagen-derived synthetic peptide, IV-H1, interacts with human melanoma CD44/chondroitin sulfate proteoglycan and inhibits invasion of basement membranes. *Proc. Am. Assoc. Cancer Res.* **36**, 68.

100. Toth, M., Hernandez-Barrantes, S., Osenkowski, P., et al. (2002) Complex pattern of membrane type I matrix metalloproteinase shedding. *J. Biol. Chem.* **277**, 26,340–26,350.

101. Osenkowski, P., Toth, M., and Fridman, R. (2004) Processing, shedding, and endocytosis of membrane type 1-matrix metalloproteinase (MT1-MMP). *J. Cell. Physiol.* **200**, 2–10.

102. Hsu, M.-Y., Meier, F., and Herlyn, M. (2002) Melanoma development and progression: a conspiracy between tumor and host. *Differentiation* **70**, 522–536.

103. Park, C. C., Bissell, M. J., and Barcellos-Hoff, M. H. (2000) The influence of the microenvironment on the malignant phenotype. *Mol. Med. Today* **6**, 324–329.

104. Li, G., Satyamoorthy, K., Meier, F., Berking, C., Bogenrieder, T., and Herlyn, M. (2003) Function and regulation of melanoma-stromal fibroblast interactions: when seeds meet soil. *Oncogene* **22**, 3162–3171.

105. Bogenrieder, T. and Herlyn, M. (2002) Cell-surface proteolysis, growth factor activation and intercellular communication in the progression of melanoma. *Crit. Rev. Oncol. Hematol.* **44**, 1–15.

106. Wandel, E., Grabhoff, A., Mittag, M., Haustein, U. F., and Saalbach, A. (2000) Fibroblast surrounding melanoma express elevated levels of matrix metalloproteinase-1 (MMP-1) and intercellular adhesion molecule-1 (ICAM-1) in vitro. *Exp. Dermatol.* **9**, 34–41.

107. Wang, T. N., Albo, D., and Tuszynski, G. P. (2002) Fibroblasts promote breast cancer cell invasion by upregulating tumor matrix metalloproteinase-9 production. *Surgery* **132**, 220–225.

108. Park, J. E., Lenter, M. C., Zimmermann, R. N., Garin-Chesa, P., Old, L. J., and Rettig, W. J. (1999) Fibroblast activation protein, a dual specificity serine protease expressed in reactive human tumor stromal fibroblasts. *J. Biol. Chem.* **274**, 36,505–36,512.

109. Basset, P., Okada, A., Chenard, M. P., et al. (1997) Matrix metalloproteinases as stromal effectors of human carcinoma progression: therapeutic implications. *Matrix Biol.* **15**, 535–541.

110. Chung, L., Shimokawa, K., Dinakarpandian, D., Grams, F., Fields, G. B., and Nagase, H. (2000) Identification of the RWTNNFREY(183–191) region as a critical segment of matrix metalloproteinase 1 for the expression of collagenolytic activity. *J. Biol. Chem.* **275**, 29,610–29,617.

111. Itoh, Y., Binner, S., and Nagase, H. (1995) Steps involved in activation of the complex of pro-matrix metalloproteinase 2 (progelatinase A) and tissue inhibitor of metalloproteinases (TIMP)-2 by 4-aminophenylmercuric acetate. *Biochem. J.* **308**, 645–651.

112. Huang, W., Suzuki, K., Nagase, H., Arumugam, S., Van Doren, S., and Brew, K. (1996) Folding and characterization of the amino-terminal domain of human tissue inhibitor of metalloproteinases-1 (TIMP-1) expressed at high yield in E. coli. *FEBS Lett.* **384**, 155–161.

113. Hurst, D. R., Schwartz, M. A., Ghaffari, M. A., et al. (2004) Catalytic- and ecto-domains of membrane type 1-matrix metalloproteinase have similar inhibition profiles but distinct endopeptidase activities. *Biochem. J.* **377**, 775–779.

114. Lauer-Fields, J. L., and Fields, G. B. (2002) Triple-helical peptide analysis of collagenolytic protease activity, *Biol. Chem.* **383**, 1095–1105.

115. Singh, A., Nelson-Moon, Z. L., Thomas, G. J., Hunt, N. P., and Lewis, M. P. (2000) Identification of matrix metalloproteinases and their tissue inhibitors type 1 and 2 in human masseter muscle. *Arch. Oral Biol.* **45**, 431–440.

116. Wong, H., Muzik, H., Groft, L. L., et al. (2001) Monitoring MMP and TIMP mRNA expression by RT-PCR, in *Matrix Metalloproteinase Protocols: Methods in Molecular Biology, Vol. 151* (Clark, I. M., ed.) Humana, Totowa, NJ: pp. 305–333.
117. Crowther, J. R. (1995) *ELISA: Theory and Practice*, Vol. 42. Humana, Totowa, NJ.
118. Lauer, J. L., Gendron, C. M., and Fields, G. B. (1998) Effect of ligand conformation on melanoma cell $\alpha 3\beta 1$ integrin-mediated signal transduction events: implications for a collagen structural modulation mechanism of tumor cell invasion. *Biochemistry* **37**, 5279–5287.

7

β-Amyloid Protein Aggregation

Marcus A. Etienne, Nadia J. Edwin, Jed P. Aucoin, Paul S. Russo, Robin L. McCarley, and Robert P. Hammer

Summary

The β-amyloid peptide aggregates via a nucleation pathway where micellar aggregates propagate to form oligomers (protofibrils), which then polymerize into insoluble fibrils. This fibrillogenic process has been linked to the pathogenesis associated with Alzheimer's disease. One purpose of this chapter is to provide a protocol for reliably producing monomeric Aβ as a starting point for physical and biological studies. Many research groups have used organic solvents to disaggregate pre-seeded Aβ in an attempt to acquire monomeric starting materials. Others have used instrumental techniques such as size exclusion chromatography to isolate monomer, structural intermediates, and fibrils and study their affects on Aβ nucleation. This chapter discusses a modified method of Aβ preparation using organic solvents followed by dissolution into aqueous phosphate buffer systems that renders monomeric Aβ starting solutions for kinetic experiments. Additionally, this chapter details a number of physical techniques such as scanning force microscopy, circular dichroism spectroscopy, transmission electron microscopy, fluorescence spectroscopy, fluorescence photobleaching recovery, and dynamic light scattering, together with physiological techniques such as cell viability assays to characterize Aβ nucleation, aggregation, and fibrillization and the potential biological activity of the various Aβ particles.

Key Words: Aβ aggregation; β-sheet breaker; $C^{\alpha,\alpha}$ disubstituted amino acids; fibril formation; micellar aggregates; peptide-based inhibitors.

1. Introduction

Alzheimer's disease (AD) is a degenerative disease most commonly associated with dementia. It is characterized by the formation of protein aggregates that assemble into fibrillar structures. Many believe that the causative factor of AD is the proliferation of amyloid plaques, mostly consisting of the

From: *Methods in Molecular Biology, vol. 386: Peptide Characterization and Application Protocols*
Edited by: G. Fields © Humana Press Inc., Totowa, NJ

β-amyloid (Aβ) peptide *(1–3)*. Under pathogenic conditions, Aβ aggregation is likely to proceed via nucleation to initiate fibril formation followed by elongation to full length fibrils. Identification of monomeric material, which consists of a mixture of monomers with lower oligomers (dimers, trimers, tetramers) in rapid equilibrium, as well as structural intermediates of fibrillogenesis (pentamers-decamers) *(4)*, and elucidation into the mechanistic roles of these intermediates are necessary for the identification of possible therapeutic targets.

A significant problem associated with the biophysical studies of synthetic Aβ is its time-dependent aggregation in aqueous solution *(5)*. Controlling the initial aggregated state of the peptide is generally a challenge as synthetic Aβ is difficult to dissolve directly into physiological buffers *(6)*. The existence of preseeded material in commercially available Aβ from different manufacturers results in various initial aggregation states. Different lot batches developed by the same company also vary; therefore; protocols obtaining monomeric solutions of Aβ for elucidation into the toxicity of Aβ aggregates and protofibrils are paramount. Many groups follow protocols that involve the dissolution of Aβ in organic solvents *(7)* such as: trifluoroacetic acid (TFA) *(8)*, 2, 2, 2-trifluoroethanol (TFE) *(9,10)*, dimethyl sulfoxide (DMSO) *(11,12)*, and hexafluoroisopropanol (HFIP) *(13,14)*, followed by dilution in aqueous buffer systems. Organic solvents are widely used because of their ability to promote the α-helix and to dissolve preseeded material. The authors' laboratory has derived a modified protocol where Aβ is dissolved in organic solvents TFA and HFIP followed by dissolution in KOH. The stock solution is then dissolved in PBS at desired concentrations to obtain working solutions.

2. Materials

2.1. Inhibitor Peptide Synthesis, Purification, and Characterization

1. 9-Fluorenylmethyloxycarbonyl (Fmoc)-amino acids, Fmoc-resins, and coupling agents/activators are purchased from Applied Biosystems (Warrington, UK).
2. Inhibitor peptides are partially synthesized (oligolysine *N*- or *C*-terminus) using a Pioneer Peptide Synthesizer System (Applied Biosystems).
3. Crude peptides are purified using reversed-phase high-performance liquid chromatography (HPLC) (Waters) with Empower software (Waters); C_4 and/or C_{18} columns operating on a linear gradient using A = H_2O + 0.1% TFA and B = CH_3CN + 0.1% TFA. Homogeneity is typically >99%.
4. Net peptide content is determined via Amino Acid Analysis (Dionex).
5. MALDI-MS data are acquired using a Bruker ProFLEX III.

2.2. Aβ$_{1-40}$ Peptide Aggregation

1. Aβ$_{1-40}$ is purchased from BioSource (Camarillo, CA). Aβ$_{1-40}$ (DLS, cat. no. 20698) and fluorescein Aβ$_{1-40}$ (FPR, cat. no. 23513) are purchased from Anaspec, Inc. (San Jose, CA).
2. Potassium hydroxide (KOH NF/FCC Pellets (P251-3) is purchased from Fisher Scientific (Fair Lawn, NJ).
3. Sodium chloride (NaCl certified for biological work, S671-3) is purchased from Fisher Scientific (Fair Lawn, NJ), whereas sodium chloride, Puratronic (CAS # 7647-14-5) is purchased from Alfa-Aesar (Ward Hill, MA).
4. Phosphoric acid (99.999%) is purchased from Aldrich (34,524-5; Milwaukee, WI).
5. An Eppendorf thermomixer (1.5 mL) and Eppendorf mastercycler (96-well plate) (Eppendorf) are used to incubate Aβ/Aβ:inhibitor samples.
6. Deionized water (18 MΩ) is prepared using reverse osmosis filter and ion-exchange filters followed by a 5-mm spiral ultrafilter (Nanopure, Barnstead).
7. HFIP is purchased from Aldrich (105228-100G, Milwaukee, WI).
8. TFA (peptide synthesis-grade) is purchased from Fisher (BP618-500, Fairlawn, New Jersey).
9. 0.02 micron Anotop filters are purchased from Whatman (cat. no. 6809-3102).
10. PBS buffer: 50 mM phosphate-buffered saline (PBS) and 150 mM NaCl, pH 7.4.

2.3. Circular Dichroism Measurements

All measurements are carried out using an Aviv Circular Dichroism Spectrometer Model 62DS with Igor plotting software.

2.4. Scanning Force Microscopy

1. A Nanoscope III Multimode scanning force microscope (tapping mode) with noncontact high frequency silicon cantilevers (NSC-15, MikroMasch) is utilized.
2. Atomically flat and hydrophilic mica sheets (ASTM V–2 Grade 3 Ruby muscovite) are from Lawrence & Co.
3. A 15-mm metal specimen disc is from Ted Pella Inc. (vendor no. 16218) with double–backed tape (Gluespot by Digital Instruments, catalog reference STKYDOT).
4. Samples not immediately imaged are stored in semiconductor wafer containers from Entegris (vendor no. H22-101-0615 tray and number H22-10-0615) under an ambient environment.

2.5. Transmission Electron Microscopy

1. All experiments are performed using JEOL 100 CX transmission electron microscope.
2. All substrates, storage boxes, lens tissue, and chemicals used for transmission electron microscopy are purchased from Electron Microscopy Sciences (EMS), Fort Washington, PA.

3. Polystyrene petri dish cover (150 × 15 mm) is from Fisher (vendor no. 08-757-14).
4. Ross optical lens tissue (vendor no. 71700), 2% collodion in amyl acetate (vendor no. 12620-00), 400 mesh Cu grids (vendor no. CF400–Cu), 2% uranyl acetate (vendor no. 22400), and a storage box (vendor no. 71137, EMS) which holds 100 2- to 3-mm specimen support grids, are from EMS.

2.6. Fluorescence Measurements

1. All 96-well plate fluorescence assays are performed on a Victor^{2}V with Wallace 1420 operating software.
2. All cuvet assays are performed using a Cary Eclipse Fluorescence Spectrophotometer with Cary Eclipse software.
3. Thioflavin-T (T-3516) is purchased from Sigma (St. Louis, MO).

2.7. Cell Viability Assay

1. Rat tail collagen, type 1 (cat. no. 354236) is purchased from BD Biosciences (Bedford, MA).
2. CellTiter 96 Non-Radioactive Cell Proliferation Assay Kit (part no. G4002) is purchased from Promega Corporation (Madison, WI).
3. RPMI medium 1640 with L-glutamine without phenol red is purchased from Gibco Invitrogen Corporation (Grand Island, NY).
4. Fetal bovine serum, certified (cat. no. 16000) is purchased from Gibco Invitrogen Corporation.
5. Horse serum (heat-activated) (cat. no. 26050) is purchased from Gibco Invitrogen Corporation.
6. PC-12 cell media is prepared using 81% RPMI media 1640, 10% horse serum, 5% fetal bovine serum, 1% amino acid, 1% nonessential amino acids, 1% vitamins, 1% glutamine, with Primocin (200 μL/100 mL of media).

3. Methods

A protocol has been developed to provide consistent sample preparation of monomeric Aβ protein and avoid ambiguity in experimental results. In addition, several physical techniques are used to characterize and monitor Aβ in its initial monomeric state as it nucleates and polymerizes to fibrils. Thioflavin-T (Th-T) fluorescence is a technique used to elucidate the aggregation kinetics of Aβ and dissolution of fibrils associated with Aβ fibrillogenesis. It is widely used because Th-T specifically binds to Aβ aggregates by inserting itself into hydrophobic pockets as Aβ monomers self-associate to form fibrils, which are predominantly β-sheet in conformation *(15)*. The secondary structure of Aβ working solutions are monitored by circular dichroism (CD) spectroscopy, where a transition from random coil (monomer material) to β-sheet (protofibrils and fibrils) can be observed. Scanning force microscopy (SFM) and transmission electron microscopy (TEM) allows direct monitoring of Aβ growth and morphology as monomers self-associate into aggregates, which then develop into mature fibrils. Dynamic light scattering (DLS) data are used to obtain quantitative information on changes in the average size of particles in solution. Fluorescence photobleaching recovery (FPR) is a useful technique for providing diffusion coefficient data which can be related to particle size through the Stokes-Einstein relation (D = $kT/6\pi\eta$ R_h, where η represents the viscosity). In vitro cellular viability assays (3-[4,5-Dimethylthiazol-2-yl]-2,5-diphenyltetrazolium bromide [MTT] proliferation assay) to determine the toxicity of Aβ mononers, intermediate structural assemblies (oligomers/protofibrils), and fibrils are performed using PC-12 cells. MTT is a yellow tetrazolium salt that is converted by mitochondrial reductoses to a purple formazan dye, which can be solubilized and measured via ultraviolet (UV) spectroscopy. Reliable methods for producing monomeric Aβ starting material, combined with use of the physical techniques as well as physiological studies, allows for optimal isolation and identification of the Aβ toxic species that contribute to amyloid plaques. Understanding the nucleation pathway of Aβ may lead to the development of compounds capable of interfering with this process, ultimately altering Aβ aggregation and toxicity.

3.1. Peptide Synthesis

From mutagenesis studies of Aβ, it has been shown that the hydrophobic core, KLVFF, serves as a key sequence in the self-assembly of Aβ *(16)*. This peptide sequence forms amorphous aggregates, do not form amyloid fibrils *(17)*, but is capable of binding to and inhibiting formations of amyloid fibrils. Thus, a number of groups have used this sequence as a recognition element for Aβ and have investigated peptides containing this hydrophobic core as possible inhibitors of aggregation and/or Aβ fibril dissolution agents. Soto used the recognition sequence, but incorporated L-proline, due to its "β-sheet breaking" capabilities, at key positions and found that his inhibitor was able to convert Aβ fibrils to amorphous aggregates and subsequently inhibit toxicity in vitro *(18)*. Murphy et al. *(19,20)* used a strategy where they designed peptides that bind to the hydrophobic core of Aβ (KLVFF) and disrupt fibril formation by altering the rate of aggregation and interfering with cellular toxicity. Their approach relied on self recognition of the hydrophobic core with inhibitor compounds, and the adaptation of specific conformational changes of Aβ with their proposed inhibitors to limit toxicity. What they ultimately found was that their specific inhibitors, mostly consisting of the hydrophobic core of Aβ with disrupting oligolysine groups at either *N* or *C*-termini, not only altered Aβ self assembly (aggregation), but also blocked toxicity in vitro *(21)*. Meredith et al. described inhibitors that contained ester bonds *(22)*, which replaced amide bonds in the peptide backbone, and inhibitors, where *N*-methylated amino acids were used *(23,24)* in alternating positions of the hydrophobic core of Aβ (residues 16–20), that were adequate for blocking fibril formation and the disassembly of pre-formed fibrils. They also found that the alternating positioning of *N*-methylated amino acids was crucial for inhibition because their peptides were designed so that when in β-strand conformations, one face would be blocked, therefore hydrogen bonding from that face could not occur.

Aβ transitions from random coil to β-sheet conformations as it aggregates in solution (CD); therefore, it has been hypothesized that β-sheet "blockers," such as proline, *N*-methylated amino acids, and D-amino acids, would inhibit protein aggregation and block amyloid fibril formation. For this reason, the use of α,α-disubstituted amino acids (ααAA) was explored. ααAAs are widely used in peptide design because of their structure-promoting effects. Incorporation of $C^{\alpha,\alpha}$-disubstituted amino acids dipropyl glycine (Dpg), diisobutyl-glycine (Dibg), and dibenzylglycine (Dbg) into the hydrophobic core of Aβ (KLVFF) are of particular interest due to their potential to adopt fully extended conformations *(25)*, which may have increased affinity for Aβ assemblies or protofibrils. Difficulties associated with the incorporation of sterically hindered $C^{\alpha,\alpha}$-disubstituted amino acids into peptide sequences by conventional methods caused for the exploration of alternative routes to synthesizing peptides containing the novel amino acids *(26–28)*. **Scheme 1** (**B** and **C**) displays inhibitor compounds synthesized by the Hammer laboratory.

3.1.1. AMY-1

Coupling: 7-azabenzotriazol-1-yloxy-tris-(pyrrolidino)phosphonium hexa-fluorophosphate (PyAOP), 4 eq; diisopropylethylamine (DIEA), 8 eq; Fmoc-amino acids, 4 eq (0.3 *M*); coupling 24 h. Fmoc removal: DMF:piperidine:DBU (93:5:2) for 30 min. Cleavage (precipitation method): TFA:phenol:H_2O:TIPS (8.2:0.5:0.5:0.2) for 2 h followed by precipitation in cold Et_2O.

1. The first six Lys residues are coupled to PAL-PEG-PS resin on a Pioneer Peptide Synthesizer using PyAOP in DMF.
2. The remaining amino acid residues are incorporated manually into the sequence. Dpg, Phe, and Dbzg are incorporated into the sequence using PyAOP/DIEA in DCE:DMF (1:1) at 50°C. The equiv of each reagent for the coupling is as described above.

A   H₂N-DAEFRHDSGYEVHHQKLVFFAEDVGSNKGAIIGLMVGGVVIAL-COOH

B

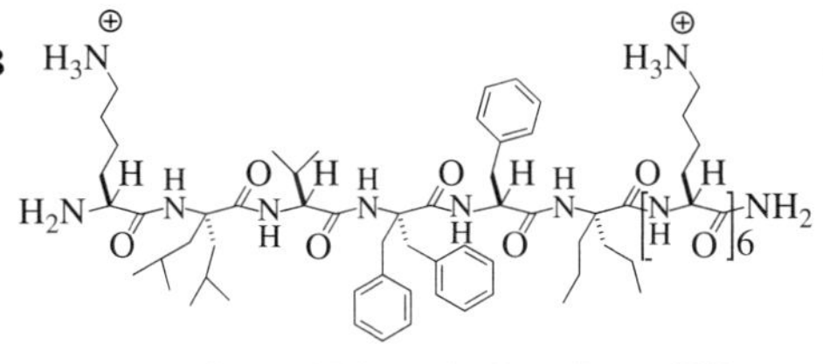

H-Lys-Dibg-Val-Dbzg-Phe-Dpg-(Lys)₆-NH₂

C

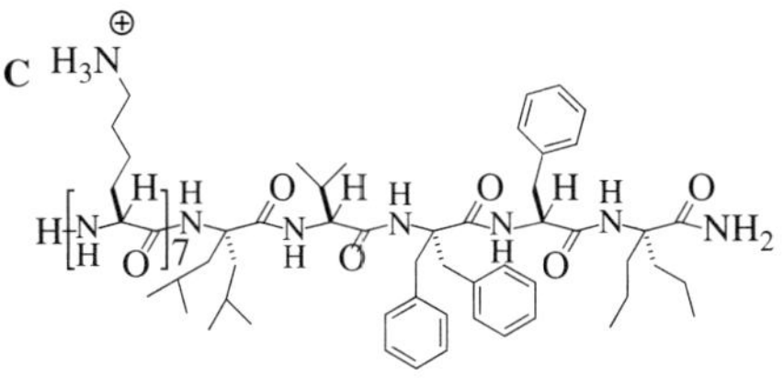

H-(Lys)₇-Dibg-Val-Dbzg-Phe-Dpg-NH₂

D

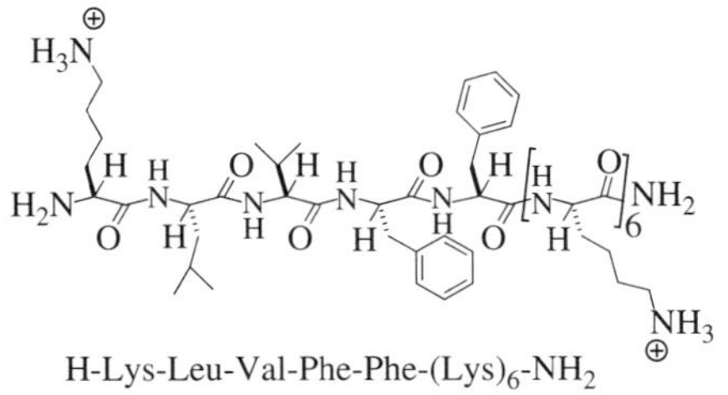

H-Lys-Leu-Val-Phe-Phe-(Lys)₆-NH₂

Scheme 1. Peptide based β-strand mimics designed to alter Aβ aggregation. (A) Sequence of Aβ$_{1-40}$. Amino acids highlighted in red (16–20) represent the hydrophobic core. Amino acids highlighted in green represent the turn portion of the β-sheet. (B) AMY-1 inhibitor. (C) AMY-2 inhibitor. (D) Murphy inhibitor (control peptide). The use of ααAAs incorporated into the hydrophobic sequence of Aβ was compared to the Murphy peptide (19–21) to elucidate the role of ααAAs on aggregation.

3. The Val residue is coupled to the *N*-terminus of Dbzg via amino acid symmetrical anhydride method. The symmetrical anhydride is prepared by treatment of 2 eq of Fmoc-Val-OH with 1 eq of DCC in CH₂Cl₂ at room temperature for 2 h followed by removal of the precipitated DCU by filtration. The symmetrical anhydride is concentrated, redissolved in the higher boiling DCE:DMF (9:1), and added to the resin-bound peptide. The reaction is carried out at 50°C for 24 h.

4. Dibg is coupled to Val using HATU (4 eq), HOAt (4 eq), DIEA (8 eq), and Fmoc-Dibg-OH (4 eq, 0.3 *M*) in DCE:DMF (1:1) at 50°C.

5. The last residue, Lys, is coupled to Dibg via symmetrical anhydride, coupling (as in **step 3** above) at 50°C.

6. A final deprotection procedure occurs followed by peptide cleavage from the resin support.

3.1.2. AMY-2

Coupling: PyAOP, 4 eq; DIEA, 8 eq; Fmoc amino acids, 4 eq $(0.3 M)$; coupling 24 h. Fmoc removal: DMF:piperidine:DBU (93:5:2) for 30 min. Cleavage (extraction method): TFA:phenol:H$_2$O:TIPS (8.2:0.5:0.5:0.2) for 2 h followed by extraction with 30% acetic acid (HOAc) and Et$_2$O.

1. Dpg is coupled to PAL-PEG-PS resin by manual coupling using PyAOP/DIEA in DCE:DMF (1:1) at 50°C. The equivalent of each reagent for the coupling is as described above.
2. Capping of resin to eliminate any unreactive active sites is performed using acetic anhydride $(0.2 M$ in $0.28 M$ DIEA).
3. Phe and Dbzg are incorporated into the sequence using PyAOP/DIEA in DCE:DMF (1:1) at 50°C.
4. The Val residue is coupled to the *N*-terminus of Dbzg via amino acid symmetrical anhydride method (*see* **Subheading 3.1.1., step 3**)
5. Dibg is coupled to Val (*see* **Subheading 3.1.1., step 4**)
6. Lys is coupled to Dibg via symmetrical anhydride at 50°C.
7. The last six Lys residues are coupled to the resin using the Pioneer Peptide Synthesizer using PyAOP in DMF. Conditions: PyAOP, 4 eq; DIEA, 8 eq; Fmoc Lys(Boc)-OH, 4 eq $(0.3 M)$.
8. A final deprotection procedure occurs followed by peptide cleavage from the resin support.

3.1.3. Murphy Inhibitor

1. The Murphy Inhibitor is synthesized using 4 eq of amino acid and activation with TBTU, HOBt, and DIEA at a final concentration of $0.2 M$.
2. A stepwise coupling of each amino acid is obtained using the standard solid-phase Fmoc coupling chemistry. Peptide cleavage is performed as that of AMY-1 peptides (section 3.1.1).

3.2. Peptide Cleavage

1. Extraction Method *(29,30)*. Resins are treated with standard cleavage solution (TFA:phenol:H$_2$O:TIPS; 8.8: 0.5: 0.5: 0.2) at room temperature for 2 h to cleave the peptide from the solid support. After filtration, the filtrate is diluted with cold 30% HOAc solution, washed with cold Et$_2$O (5 × 40 mL), and lyophilized to yield crude peptides.
2. Precipitation Method *(31)*. Resins are treated with standard cleavage solution (TFA:phenol:H$_2$O:TIPS; 8.8: 0.5: 0.5: 0.2) at room temperature for 2 h to cleave the peptide from the solid support. After filtration, the filtrate is diluted with cold Et$_2$O and allowed to co-precipitate for 24 h. The precipitate is then washed with cold

Et$_2$O and centrifuged at 1600 g for 10 min (three times). The supernatant is decanted and the remaining pellet was allowed to dry 12 h yielding the crude product.

3.3. Peptide Purification and Characterization

Linear gradients of 0.1% TFA in H$_2$O (v/v) (Buffer A) and 0.1% TFA in CH$_3$CN (v/v) (Buffer B) are utilized in all HPLC experiments.

1. AMY-1; Analytical HPLC is performed using reversed-phase HPLC with a C$_{18}$ column. Analytical conditions: 10% to 70% B over 60 min at 1 mL/min. HPLC retention time t$_R$: 36.64 min. Semipreparative conditions: four preps (30–35 mg). Gradient conditions: ramp from 10% to 40% B over 15 min at 15 mL/min followed by a change from 40% to 60% B at 55 min, which is then eluted to 70% B over 60 min to ensure no residue on column. Average HPLC retention time t$_R$: 25.88 min. Analytical conditions for purity testing (Homogeneity >99%): 10% to 35% B over 25 min at 1 mL/min with t$_{Ravg}$: 22.95 min. Characterization of peptide using Matrix-assisted laser desorption/ionization (MALDI)-mass spectrometry (MS) reveals AMY-1 [M+H]$^+$ at 1709.54, with m/z calculated peaks indicating [M+Na]$^+$ at 1731. Expected molecular weight of AMY-1 is 1708 Da. (*see* **Note 1**)
2. AMY-2; Analytical HPLC is performed using reversed-phase HPLC with a C$_{18}$ column. Analytical conditions: 10% to 70% B over 60 min at 1 mL/min. HPLC retention time t$_R$: 50.98 min. Semipreparative conditions: three preps (30 mg). 10% to 70% B over 60 min at 20 mL/min. Average HPLC retention time t$_R$: 33.06 min. Homogeneity (>99%) is determined by analytical HPLC. Analytical conditions for purity testing: 10% to 30% B over 25 min at 1 mL/min with t$_{Ravg}$: 12 min. Characterization of peptide using MALDI-MS reveals the molecular weight of AMY-2 aq at 1707 Da, with m/z calculated peaks indicating [M+Na]$^+$ at 1730. Expected molecular weight of AMY-2 is 1708 Da.
3. Murphy Inhibitor: analytical HPLC is performed using reversed-phase HPLC with a C$_4$ column. Analytical conditions: 10% to 70% B over 60 min at 1 mL/min. HPLC retention time t$_R$: 29.84 min. Semipreparative conditions: three preps (45 mg). 5% to 18% B over 40 min at 15 mL/min, which remains unchanged for 15 additional minutes, then ramping to 50% B at 60 min. Average HPLC retention time t$_R$: 43.69–52.76 min. Analytical conditions for purity testing: 10% to 25% B over 15 min at 1 mL/min with t$_{Ravg}$: 13.65 min. Characterization of peptide using MALDI-MS reveals the molecular weight of Murphy inhibitor at 1420.9 Da.

3.4. Aβ$_{1-40}$ Peptide Aggregation (Monomeric Starting Material) (see Notes 2 and 3)

1. Lyophilized Aβ is dissolved in neat TFA at 1 mg/mL and sonicated for 10–15 min. TFA is then evaporated using centrivac to obtain a dark yellow oil.
2. The resulting oil is dissolved in HFIP at 1 mg/mL and incubated at 37°C for 1 h. HFIP is removed via centrivac to obtain a white powder.

3. The white powder is dissolved in HFIP at 1 mg/mL and incubated at 37°C for 1 h. HFIP is then removed via centrivac and the resulting white powder is lyophilized for 24 h (*see* **Note 4**).
4. The white powder is then redissolved in 2 mM KOH and 2X PBS at 1:1 and centrifuged for 10 min at 13,000g (*see* **Note 5**).
5. Amino acid analysis is performed on the supernatant to verify peptide concentration.
6. Sample is stored in the freezer at −80°C until ready for usage.
7. For co-incubation experiments, 1:1 molar ratio mixtures of Aβ:inhibitor (500 μM: 500 μM; final concentration 250 μM) are diluted 10-fold and allowed to co-incubate at 37°C at various time intervals.

3.5. Aβ$_{1-40}$ Peptide Aggregation (Protofibril/Fibril Formation)

1. A 500-μM stock solution of Aβ is freshly prepared by dissolving the lyophilized peptide in filtered 10 mM KOH at pH 10.7.
2. Aliquots of Aβ stock are then dissolved in PBS to acquire a 50-μM working solution.
3. Incubations of the working solutions are performed using an Eppendorf thermomixer set at 37°C at various time intervals to allow fibril growth.
4. Inhibitor peptides are dissolved in filtered ultra pure deionized H$_2$O to yield a concentration of 500 μM.
5. For co-incubation experiments, 1:1 molar ratio mixtures of Aβ:inhibitor (500 μM: 500 μM; final concentration 250 μM) are diluted 10-fold and allowed to co-incubate at 37°C at various time intervals.

3.6. Circular Dichroism Spectroscopy Measurements

1. Each peptide stock solution, buffer, and ultra pure deionized H$_2$O is filtered using a 0.02-μm Anatop filter (Whatman).
2. Aβ samples are prepared by diluting 10X stock solution of Aβ$_{1-40}$ (500 μM) and 10X PBS in filtered H$_2$O to acquire a working solution of 50 μM Aβ in PBS.
3. Inhibitor peptides are dissolved in filtered H$_2$O to yield a concentration of 500 μM.
4. For co-incubation experiments, 1:1 molar ratio mixtures of Aβ:inhibitor are acquired yielding 250 μM stock of Aβ:inhibitor. Working stock solutions are then diluted 10-fold and incubated at 37°C over various time intervals.
5. CD spectra are the average of three scans taken at 1.00-nm intervals acquired from $\lambda = 260$ nm to 190 nm (UV absorbance range) recorded at 25°C **(Fig. 1)**.
6. The spectrum of PBS buffer is used as the background subtraction in all experiments.
7. The CD instrument is calibrated using a 1.0 mg/mL solution of camphorsulfonic acid (CSA)(*see* **Note 6**).

3.7. Scanning Force Microscopy

1. Cleavage and exposure of interior mica planes (which are atomically flat over large areas and ideal for SFM imaging) are achieved by placing a razor blade in the

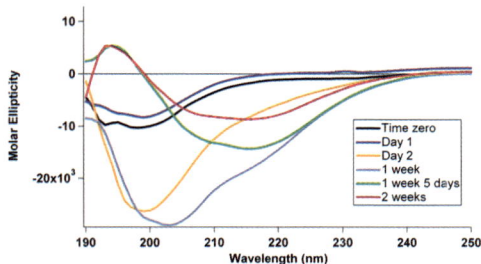

Fig. 1. Circular dichroism (CD) spectrum of Aβ (50 μ*M*) in phosphate-buffered saline (50 m*M* 150 NaCl, pH 7.4). Monomeric starting material orients in random coil conformations while aggregated intermediates (protofibrils) and fibrils orient in β-sheet conformation. Molar Ellipticity = [θ] units: deg cm^2 dmol^{-1}.

middle of the sheet edge layers of a 1-cm^2 mica sheet and separating the layers by gripping one side with a pair of tweezers.

2. A sample aliquot of 5 μL is then placed on the freshly exposed mica sheet and allowed to dry for 25 min, unless otherwise noted.

3. After a 25-min adsorption step, the sample/substrate is rinsed with 400 μL of filtered deionized H$_2$O followed by tilting the substrate and placing its edge on a Kimwipe tissue to wick away the H$_2$O (*see* **Note 9**).

4. The mica samples are then placed sample-exposed face up on a 15-mm metal specimen disc.

5. When ready for imaging, each specimen disc is placed on top of the piezoelectric scanner of the SFM instrument and imaged (**Fig. 2**) using SFM tapping mode. Samples not immediately imaged are stored in semiconductor wafer containers under an ordinary lab ambient environment (*see* **Note 10**).

3.8. Transmission Electron Microscopy (see Fig. 3)

1. For electron microscopy, the sample solutions are adsorbed onto a thin carbon-coated collodion layer placed on a copper grid.

2. The collodion coating is used to keep the sample on the grid; deposition of carbon onto this grid is done to ensure thermal stability of the film.

   a. In order to make the collodion film, a pool of filtered deionized H$_2$O is placed in a 150 × 15 mm polystyrene Petri dish cover.

   b. The water is added until almost spilling over the sides of the Petri dish.

   c. A sheet of Ross optical lens tissue is placed on top of the pool of water and gently pulled across the surface to remove any particles.

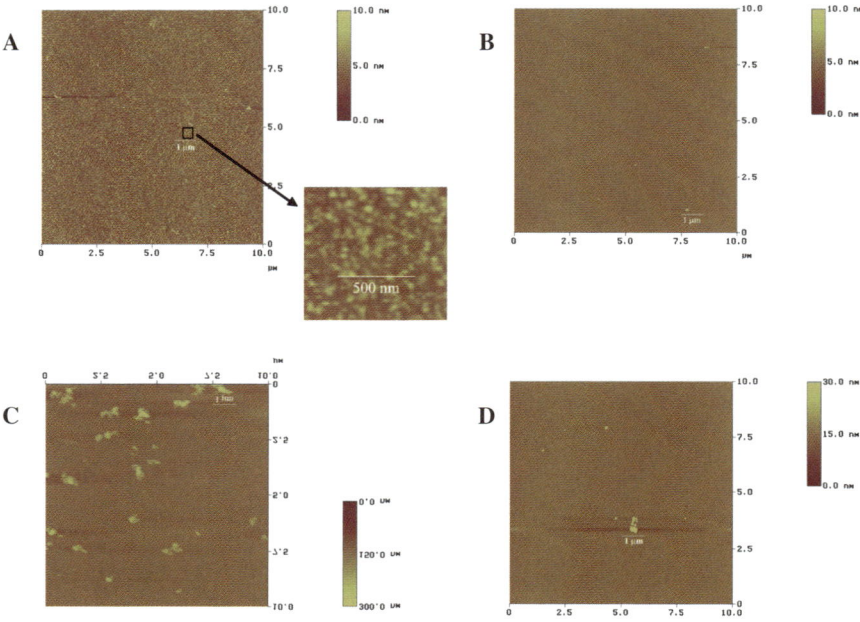

Fig. 2. Scanning force microscopy (SFM) images of inhibitor assays with Aβ$_{1-40}$. Samples were incubated at 37°C for 1.5 h followed by incubation at room temperature for 1 wk. A) Aβ absorbed on mica, with magnified image of Aβ. This sample was used as the control in the aggregation studies (shown in **Fig. 2B-D**). Fibrils are approx 3 nm in height and approx 480 nm in length. pH was 7.38. (**B**) SFM image of Aβ in the presence of AMY-1. Results show minimum amount of aggregates (8–17 nm in height). (**C**) SFM image of Aβ in the presence of AMY-2. Results show large amount of aggregates (100 nm in height), yet no fibril formation. (**D**) SFM image of Aβ in the presence of Murphy inhibitor. Minimal amount aggregates (30–80 nm in height) were formed with no indication of fibrillic assemblies. Aggregate height was larger than that of AMY-1.

   d. A thin film of collodion is prepared by dropping roughly 20 μL of nonfiltered 2% collodion in amyl acetate until the solution began to solidify and thicken on the surface.
   e. After the collodion film is formed, several 400-mesh Cu grids were placed onto the film and allowed to stand.
   f. Removal of the coated grids is accomplished by placing the paper used to separate Parafilm layers over the collodion membrane until the entire paper is wet; this is to ensure proper adhesion of the grid to the Parafilm paper.

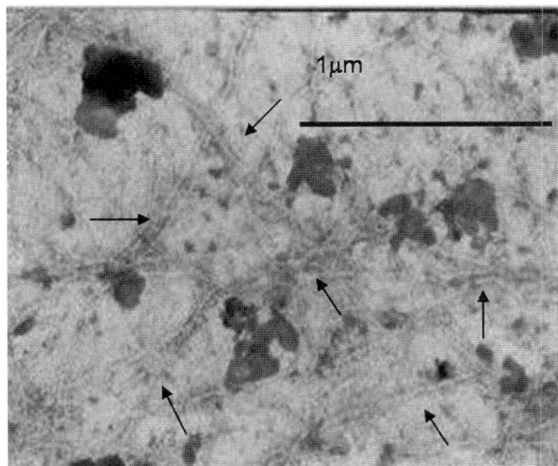

Fig. 3. Transmission electron microscopy (TEM) image of Aβ (50 μ*M*) incubated at 37°C for 20 d. Fibrils range from 7 to 10 μm in length. Circular dichroism measurements of the solution containing fibrils (shown here) were analyzed and contained β-sheet conformations. Arrows denotes bundles of fibrils.

    g. The paper is then removed causing the grids and adsorbed collodion film to be removed from the water/collodion interface.

    h. After adsorbing the film onto the grid, the grids are placed in Petri dishes for 24 h under ambient conditions to dry.

    i. The grids are coated with carbon (bench top Turbo Denton vacuum evaporator) to keep the collodion from being vaporized by the thermal energy generated from the bombardment of incident electrons.

3. Samples for TEM analysis are prepared by inverting the carbon/collodion exposed grid on a 5-μL droplet of sample for approximately 30 sec.

4. Excess solvent is wicked away and the grid is then rinsed by inverting the grid on a 5 μL droplet of water for approximately 5–10 sec.

5. The grid is stained with 2% uranyl acetate in 0.05 *M* HCl by inverting on a 5 μL droplet of filtered 2% uranyl acetate for 5–10 sec.

6. Excess liquid is wicked away from the original grid using filter paper and placed in a Petri dish where each sample is properly labeled.

3.9. Fluorescence Measurements

1. All experiments are performed using an excitation wavelength filter of 450 nm, with a slit of 5 nm, and an emission wavelength filter of 480 nm, with a slit of 10 nm.

2. Averaging time for scans is 0.1 nm/s.

3. Th-T is dissolved in 10X PBS to acquire a 10X stock solution (100 μ*M*) of dye (**Note 11**).
4. 90 μL of peptide sample stock (peptide at desired concentrations) is diluted with 10 μL of dye to acquire fluorescence spectroscopic working solutions (**Fig. 4**).
5. Aβ and co-incubation dilutions are made as previously described under **Subheading 3.6.**, and were allowed to incubate in a 96-well plate or quartz micro-cuvettes at various time intervals (**Note 12**).

3.10. Cell Viability Assay

Coating Plate (coat plates with collagen for cells to better adhere to plate): thoroughly mix 45 μL of collagen with 1 mL of 0.02 *N* sterile HOAc in a sterile reservoir. Tritulate to ensure proper mixing. Aliquot 1 mL of solution into culture plate. Be sure to allow the solution to coat the entire surface of the culture plate where cells are to attach. Let solution sit in the plate for 1 h, with periodic shaking, followed by removal of collagen coating solution.

1. Calculate number of cells need for an experiment.
   a. For a typical experiment, 500,000 cells/mL is desired.
   b. Example: 8 (rows) × 5 (repeats) (from 96 well plate) = 40 (100 μL [total volume per well]) = 4 mL total volume.
   c. 4 mL × 500,000 cells/mL = 2 million cells needed for the experiment
   d. 500,000 cells/mL because each well requires 100 μL at 50,000 cells/100 μL.

2. Warm the media in aquabath at 37°C.
3. Dislodge cells from culture plates.

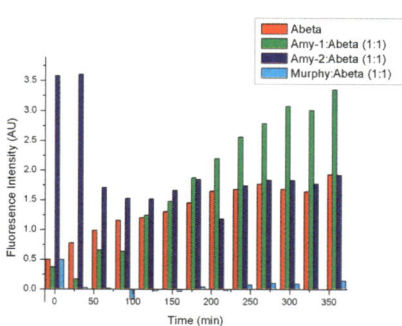

Fig. 4. Thioflavin-T fluorescence assays. Aggregation of Aβ in the presence of inhibitor compounds. Results indicate that AMY-1 increases Aβ aggregation rate as well as AMY-2. The Murphy:Aβ (1:1) sample exhibited no fluorescence signal as a result of short experimental time and no fibril formation.

a. Pipet off old media from cells using a vacuum line.

b. Add 3–5 mL of fresh media to culture plate and tritulate to dislodge cells from the plate forming a cell suspension, and place in a sterile 10- to 15-mL centrifuge tube.

4. Set up counting tube.

a. Place 1 mL of media, 0.5 mL of trypsin blue, 0.5 mL of cell suspension (**step 3b**) in a sterilized tube.

b. Mix well and pipet 10 μL into counting apparatus (hemocytometer) and count cells in each quadrant (four quadrants total).

c. Example: count 1–17, 18, 19, 20 are the number of cells counted from each quadrant. Count 2–20, 21, 22, 23; find the sum of all quadrants, count 1–74, and count 2–86. Take the average of count 1 and count 2, giving a total of 80 cells averaged (**Note 13**).

d. Average = 800,000 cells per mL (remember 100 μL volume) (**Note 13**).

5. 2.0 million cells (number of cells needed for experiment; step 1c)/800,000 cells mL^{-1} (cells counted from cell suspension; step 4d) = 2.5 mL of cell suspension (step 3b) needed in order to obtain 500,000 cells/mL needed for plating (place in a separate reservoir).

6. 4 mL total volume needed to fill well plate (**step 1b**); 4 mL − 2.5 mL = 1.5 mL of media needed and will be added to 2.5 mL of cell suspension (in reservoir) to obtain 500,000 cells/mL.

7. Plate cells by adding 100 μL of the 4 mL cell solution (**step 6**) to each mapped well.

8. Allow cells to adhere to the wells for 24 h at 37°C, 5% CO_2.

9. Option: Split remaining cells in counting tube or discard.

10. Cells adhere to plate for 24 h followed by co-incubation with analyte at desired concentrations with analyte at desired concentrations and time intervals (**Note 14**).

11. Following co-incubation at desired concentrations (preferably for 24 h) or at desired incubation intervals, 15 μL of Dye substrate (Promega) is added to each well and incubated for 4 h at 37°C, 5% CO_2.

12. Add 100 μL of Stop/Solubilizing solution is added to each well and incubated at room temp for 1 h.

13. The 96 well plate is read using UV absorbance at 570 nm (**Fig. 5**).

3.11. Dynamic Light Scattering Measurements

1. The DLS instrument has been described previously (*32*); it is built in house and consists of a Coherent Innova 90 argon ion laser and a copper sample cell with a filtered toluene bath. A Hamamatsu R955P photomultiplier tube is mounted on a rotating arm. A Pacific Precision Instruments model 126 pulse amplifier/discriminator feeds data to a computer with an ALV 5000 correlator and

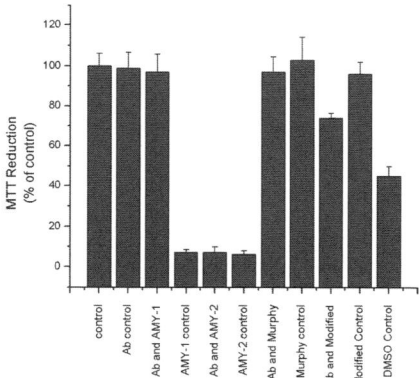

Fig. 5. Cell viability assays of inhibitor peptides. Samples were incubated for 1 wk at 37°C and diluted in RPMI media (1:10 dilution). The AMY-1 (toxic alone) inhibitor, when in the presence of β -amyloid (Aβ), exhibited no cytotoxicity while the AMY-2 inhibitor caused cell death. The Murphy and Modified A-1 inhibitors were nontoxic to cells, suggesting that they are capable of inhibiting cell death caused by Aβ fibrilloge-nesis. Experimental results indicate that the AMY inhibitors in the absence of Aβ are toxic to PC-12 cells whereas the Murphy and Modified inhibitors are nontoxic.

associated software installed. The sample holder temperature is controlled using a Lauda RM6 water bath to within 0.1°C (**Note 15**).

2. Stock Aβ is prepared by diluting $A\beta_{1-40}$ in 100% DMSO at a concentration of 1 mg/mL.
3. Stock solutions of $A\beta_{1-40}$ in 100% DMSO are incubated for 2 h at room temperature.
4. Samples are prepared by diluting the stock solution into filtered 50 mM PBS, pH 7.4 to a final peptide concentration of 100 μM.
5. The samples are then loaded into a dust-free, disposable borosilicate glass cell (6 × 13 mm), Kimble.
6. DLS measurements are taken at 90° scattering angle. Custom software is used to analyze each of the many sequential runs of the experiment (**Fig. 6**, *see* **Note 16**).

3.12. Fluorescence Photobleaching Recovery Measurements

1. A brief summary of the FPR apparatus follows; details have appeared elsewhere *(33)*. A fringed pattern is bleached into the sample by intense laser illumination (at λ = 488 nm, 3–7 W/cm^2) of a coarse diffraction grating (50, 100, 150, and 300 lines/inch), referred to as a Ronchi ruling (Edmund Scientific), held in the rear image plane of an objective mounted on an epifluorescence microscope. The stripe pattern is described by the grating constant, $K = 2\pi/L$, where L is the period of the repeat pattern in the sample. After bleaching, the laser beam is shifted

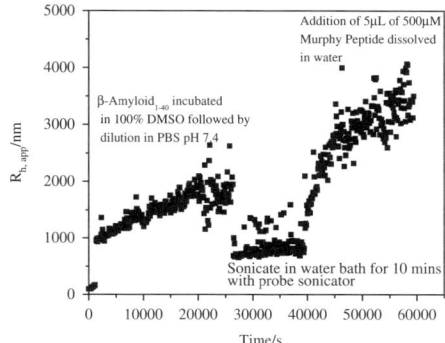

Fig. 6. Growth kinetics of Aβ at physiological pH. β-amyloid was dissolved in neat dimethyl sulfoxide at a concentration of 1 mg/mL followed by dilution in 50 m*M* phosphate-buffered saline, 150 m*M* NaCl, pH 7.4, yielding a concentration of 100 μ*M*. The apparent hydrodynamic radius was determined from cumulant analysis of dynamic light scattering data taken at 90° scattering angle.

to lower intensity to detect fading of the pattern. The detection is accomplished by a modulation system *(34)*. After the shutter, which protects the RCA 7265 photomultiplier tube (PMT) during the photobleaching step, reopens, the Ronchi ruling is vibrated electromechanically in a direction perpendicular to its stripes to produce a modulated signal. This signal is buffered by a Stanford Research Systems SR560 low-noise preamplifier then fed to a tuned amplifier to isolate the effects of shallow (typically 5%) photobleaching from random noise. The modulation detection system not only enables shallow, weakly perturbing photobleaches to be used, but it also eliminates decay terms due to higher Fourier components of the stripe pattern. Thus, each diffuser gives rise to a single decay term *(35)*. A zero-crossing detector/peak voltage detection circuit triggers an analog-digital card (National Instruments #AT-MIO-16D, part no. 320489-01) to acquire the peak voltage of the modulated signal, or contrast $C(t)$, which decays exponentially due to diffusion: $C(t) = baseline + C(0)e^{-K^2 Dt}$.

2. Samples are prepared by dissolving 500-μ*M* stock 5-carboxyfluorescein-labeled peptide in filtered 10 m*M* KOH. The sample is difficult to solubilize, therefore sonication in a Branson model no. 2510 bath sonicator for short cycles of 10 sec to prevent heat from affecting the sample is performed.

3. A 50-m*M* phosphate-buffered solution, pH 7.4 is prepared and filtered through a 0.02-μm Anatop filter, after which the desired amounts of sodium chloride are added.

4. Aliquots of the 500-μ*M* stock solution (**step 2**) are diluted to the desired concentration with 50 m*M* PBS.

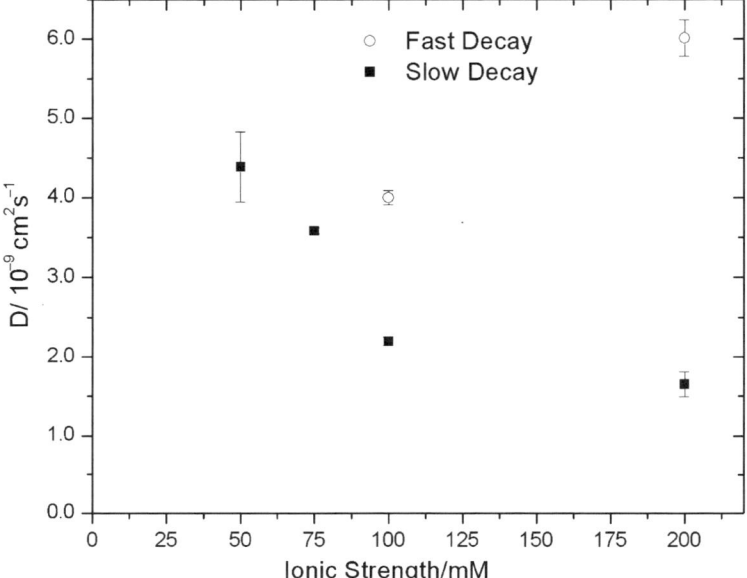

Fig. 7. Fluorescence photobleaching recovery measurements on samples containing 100% 5-carboxyfluorescein-labeled peptide in 50 mM phosphate buffer, at pH 7.4 with increasing sodium chloride concentrations.

5. The samples are loaded into rectangular microslides (0.2-mm pathlength, VitroCom) by capillary action, and the microslides are flame-sealed.
6. Following FPR measurements, the recovery profiles are analyzed with homemade software and the plots are best fitted with a single and double exponential function (**Fig. 7**).

4. Notes

1. For HPLC purification, columns are allowed to re-equilibrate for 30 min after each analytical run and 45 min after each preparatory run. Re-equilibration is performed to ensure column starting conditions.
2. To ensure monomeric starting solutions, tips, tubes, and bottles that are to house stock solutions are autoclaved to remove any possible contamination by bacteria. Impurities such as metals are minimized by the use of salts with purity contents ≥99.99%. The removal of dust and preformed aggregates are obtainable by the use of 0.02-μm Anotop filters.
3. Surface roughness affects aggregation rates of Aβ. It was determined that Dot Scientific Microcentrifuge tubes (509-FTG) were optimal for dissolution protocols

(36). These specific tubes have smooth surface textures which prolong the initiation of Aβ aggregation compared to rough textured tubes.

4. Lyphilization of Aβ following HFIP evaporation is performed to ensure that all residual organic solvent is removed. Any trace amounts of organic solvent will distort CD images giving rise to α-helical conformations and will affect overall aggregation kinetics.

5. Following centrifugation of Aβ at 13,000g, protofibrillar material sediments out forming a pellet. The resulting supernatant is the monomeric material needed. Carefully pipette off the supernatant from the top of the solution. Be sure not to allow the tip of the pipette tip to remove any of the pellet. To alleviate the possibility of removing sedimented material, the pellet remains submerged in approx 3–5 μL of supernatant.

6. CSA calibration can be performed on the CD instrument or using an UV instrument.

   a. UV: 1 mg/mL CSA, wavelength (λ = 320–240 nm), 1-nm bandwidth, 1 s average time, 0.5 nm step.

      i. Run water blank, record baseline at λ = 320 nm.
      ii. Run CSA standard, record baseline at λ = 320 nm.
      iii. Baseline at λ = 320 nm should be the relatively same value for both samples.
      iv. Correct spectrum (subtract blank).
      v. Find AU using Beer's Law: OD_{285} = 0.1486 at 1.00 mg/mL.; mDeg = $mDeg_0 \times$ (length × concentration) (mDeg = $mDeg_0 \times$ [Abs/ε].

   b. CD ratio test: 1 mg/mL CSA, 0.1 cm cell, wavelength (λ = 320–192.5 nm), 0.5 nm step, 1.0 nm bandwidth.

      i. Measure CSA standard.
      ii. Measure air baseline.
      iii. Subtract air baseline from CSA standard.
      iv. Find peak values near λ = 290 nm and λ = 192.5 nm.
      v. Calculate peak ration 290/195.2.
      vi. Ratio should be between 1.9 and 2.2.

7. Monitor the signal dynode voltage of the CD sample as spectra is being acquired; it should not exceed 1000 mV. If this occurs, the sample is contaminated, too aggregated, or contains too high of a salt content and should be reprepared and tested.

8. For CD measurements, the buffer blank should "flatline" in the data collection screen. If this does not occur, cell contamination or buffer contamination has occurred. It is important to notice this due to the negative impact baseline fluctuations will have on data when converting mDeg to molar ellipticity.

9. It is imperative that the sample adsorb to the mica for at least 25 min. Many groups report adsorption times of less than 10 min. This is problematic because a rinsing

step follows the adsorption. Rinsing is important for the removal of inorganic salts, but if the sample is assumed to have been absorbed to mica and is partially adsorbed, rinsing of partially absorbed material can rinse aggregates or fibrils from mica leading to the illusion of only monomeric, nonaggregated materials. This results in inconsistent interpretation of data and inconsistent and uncomparable results relative to other experimental techniques.

10. SFM tapping mode eliminates unwanted contact forces (van der Waals) created by SFM contact mode. With tapping mode, as the sample moves under the cantilever, the cantilever taps the sample rather than being in constant contact with the sample. Tapping mode is done due to peptide samples being "soft" and extremely fragile. Contact mode will disrupt morphology of samples.

11. The use of PBS buffer at pH 7.4 to dissolve Th-T (dye working solution) yields similar results compared to the conventional method of Th-T dissolution in 50 mM glycine-NaOH buffer at pH 8.5.

12. 100 µL is the maximum volume per well. Adhesive sealing films for microplates (Thomas Scientific, cat. no. 6980A13) are used to prevent evaporation of fluorescence sample. The non-usage of thermal seals causes for Th-T and NaCl salts to precipitate as a result of evaporation, disrupting the fluorescence signal.

13. The standard hemocytometer has four quadrants. A specific number of live cells will reside in each quadrant. The number of live cells are counted and totaled. Note that the live cells will appear green and dead cells will appear blue. Live cells can resist trysin blue treatment while dead cells will just absorb the dye. The total number of living cells (green) is averaged by the number of times counted. The total number of cells counted is based on the volume underneath the hemocytomer's coverslip. The standard apparatus has four basic quadrants with dimensions that equals 1.0×10^4 cm^3, which can hold 1.0×10^4 mL; therefore, if one has x number of cells multiplied by 1×10^4 mL gives rise to x^5 or x^6 number of cells per mL relative to the cell suspension of the counting tube.

14. It is stated to co-incubate analyte at desired concentration and time intervals. It is helpful, if doing concentration studies with cells, to first prepare concentration dilutions then add to cells. Make sure to take into consideration the dilution that will occur when adding analyte to cells, so 2–10X stocks of each desired concentration are preferable. When performing time dependent experiments, it is advisable to pre-incubate all samples in descending order from long to short times.

15. DLS is a technique whose sensitivity depends on the size of the scattering particle. Thus, DLS will be able to identify large aggregates more readily as compared to smaller aggregates. The limitation of this technique is particularly severe at the early stages of aggregation. Heterogeneous samples are difficult to measure, as small particles scatter less than larger ones; thus, it would be difficult to conclude whether both large and small species are present in solution. For this reason, FPR, which allows for the differentiation of large and small particles based on order of magnitude differences associated their respective diffusion coefficients, is used.

16. The DLS apparatus is set to make multiple measurements. **Figure 6** shows that immediately upon dilution of Aβ from neat DMSO stock solution into PBS at pH 7.4, large aggregates are present with an initial hydrodynamic radius of about 90 nm. The average size of the aggregates increases linearly with time as evidenced by the larger hydrodynamic radius values. After disrupting the large and slowly growing fibrils by sonication, the system remains stable. Addition of the Murphy peptide accelerates growth as expected *(37)*. The occasional spikes signify the presence of exceptionally large aggregates in a particular acquisition, but the trend is clear.

References

1. Selkoe, D. J. (1991) Amyloid protein and Alzheimer's disease. *Sci. Am.* **265(5)**, 68–78.
2. Hardy, J., and Selkoe, D, J. (2002) The Amyloid hypothesis of Alzheimer's disease: progress and problems on the road to therapeutics. *Science* **297**, 353–356.
3. Orpiszewski, J., Schormann, N., Kluve-Beckerman, B., Liepnieks, J. J., and Benson, M. D. (2000) Protein aging hypothesis of Alzheimer's disease. *FASEB J.* **14**, 1255–1263.
4. Bitan, G., Lomakin, A., and Teplow, D. B. (2001) Amyloid β-protein oligomerization: prenucleation interactions revealed by photo-induced crosslinking of unmodified proteins. *J. Biol. Chem.* **276**, 35,176–35,184.
5. Shao, H., Jao, S., Ma, K., Zagorski, M.G. (1999) Solution structure of micelle-bound amyloid β–(1–40) and β–(1–42) peptides of Alzheimer's disease. *J. Mol. Biol.* **285**, 755–773.
6. Shen, C. and Murphy, R.M. (1995) Solvent effects on self-assembly of β–amyloid peptide. *Biophys. J.* **69**, 640–651.
7. Zagorski, M. G., Yang, J., Shao, H., Ma, K., Zeng, H., and Hong, A. (1999) Methodological and chemical factors affecting amyloid β-peptide amyloidogencity. *Methods Enzymol.* **309**, 189–204.
8. Joa, S. C., Ma, K., Talafous, J., Orlando, R., and Zagorski, M. G. (1997) Trifluoroacetic acid pre-treatment reproducibly disaggregates the amyloid β-peptide. *Amyloid, Int. J. Clin. Exp. Invest.* **4**, 240–252.
9. Huang, J. T. H., Fraser, P. E., and Chakrabartty, A. (1997) Fibrillogenesis of Alzheimer Aβ Peptides Studied by Fluorescence Energy Transfer. *J. Mol. Biol.* **269**, 214–224.
10. Fezoui, Y., and Teplow, D. B. (2002) Kinetic studies of amyloid β-protein fibril assembly: differential effects of α-helix stabilization. *J. Biol. Chem.* **277**, 36,948–36,954.
11. Harper, J. D., Leiber, C. M., and Lansbury, P. T., Jr. (1997) Atomic force microscopic imaging of seeded fibril formation and fibril branching by Alzheimer's disease amyloid-β protein. *Chem. Biol.* **4**, 951–959.

12. Harper, J. D., Wong, S. S., Leiber, C. M., and Lansbury, P. T., Jr. (1997) Observation of metastable Aβ amyloid protofibrils by atomic force microscopy. *Chem. Biol.* **4**, 119–125.

13. Nichols, M. R., Moss, M. A., Reed, D. K., Cratic-McDaniel, S., Hoh, J. H., and Rosenberry, T. L. (2005) Amyloid-β protofibrils differ from amyloid -β aggregates induced in dilute hexafluoroisopropanol in stability and morphology. *J. Biol. Chem.* **280**, 2471–2480.

14. Wood, S. J., Maleeff, B., Hart, T., and Wetzel, R. (1996) Physical, morphological and functional differences between pH 5.8 and 7.4 aggregates of the Alzheimer's amyloid peptide Aβ. *J. Mol. Biol.* **256**, 870–877.

15. Khurana, R., Coleman, C., Ionescu-Zanetti, C., Carter, S. A., Krishna, V., Grover, R. K., Roy, R., and Singh, S. (2005) Mechanism of thioflavin T binding to amyloid fibrils. *J. Struct. Biol.* **151(3)**, 229–238.

16. Tjernberg, L. O., Naslund, J., Lindqvist, F., et al. (1996) Arrest of β-amyloid fibril formation by a pentapeptide ligand. *J. Biol. Chem.* **271**, 8545–8548.

17. Tjernberg, L. O., Callaway, D. J. E., Tjernberg, A., et al. (1999) A molecular model of Alzheimer amyloid β-peptide fibril formation. *J. Biol. Chem.* **274**, 12,619–12,625.

18. Soto, C., Kindy, M. S., Baumann, M., and Frangione, B. (1996) Inhibition of Alzheimer's amyloidosis by peptides that prevent β-sheet conformation. *Biochem. Biophys. Res. Commun.* **226**, 672–680.

19. Lowe, T. L., Strzelec, A., Kiessling, L. L., and Murphy, R. M. (2001) Structure-function Relationships for inhibitors of β-amyloid toxicity containing the recognition sequence KLVFF. *Biochemistry* **40**, 7882–7889.

20. Pallitto, M. M., Ghanta, J., Heinzelman, P., Kiessling, L. L., and Murphy, R. M. (1999) Recognition sequence design for peptidyl modulators of β-amyloid aggregation and toxicity. *Biochemistry* **38**, 3570–3578.

21. Cairo, C. W., Strzelec, A., Murphy, R. M., and Kiessling, L. L. (2002) Affinity-based inhibition of β-amyloid toxicity. *Biochemistry* **41**, 8620–8629.

22. Gordon, D. J. and Meredith, S. C. (2003) Probing the role of backbone hydrogen Bonding in β-amyloid fibrils with inhibitor peptides containing ester bonds at alternate positions. *Biochemistry* **42**, 475–485.

23. Gordon, D. J., Tappe, R., and Meredith, S. C. (2002) Design and characterization of a membrane permeable N-methyl amino acid-containing peptide that inhibits Aβ$_{1-40}$ fibrillogenesis. *J. Peptide Res.* **60**, 37–55.

24. Gordon, D. J., Sciarretta, K. L., and Meredith, S. C. (2001) Inhibition of β-amyloid(40) fibrillogenesis and disassembly of β-amyloid(40) fibrils by short β-amyloid congeners containing *N*-methyl amino acids at alternate residues. *Biochemistry* **40**, 8237–8245.

25. Toniolo, C., Crisma, M., Formaggio, F., and Peggion, C. (2001) Control of peptide conformation by the Thorpe-Ingold effect (C-alpha-tetrasubstitution). *Biopolymers* **60(6)**, 396–419.

26. Fu, Y. and Hammer, R. (2002) Efficient acylation of the N-terminus of highly hindered $C^{\alpha,\alpha}$-disubstituted amino acids via amino acid symmetrical anhydrides. *Org. Lett.* **4**, 237–240.

27. Fu, Y., Hammarstrom, L. G. J., Miller, T. J., Fronczek, F. R., McLaughlin, M. L., and Hammer, R. (2001) Sterically hindered $C^{\alpha,\alpha}$-disubstituted amino acids: synthesis from α-nitroacetate and incorporation into peptides. *J. Org. Chem.* **66**, 7118–7124.

28. Fu, Y., Etienne, M. A. and Hammer, R. (2003) Facile synthesis of α, α-diisobutylglycine and anchoring its derivatives onto PAL-PEG-PS resin. *J. Org. Chem.* **68**, 9854–9857.

29. (2002)*Synthetic Peptides: A User's Guide*, 2nd ed. (Grant, G. A., ed.). Oxford University Press, New York.

30. "Cleavage, Deprotection, and of Peptides after Fmoc Synthesis," Technical Bulletin, PerSeptive Biosystems.

31. Sole, N. A. and Barany, G. (1992) Optimization of solid-phase synthesis of [Ala8]-dynorphin A. *J. Org. Chem.* **57(20)**, 5399–5403.

32. Russo, P. S., Saunders, M. J., Delong, L. M., Kuehl, S. K., Langley, K. H., Detenbeck, R. W. (1986) Zero-angle depolarized light scattering of a colloidal polymer. *Anal. Chim. Acta* **189**, 69–87.

33. Bu, Z., Russo, P.S., Tipson, D.L., and Negulescu, I.I. (1994) Self-diffusion of rodlike polymers in isotropic solutions. *Macromolecules* **27**, 1187–1194.

34. Lanni, F. and Ware, B.R. (1982) Modulation detection of fluorescence photo-bleaching recovery. *Rev. Sci. Instrum.* **53**, 905.

35. Ware, B. R. (1984) Fluorescence photobleaching recovery. *Am. Lab.* **16**, 16.

36. Aucoin, J. (2004) Protein aggregation studies: inhibiting and encouraging β-amyloid aggregation. Ph.D. Dissertation, Louisiana State University Department of Chemistry.

37. Kim, J. R. and Murphy, R. M. (2004) Mechanism of accelerated assembly of β-amyloid filaments into fibrils by KLVFFK$_6$. *Biophys. J.* **86**, 3194–3203.

8

Radiometal-Labeled Somatostatin Analogs for Applications in Cancer Imaging and Therapy

Jason S. Lewis and Carolyn J. Anderson

Summary

The use of radiolabeled peptides for the diagnosis and therapy of cancer has increased greatly over the last few decades. Skillfully crafted peptide systems, which have high affinity for receptors that are overexpressed in human tumors, offer the potential to improve the characterization, grading, and eventual therapy of human cancer. Robust peptide systems can be labeled with radioactive atoms for imaging purposes using single-photon emission computed tomography and positron emission tomography technologies, or can be labeled with therapeutic nuclides for the efficient killing of tumor cells. This method-based review discusses one such class of receptor-targeted peptides and their radiolabeling with radioactive metals. The somatostatin receptor is upregulated in many types of cancer, and when labeled with a radiometal atom via a bifunctional chelate, can be employed as an agent for the imaging and radiotherapy of cancer. This review will discuss the methods used in the synthesis of the somatostatin peptides, conjugation with bifunctional chelators, and radiolabeling with metal radionuclides. Methods will also be presented for the in vitro and in vivo evaluation of the compounds produced.

Key Words: Radiometal; peptide; bifunctional chelator; PET; somatostatin; imaging; therapy.

1. Introduction

Somatostatin, a 14-amino acid peptide, is involved in the regulation and release of a number of hormones (e.g., growth hormone, thyroid-stimulating hormone, and prolactin) *(1,2)*. Somatostatin receptors (SSR) have been found in many different normal organ systems such as the central nervous system, the gastrointestinal tract, and the exocrine and endocrine pancreas *(1–3)*. A large number of human tumors also overexpress SSRs *(4)*, which makes SSR

From: *Methods in Molecular Biology, vol. 386: Peptide Characterization and Application Protocols*
Edited by: G. Fields © Humana Press Inc., Totowa, NJ

a good target for diagnostic imaging and therapeutic purposes. Because somatostatin has a very short biological half-life, the analog octreotide (OC) was developed with a longer biological half-life than the native somatostatin *(5)*. In more recent years, other analogs have been developed such as Tyr[3]-octreotide (Y3-OC), octreotate (TATE), and Tyr[3]-octreotate (Y3-TATE) (**Fig. 1**) *(6–8)*. ^{111}In-diethylenetriaminepentaacetic acid (DTPA)-OC was one of the first clinically approved peptide-based tumor receptor imaging agents for neuroendocrine tumors *(9)*. Our studies and those of other groups have shown the potential of these somatostatin analogues to be utilized for both positron emission tomography (PET) diagnostic imaging and targeted radiotherapy of cancer *(10–17)*.

The choice of radiometal to be used for the radiolabeling of peptides is dependent on the application. In selecting the ideal radionuclide for a particular application, whether it be for therapy or imaging, the most important factors apart from the half-life of the radionuclide are the energy of the radioactive emissions and the cost and availability of the isotope. Metal positron-emitting radionuclides are generally produced by nuclear reactions on enriched parent materials *(18)*. Metallic positron-emitting radionuclides available for PET imaging include copper radionuclides (^{64}Cu, [61]Cu, [60]Cu) *(19–22)*, [45]Ti *(23)*,

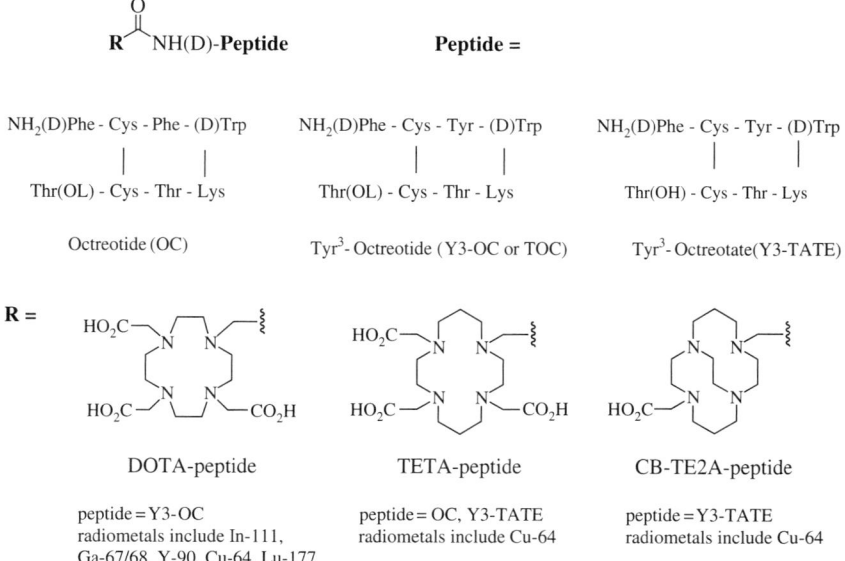

Fig. 1. Somatostatin analogs for labeling with metal radionuclides.

Table 1
Decay Characteristics of Metal Radionuclides

Isotope	$T_{1/2}$	β^- MeV (%)	β^+ MeV (%)	EC MeV (%)	γ MeV (%)
^{64}Cu	12.7 h	0.573 (39.6%)	0.655 (17.4%)	41%	0.51 (38.6%)
					1.35 (0.6%)
^{67}Ga	3.26 d	–	–	99 + %	0.093 (40%)
					0.184 (20%)
					0.296 (22%)
					0.288 (7%)
^{68}Ga	68.3 min	–	1.88 (86%)	12%	0.511 (176%)
			0.77 (2%)		1.078 (3.5%)
^{90}Y	64 h	2.27 (100%)	–	–	–
^{111}In	67 h	–	–	99+%	0.173 (89%)
					0.247 (94%)
^{177}Lu	6.7 d	0.497 (90%)	–	–	0.208 (11%)

^{66}Ga *(24,25)*, ^{86}Y *(26)*, and ^{94m}Tc *(27)*. The most widely use positron-emitting radiometal is ^{64}Cu and can be produced as previously described *(20,21)*. This method-based review will focus on the production of ^{64}Cu-labeled somatostatin peptides for the diagnosis and therapy of cancer, although many of the methods discussed are applicable for use with other metal radionuclides. ^{64}Cu has a half-life of 12.7 h and can be considered the most versatile copper radionuclide owing to its unique decay scheme (β^+ [17.4%]; β^- [40%]). Other radiometals commonly labeled to somatostatin analogs for imaging and therapy applications include ^{111}In, ^{67}Ga, ^{68}Ga, ^{90}Y and ^{177}Lu (**Table 1**), and methods for labeling somatostatin analogs with these radionuclides are also discussed.

2. Materials
2.1. Chelator-Peptide Synthesis

1. All chemicals unless otherwise stated are purchased from Sigma-Aldrich Chemical Co. (St. Louis, MO).
2. Water is distilled and then deionized (18 MΩ/cm^2) by passing through a Milli-Q water filtration system (Millipore Corp., Milford, MA).
3. All glass and plastic-ware is acid washed prior to use to reduce metal contaminants.
4. Solid-phase peptide synthesis (SPPS) is performed on commercially available peptide synthesizers employing the 9-fluorenylmethoxycarbonyl (Fmoc) method *(8,28–30)*. The Fmoc-protected amino acids can be purchased commercially.

5. The peptide conjugates are purified by reversed-phase high-performance liquid chromatography (HPLC) and the peptides are analyzed by high-resolution matrix-assisted laser desorption/ionization (MALDI) Fourier transform mass spectrometry (FTMS) to confirm identity.

6. The chelators 1,4,8,11-tetraazacyclotetradecane-1,4,8,11-tetraacetic acid (TETA) and 1,4,7,10-tetraazacyclododecane-1,4,7,10-tetraacetic acid (DOTA) are commercially available from Sigma-Aldrich (St. Louis, MO), Fluka Chemie AG (Buchs, Switerland), Strem (Newburyport, MA), and Macrocyclics Inc. (Dallas, TX) at reasonable prices. Macrocyclics Inc. also provides other bifunctional chelators (BFCs), such as DOTA-N-hydroxysuccinimide (NHS)-ester and DOTA-tris(t-butyl ester), which can be readily used for the conjugation to biomolecules. **Figure 2** depicts a series of BFCs that can be used for the attachment of ^{64}Cu and other radiometals to a biomolecule.

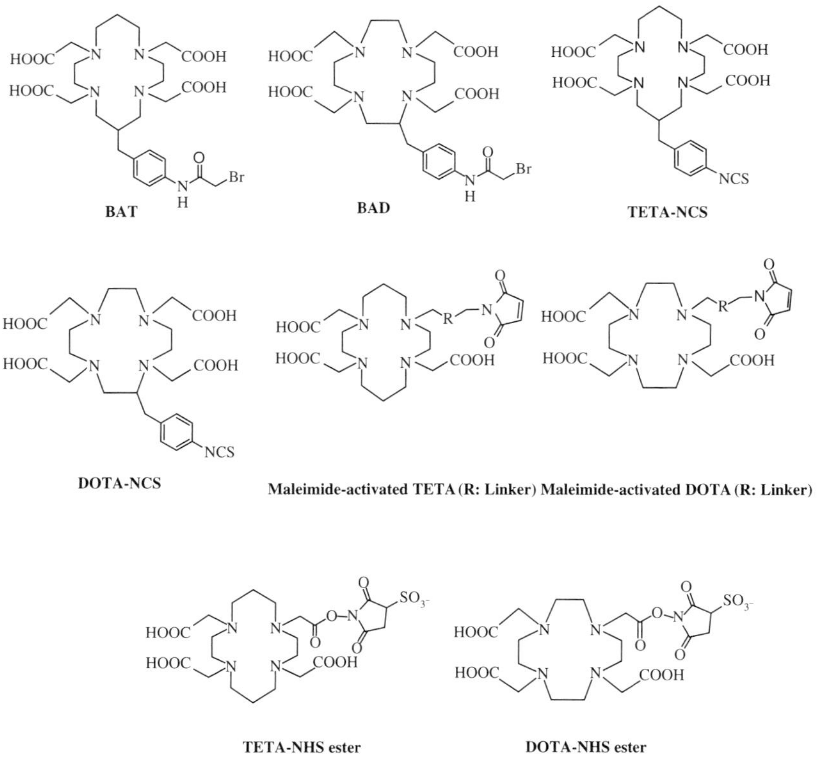

Fig. 2. Bifunctional chelates for the attachment of the radiometal to the peptide.

2.2. Radiolabeling of Peptides

The use of radioactive materials requires careful handling to ensure as low as reasonably achievable radiation exposure. Handling of radioactivity should only be done by fully trained personnel under the guidance of the local Radiation Safety office.

1. At Washington University, ^{64}Cu is produced by the nuclear reaction ^{64}Ni(p,n)^{64}Cu on the CS15 cyclotron. Ten to fifty milligrams of enriched ^{64}Ni (obtained from Department of Energy or ISOFLEX, San Francisco, CA) is plated on a gold disc. The target is generally irradiated for 0.4 to 4 h in a specially designed target holder *(20)*. Generally, for ^{64}Cu production, enriched ^{64}Ni targets are bombarded with 15–45 µA current.

2. The ^{64}Cu is isolated in a one-step procedure; ^{64}Ni (containing the ^{64}Cu) is dissolved in 4.0–6.0 mL of 6.0 M HCl (Alfa Aesar, Ward Hill, MA) under reflux. The solution is then transferred to a 4×1 cm AG1-X8 (BIO-RAD, Hercules, CA) column that separates the ^{64}Cu from the nickel. Elution is carried out with the nickel and other reaction byproducts collected in a 15–30 mL fraction of the 6.0 M HCl. The ^{64}Cu is then collected in approx 2–12 mL of 0.5 M HCl.

3. The ^{64}Cu fraction is then evaporated to dryness and redissolved in approx 150 µL 0.1M HCl (HCl from Aldrich, 99.999999%). Typically, 150–600 mCi of ^{64}Cu are produced by this process.

4. Radionuclidic purity is confirmed by gamma spectroscopy (Canberra multi-channel gamma analysis instrument system 96-6922). Utilizing the cyclotron conditions described above, the yields of ^{64}Cu are dependent on the amount of ^{64}Ni on the target and the length of the bombardment. Typically, 18.5 GBq (500 mCi) ^{64}Cu are produced with a 40 mg ^{64}Ni target and a bombardment time of 4 h. The specific activity of the ^{64}Cu ranges from 47.4 to 474 GBq/µmol (1280 to 12,800 mCi/µmol). The typical yields for ^{64}Cu productions are 0.2 mCi/µA·h per mg ^{64}Ni. It is challenging to produce high specific activity copper radionuclides due to the extensive presence of copper in the environment. Extra care is taken to use highly pure reagents and metal-free glassware and plastic ware.

5. Other metal radionuclides are commercially available. Indium-111 can be purchased from Mallinckrodt, Inc. (St. Louis, MO). Lutetium-177 is available from Missouri University Research Reactor (MURR; Columbia, MO), and also from Perkin-Elmer (Boston, MA). Gallium-68 is produced from a ^{68}Ge/^{68}Ga generator, commercially available from TCI Medical (Albuquerque, NM) with the source of this generator being Cyclotron Co. Ltd. (Obninsk, Russia). Germanium-68 has a long half-life (275 d), making this generator a convenient source of a positron-emitting radionuclide. Yttrium-90 can be purchased from Perkin-Elmer (Boston, MA).

2.3. In Vitro Receptor-Binding Assays

1. Receptor-binding buffer: 50 mM Tris-HCl, pH 7.4, 5.0 mM MgCl$_2$, 0.5 μg/mL aprotinin, 200 μg/mL bacitracin, 10 μg/mL leupeptin, and 10 μg/mL pepstatin A. It can be stored for one day at 4 °C
2. Assays are performed using Millipore MultiScreen™ Assay System 96-Well Filtration Plates (#MAFBN0B10, Millipore Bedford, MA)
3. Radiotracer and competitor ligand are prepared with the desired radionuclide and its stable counterpart, respectively, as described in Section 3.2.

3. Methods

3.1. Preparation of Chelator-Peptide Conjugates

In the method described below, somatostatin analogues are used as examples of how BFC–peptide conjugates are constructed in solid-phase as previously described *(8,28–30)*. Generally, it involves three steps: (1) preparation of the linear protected peptide, Fmoc-D-Phe-Cys(Acm)-Tyr (OtBu)-D-Trp(Boc)-Thr(OtBu)-Cys(Acm)-Thr(OtBu), using the Fmoc strategy; (2) cyclization of the peptide; and (3) conjugation of the chelator to the peptide.

1. The peptide is synthesized by an automated peptide synthesizer (Applied Biosystems Model 432A "Synergy" Peptide Synthesizer, Foster City, CA). The instrument protocol requires 25 mmol of subsequent Fmoc-protected amino acids activated by a combination of N-hydroxybenzotriazole (HOBt) and 2-(1-H Benzotriazol-1-yl)-1,1,3,3-tetramethyluronium hexafluorophosphate (HBTU). The prepacked amino acids are available from Perkin-Elmer (Norwalk, CT), BACHEM Bioscience (King of Prussia, PA), or Novabiochem (San Diego, CA).
2. At the end of the synthesis, thallium(III) trifluoroacetate is used to remove the Acm-protecting groups on the cysteines, concomitant with disulfide formation to generate the cyclic peptide.
3. The resin with the cyclic peptide is put back into the synthesizer for the conjugation of chelator. In the case of TETA and DOTA, a derivative with a monofunctional group is used, *tri-tert*-butyl TETA, or *tri-tert*-butyl DOTA. The *tri-tert*-butyl DOTA is commercially available from Macrocyclics. The TETA derivative must be synthesized *(31)*. **Note:** To prevent chelation of the thallium, the thallium-mediated peptide cyclization must be done before the DOTA or TETA conjugation.
4. The resulting peptide is cleaved from the resin and deprotected with trifluoroacetic acid (TFA)–thioanisole–phenol–water (85:5:5:5) for 8–10 h.
5. Final purification is accomplished by C-18 reversed-phase HPLC (solvent A: 0.1% TFA [*(32)*]/water, solvent B: 0.1% TFA/10% H$_2$O/acetonitrile; gradient: 90% A/10% B to 30% A/70% B in 40 min; detection mode: ultraviolet [UV] at $\lambda = 230$ nm).

6. Fractions are analyzed by analytical HPLC (detection mode: UV at $\lambda = 214$ nm) and mass spectrometry (MS) prior to final lyophilization.
7. Other chelators that contain carboxylic acid moieties, such as CB-TE2A, can be conjugated to peptides in the same manner as the *tri-tert*-butyl esters of TETA or DOTA *(14)*.

3.2. Radiolabeling Methods

3.2.1. Radiolabelinog DOTA– and TETA–Peptide Conjugates

The DOTA– and TETA–peptide conjugates are directly radiolabeled with ^{64}Cu in ammonium acetate (or ammonium citrate) buffer and purified by previously published protocols *(8,10,11,30,33)* (*see* **Note 1**).

1. In a typical labeling of TETA-Y3-TATE as an example, 1.0 μg of TETA-Y3-TATE is radiolabeled with more than 1000 μCi (37 MBq) of ^{64}Cu-acetate in 0.1 M ammonium acetate, pH 5.5 in 30 min incubation at room temperature (*see* **Note 2**)
2. For labeling CB-TE2A-Y3-TATE, 1–15 mCi (37-555 MBq) of ^{64}Cu in 0.1 M ammonium acetate (pH 8.0) is added to 1–15 μg of the peptide in 0.1 M ammonium acetate (pH 8.0). The reaction mixture is heated for 1 h at 95°C.
3. The ^{64}Cu-TETA-Y3-TATE purification is accomplished using a C_{18} SepPak cartridge (Waters Corp. Milford, MA); however, C_{18} SepPak purification is not used for ^{64}Cu-CB-TE2A-Y3-TATE, since this compound is difficult to elute from the SepPak, most likely due to its increased lipophilic character. For a 1 mL ethanol elution, >90% of the product is in the fractions between 150 and 500 μL.
4. The radiochemical purity is confirmed by radio-thin layer chromatography (TLC) and/or radio-HPLC. The TLC conditions are C_{18}-coated plates eluted in 7:3 (v/v) MeOH:10% NH_4OAc. Analytical reversed-phase HPLC is performed on a Waters 600E (Milford, MA) chromatography system with a Waters 996 photodiode array detector and an Ortec Model 661 (EG & G Instruments, Oak Ridge, TN) radioactivity detector. HPLC samples are analyzed on a Vydac diphenyl column (4.6 × 100 mm) (Hesperia, CA). The mobile phase is H_2O (0.1% TFA) (solvent A) and 90% acetonitrile (ACN) (0.1% TFA) (solvent B). The gradient is 5% B to 70% B in 20 min (1.0 mL/min flow rate). The radiochemical purity is typically >95%.
5. The ^{64}Cu-TETA-Y3-TATE in ethanol may need to be evaporated under nitrogen or argon prior to dilution with saline (ethanol content must be less than 5%) for PET imaging and/or radiotherapy in tumor-bearing rodent models or humans.

3.2.2. Labeling $^{67/68}$Ga, ^{111}In, ^{90}Y, and ^{177}Lu to Somatostatin Analogs

For labeling $^{67/68}$Ga, ^{111}In, ^{90}Y, or ^{177}Lu to somatostatin analogs, the DOTA conjugates are preferred, since DOTA stably binds these +3 metal ions. Breeman et al. optimized conditions for radiolabeling with the four radionuclides, and determined that lower pH (~ 4.0) and high temperatures (80–100°C) are required for 30 min *(34)*. The labeling buffer used in these reactions is 25 mM sodium ascorbate in 50 mM sodium acetate.

Breeman et al. have developed labeling conditions for preparing ^{68}Ga-DOTA-Y3-OC (also called DOTA-TOC) and ^{68}Ga-DOTA-Y3-TATE (also called DOTA-TATE) *(35)*.

1. DOTA-Y3-OC and DOTA-Y3-TATE are dissolved in 0.01 M acetic acid in Milli-Q water with a final concentration of 1 mM.
2. Gallium-68 is eluted from the [68]Ge/[68]Ge generator in 0.1 M HCl, with approx 80% of the activity eluted in a 1-mL volume.
3. The peptide and ^{68}Ga activity are mixed with 1.25 M sodium acetate to achieve a pH of 3.5–4.0. Reaction volumes may vary from 130 µL to 1.5 mL. Under these conditions, complete incorporation of ^{68}Ga occurs in 5 min at 80°C.
4. For rapid quality control, radio-TLC is performed using instant thin-layer chromatography (ITLC)-silica gel (SG) strips (Gelman Sciences, Ann Arbor, MI) eluting in HCl-acidified saline, pH 3.5. Under these conditions, ^{68}Ga-labeled peptide remains at the origin, while free ionic ^{68}Ga migrates with an R_f of 0.5–0.7.
5. To distinguish between ^{68}Ga-colloid and ^{68}Ga-peptide, both of which will remain at the origin, ethylenediamine tetraacetic acid (EDTA) can be added to the reaction mixture. ^{68}Ga-EDTA migrates with the solvent front ($R_f = 1.0$). HPLC conditions as described above can also be employed.

3.3. In Vitro Receptor-Binding Assays

After the radiometal-labeled bioconjugate is prepared, its biological activity and effectiveness as a radiopharmaceutical are determined by several methods. Determining that the radiometal-labeled BFC–biomolecule conjugate retains its biological activity and evaluation of its biodistribution and/or therapeutic efficacy are essential prior to human trials.

The binding affinity is measured using an appropriate cell line that expresses the desired receptor to which the peptide conjugate is designed to bind. Receptor binding assays can be performed using membranes from receptor-rich tumors such as those from CA20948 *(36)* or AR42J *(37,38)* tumors harvested from euthanized animals. For homologous competitive assays, the competing ligands are the natural metal complexes of the BFC-somatostatin analogs prepared using high purity metal chloride salts using the procedure described above for preparation of the radiometal-labeled peptides. Complex formation is confirmed by HPLC and electrospray (ES)-MS. To reduce nonspecific binding, 0.1% bovine serum albumin (BSA) is added to the receptor-binding buffer.

1. The radiotracer (0.05 nM) is diluted in binding buffer.
2. Aliquots of 10–30 µg of AR42J or CA20948 membrane protein are incubated in triplicate on 96-well filtration plates along with 0.05 nM radioligand and 13 doses of the competitor ligand, covering a concentration range from 0 to 400 nM.

3. Samples are mixed gently and allowed to incubate for 2 h at room temperature.
4. Following incubation, supernatants are aspirated through the filters using a vacuum manifold.
5. Filters are washed twice with 250 µL of ice-cold binding buffer, and then air-dried.
6. A filter-punch apparatus (Millipore, Inc.) is used to extricate the filters from the plate and dispense them into tubes for counting.
7. The resulting filter samples along with samples representing total counts added were counted for radioactivity.
8. The best-fit IC_{50} value for the cell line is calculated according to published methods (10) using PRISM (Graphpad, San Diego, CA).

3.4. In vivo Evaluation of Radiolabeled Peptides

3.4.1. Biodistribution Studies

All animal studies should be done under the auspices of the respective Institutional Guidelines for the Care and Use of Research Animals.

1. The somatostatin receptor-positive rat pancreatic tumors AR42J *(30)* and CA20948 *(7)* are maintained by serial passage in male Lewis rats.
2. A 16G cancer implant needle is used to subcutaneously implant 4–8 mm^3 pieces of tumor.
3. AR42J tumor is implanted in the lower leg of immature (21 d old; $\sim$ 40 g) rats and CA20948 into the nape of the neck of mature (150 g) rats.
4. Tumors are allowed to grow for 12–14 d, until approx 0.2 to 0.5 g in size.
5. The *in vivo* evaluation of radiometal-labeled bioconjugates is typically carried out using one of the tumor-bearing animal models described above. The radiometal-labeled somatostatin analogs are injected into the tumor-bearing rats, and then at several time points postinjection (usually between 1 and 48 h), the rats are sacrificed.
6. Selected organs (tumor, blood, lung, liver, spleen, kidney, muscle, fat, heart, brain, pituitary, bone, adrenals, pancreas, stomach, small intestine, upper large intestine, and lower large intestine) are dissected, drained of blood, weighed, and counted in a γ counter.
7. By comparison with a standard representing the injected dose per animal, the samples are corrected for radioactive decay, in order to calculate percent injected dose per gram (% inner diameter [ID]/g) of tissue and percent injected dose per organ (% ID/organ).

Competitive binding studies are performed in vivo by having an additional group of animals in which a "blockade" of nonradioactive peptide analog is co-injected to observe receptor mediated uptake of the radiolabeled peptide conjugate. Typically, the amount used to block the receptors is 100- to 1000-fold higher than the concentration of somatostatin analog in the injected tracer.

The targeting capability of the ^{64}Cu-labeled compound is evaluated as the ratio of tumor to surrounding normal tissue (e.g., tumor:muscle), or tumor:blood. In addition to determining the target tissue uptake, clearance of the agent through the blood, liver, and kidneys is also measured, with more rapid clearance through the non-target tissues being desirable. To compare differences between the data sets, a student's t-test can be performed. Differences at the 95% confidence level ($p < 0.05$) are generally considered significant.

3.4.2. Small Animal PET/Computed Tomography Imaging

Small animal PET imaging has recently become a standard method to evaluate PET radiopharmaceuticals using animal models of disease *(30,39–42)*. For studies involving expensive transgenic rodent models, this non-invasive technique is greatly preferred, because repeat studies can be performed without sacrificing the animals. Somatostatin analogs labeled with positron-emitting radionuclides, such as ^{64}Cu, are quickly evaluated by both static and dynamic imaging.

1. For imaging studies on rats and mice, the rodents are anesthetized with 1–2% isoflurane prior to scanning and positioned supine and immobilized in a custom prepared cradle.
2. Two mice are imaged side by side and remain in the same bed position for all time points.
3. All imaging is performed in a temperature-controlled imaging suite with close monitoring of the physiological status of the animals.
4. At Washington University, co-registration of the PET images is achieved in combination with a microCAT-II camera (Imtek Inc., Knoxville, TN), which provides high-resolution X-ray computed tomography (CT) anatomical images. The image registration between microCT and PET images is accomplished by using a landmark registration technique AMIRA image display software (AMIRA, TGS Inc., San Diego, CA.). The registration method proceeds by rigid transformation of the microCT images from landmarks provided by fiducial markers directly attached to the animal bed.
5. The activity concentrations in the regions of interest (ROIs) are quantified by measuring the radioactivity in the selected volumetric regions and averaging the activity concentration over these regions. For the microPET rodent scanners (Concorde Microsystems, Knoxville, TN) *(40,43,44)*, the ROI analysis software consists of two programs: Analyze AVW 3.0 (Biomedical Imaging Resource, Mayo Foundation, Rochester, MN) and a viewing application program developed by R. Laforest at Washington University School of Medicine in St. Louis using International Data Language (Research Scientific, Boulder, CO).

4. Notes

1. For ^{64}Cu, it has been shown that the use of the DOTA bifunctional chelating system may not stably complex the Cu(II), resulting in elevated uptake in non-target tissues such as the blood and liver *(45,46)*. This background activity could limit the use of ^{64}Cu-DOTA-peptides due to high absorbed radiation dose levels and poorer tumor/background tumor ratios. Other chelates such as TETA may improve the biodistribution, and new chelates for ^{64}Cu based on a cross-bridged design that have shown significant improvement in metal-chelate stability and subsequent lowering in nontarget and clearance tissue accumulation *(14,45)*.

2. It is of great importance that receptor-targeting peptides be radiolabeled with metal radionuclides in high specific activity ($>27,750\,MBq/\mu mol$ [$750\,mCi/\mu mol$]). Low capacity tumor receptors, such as somatostatin, are easily saturated, and the presence of significant amounts of unlabeled peptide will affect the radio-labeled peptide uptake in the tumor. It is challenging to produce high specific activity copper radionuclides because of the extensive presence of copper in the environment. Extra care is taken to use highly pure reagents and metal-free glassware and plastic ware.

References

1. Reichlin, S. (1983) Somatostatin (part 1). *New Engl. J. Med.* **309,** 1495–1501.
2. Reichlin, S. (1983) Somatostatin (part 2). *New Engl. J. Med.* **309,** 1556–1563.
3. Guillemin, R. (1978) Peptides in the brain: the new endocrinology of the neuron. *Science* **202,** 390–402.
4. Reubi, J. C., Kvols, L. K., Krenning, E. P., and Lamberts, S. W. J. (1990) Distribution of somatostatin receptors in normal and tumor tissue. *Metabolism* **39(Suppl 2),** 78–81.
5. Bauer, W., Briner, U., Doepfner, W., et al. (1982) SMS 201-995. *Life Sci.* **31,** 1133–1140.
6. de Jong, M., Bernard, B. F., de Bruin, E., et al. (1998) Internalization of radiolabelled [DTPA⁰]octreotide and [DOTA⁰, Tyr³]octreotide: peptides for somatostatin receptor-targeted scintigraphy and radionuclide therapy. *Nucl. Med. Commun.* **19,** 283–288.
7. de Jong, M., Breeman, W. A. P., Bakker, W. H., et al. (1998) Comparison of ^{111}In-labeled somatostatin analogues for tumor scintigraphy and radionuclide therapy. *Cancer Res.* **58,** 437–441.
8. Lewis, J. S., Lewis, M. R., Srinivasan, A., Schmidt, M. A., Wang, J., and Anderson, C. J. (1999) Comparison of four ^{64}Cu-labeled somatostatin analogs *in vitro* and in a tumor-bearing rat model: Evaluation of new derivatives for PET imaging and targeted radiotherapy. *J. Med. Chem.* **42,** 1341–1347.

9. Krenning, E. P., Bakker, W. H., Kooij, P. P. M., et al. (1992) Somatostatin receptor scintigraphy with indium-111-DTPA-D-Phe-1-octreotide in man: metabolism, dosimetry and comparison with iodine-123-Tyr-3-octreotide. *J. Nucl. Med.* **33,** 652–658.
10. Anderson, C. J., Pajeau, T. S., Edwards, W. B., Sherman, E. L. C., Rogers, B. E., and Welch, M. J. (1995) In vitro and in vivo evaluation of copper-64-labeled octreotide conjugates. *J. Nucl. Med.* **36,** 2315–2325.
11. Anderson, C. J., Jones, L. A., Bass, L. A., et al. (1998) Radiotherapy, toxicity and dosimetry of copper-64-labeled TETA-octreotide in tumor-bearing rats. *J. Nucl. Med.* **39,** 1944–1951.
12. Lewis, J. S., Srinivasan, A., Schmidt, M. A., Schwarz, S. W., Jones, L. A., and Anderson, C. J. (1998) Radiotherapy and dosimetry of copper-64-TETA-Tyr3-octreotate in a somatostatin receptor positive tumor bearing animal model [abstract]. *J. Nucl. Med.* **39,** 104P.
13. Anderson, C. J., Dehdashti, F., Cutler, P. D., et al. (2001) Copper-64-TETA-octreotide as a PET imaging agent for patients with neuroendocrine tumors. *J. Nucl. Med.* **42,** 213–221.
14. Sprague, J. E., Peng, Y., Sun, X., et al. (2004) Preparation and biological evaluation of copper-64–labeled Tyr^3-octreotate using a cross-bridged macrocyclic chelator. *Clin. Cancer Res.* **10,** 8674–8682.
15. de Jong, M., Valkema, R., Kwekkeboom, D. J., and Krenning, E. P. (2004) Somatostatin receptor targeted-radio-ablation-of tumors. *Endocrine Updates* **24,** 233–249.
16. Kwekkeboom, D. J., Mueller-Brand, J., Paganelli, G., et al. (2005) Overview of results of peptide receptor radionuclide therapy with 3 radiolabeled somatostatin analogs. *J. Nucl. Med.* **46(suppl. 1),** 62S–66S.
17. Maecke, H. R., Hofmann, M., and Haberkorn, U. (2005) ^{68}Ga-labeled peptides in tumor imaging. *J. Nucl. Med.* **46(suppl. 1),** 172S–178S.
18. McQuade, P., Rowland, D. J., Lewis, J. S., and Welch, M. J. (2005) Positron-emitting isotopes produced on biomedical cyclotrons. *Curr. Med. Chem.* **12,** 807–818.
19. Blower, P. J., Lewis, J. S., and Zweit, J. (1996) Copper radionuclides and radiopharmaceuticals in nuclear medicine. *Nucl. Med. Biol.* **23,** 957–980.
20. McCarthy, D. W., Shefer, R. E., Klinkowstein, R. E., et al. (1997) Efficient production of high specific activity ^{64}Cu using a biomedical cyclotron. *Nucl. Med. Biol.* **24,** 35–43.
21. McCarthy, D. W., Bass, L. A., Cutler, P. D., et al. (1999) High purity production and potential applications of copper-60 and copper-61. *Nucl. Med. Biol.* **26,** 351–358.
22. Sun, X., and Anderson, C. J. (2004) Production and applications of copper-64 radiopharmaceuticals. *Meth. Enzymol.* **386,** 237–261.
23. Vavere, A. L., and Welch, M. J. (2005) Preparation, biodistribution, and small animal pet of [45]Ti-transferrin. *J. Nucl. Med.* **46,** 683–690.

24. Lewis, M. R., Reichert, D. E., Laforest, R., et al. (2002) Production and purification of gallium-66 for preparation of tumor-targeting radiopharmaceuticals. *Nucl. Med. Biol.* **29**, 701–706.

25. Szelecsenyi, F., Boothe, T. E., Tavano, T., Plitnikas, M. E., and Tarkanyi, F. (1994) Compilation of cross sections/thick target yields for [66]Ga, ^{67}Ga and ^{68}Ga production using Zn targets up to 30 MeV proton energy. *Appl. Radiat. Isot.* **45**, 473–500.

26. Reischl, G., Rosch, F., and Machulla, H. J. (2002) Electrochemical separation and purification of yttrium-86. *Radiochim. Acta* **90**, 225–228.

27. Roesch, F., and Qaim, S. M. (1993) Nuclear data relevant to the production of the positron emitting technetium isotope [94m]Tc via the [94]Mo(p,n)-reaction. *Radiochim. Acta* **62**, 115–121.

28. Edwards, W. B., Fields, C. G., Anderson, C. J., Pajeau, T. S., Welch, M. J., and Fields, G. B. (1994) Generally applicable, convenient solid-phase synthesis and receptor affinities of octreotide analogs. *J. Med. Chem.* **37**, 3749–3757.

29. Achilefu, S., Jimenez, H. N., Dorshow, R. B., et al. (2002) Synthesis, in vitro receptor binding and in vivo evaluation of fluorescein and carbocyanine peptide-based optical contrast agents. *J. Med. Chem.* **45**, 2003–2015.

30. Li, W. P., Lewis, J. S., Kim, J., et al. (2002) DOTA-D-Tyr1-octreotate: a somatostatin analog for labeling with halogen and metal radionuclides for cancer imaging and therapy. *Bioconjug. Chem.* **13**, 721–728.

31. Mishra, A. K., Draillard, K., Faivrechauvet, A., Gestin, J. F., Curtet, C., and Chatal, J. F. (1996) A convenient, novel approach for the synthesis of polyaza macrocyclic bifunctional chelating agents. *Tetrahedron Lett.* **37**, 7515–7518.

32. Yorke, E. D., Williams, L. E., Demidecki, A. J., Heidorn, D. B., Roberson, P. L., and Wessels, B. W. (1993) Multicellular dosimetry for beta-emitting radionuclides: autoradiography, thermoluminescent dosimetry and three-dimensional dose calculations. [review]. *Med. Phys.* **20**, 543–550.

33. Lewis, J. S., Laforest, R., Lewis, M. R., and Anderson, C. J. (2000) Comparative dosimetry of copper-64 and yttrium-90-labeled somatostatin analogs in a tumor-bearing rat model. *Cancer Biothet. Radiopharm.* **15**, 593–604.

34. Breeman, W. A. P., de Jong, M., Visser, T. J., Erion, J. L., and Krenning, E. P. (2003) Optimising conditions for radiolabelling of DOTA-peptides with ^{90}Y, ^{111}In and ^{177}Lu at high specific activities. *Eur. J. Nucl. Med. Mol. Imag.* **30**, 917–920.

35. Breeman, W. A. P., de Jong, M., de Blois, E., Bernard, B. F., Konijnenberg, M., and Krenning, E. P. (2005) Radiolabelling DOTA-peptides with [68]Ga. *Eur. J. Nucl. Med. Mol. Imag.* **32**, 478–485.

36. Longnecker, D. S., Lilja, H. S., French, J., Kuhlmann, E., and Noll, W. (1979) Transplantation of azaserine-induced carcinomas of pancreas in rats. *Cancer Lett.* **7**, 197–202.

37. Rosewicz, S., Vogt, D., Harth, N., et al. (1992) An amphicrine pancreatic cell line: AR42J cells combine exocrine and neuroendocrine properties. *Eur. J. Cell Biol.* **59**, 80–91.

38. Christophe, J. (1994) Pancreatic tumoral cell line AR42J: An amphicrine model. *Am. J. Physiol.* **266(6 pt 1),** G963–G971.

39. Wipke, B. T., Wang, Z., Kim, J., McCarthy, T. J., and Allen, P. M. (2002) Dynamic visualization of a joint-specific autoimmune response through positron emission tomography. *Nat. Immunol.* **3,** 366–372.

40. Cherry, S. R., Shao, Y., Silverman, R. E., et al. (1997) Micropet: a high resolution pet scanner for imaging small animals. *IEEE. Trans. Nucl. Sci.* **44,** 1161–1166.

41. Lewis, J. S., Achilefu, S., Garbow, J. R., Laforest, R., and Welch, M. J. (2002) Small animal imaging: current technology and perspectives for oncological imaging. *Eur. J. Cancer* **38,** 2173–2188.

42. Rowland, D. J., Lewis, J. S., and Welch, M. J. (2002) Molecular imaging: the application of small animal positron emission tomography. *J. Cell. Biochem.* **Suppl 39,** 110–115.

43. Knoess, C., Siegel, S., Smith, A., et al. (2003) Performance evaluation of the microPET R4 pet scanner for rodents. *Eur. J. Nucl. Med. Mol. Imag.* **30,** 737–747.

44. Tai, Y. C., Ruangma, A., Rowland, D. J., et al. (2005) Performance evaluation of the microPET FOCUS: a third-generation microPET scanner dedicated to animal imaging. *J. Nucl. Med.* **46,** 455–463.

45. Boswell, C. A., Sun, X., Niu, W., et al. (2004) Comparative in vivo stability of copper-64-labeled cross-bridged and conventional tetraazamacrocyclic complexes. *J. Med. Chem.* **47,** 1465–1474.

46. Sun, X., Wuest, M., Weisman, G. R., et al. (2002) Radiolabeling and in vivo behavior of copper-64-labeled cross-bridged cyclam ligands. *J. Med. Chem.* **45,** 469–477.

9

Cell-Penetrating Proline-Rich Peptidomimetics

Josep Farrera-Sinfreu, Ernest Giralt, Miriam Royo,
and Fernando Albericio

Summary

Cell-penetrating peptides (CPPs) offer potential as delivery agents for the cellular administration of drugs. However, the pharmacological utility of CPPs that are derived from natural
amino acids is limited by their rapid metabolic degradation, low membrane permeability,
and toxicity. Various peptidomimetics able to overcome these problems have been described,
including peptides formed by D-amino acids and β-peptides. This chapter summarizes the
synthesis of γ-proline-derived peptides and polyproline dendrimers for drug delivery applications, and includes descriptions of several modifications in the γ-peptides (mimicking the side
chains of the α-amino acids) or modulating the dendrimer surface. 5(6)-Carboxyfluorescein
labeling of the aforementioned peptidomimetics for use in cell translocation studies is also
described. Furthermore, different protocols for the study of the drug delivery capabilities of these
compounds are reviewed, including enzymatic stability studies, cellular uptake measurements
by plate fluorimetry and flow cytometry, confocal laser scanning microscopy, and cytotoxicity
assays.

Key Words: Cellular uptake; drug delivery; foldamers; γ-peptides; solid-phase; dendrimers.

1. Introduction

In the past few years, several peptides capable of crossing cell membranes
have been described (*1,2*), namely cell-penetrating peptides (CPP). CPPs have
thus been proposed for the cellular delivery of biomolecules as an alternative
to inefficient and often toxic approaches such as viral delivery systems (*3*),
liposomes (*4,5*), polymeric encapsulation of drugs (*6*), electroporation (*7*), or
receptor mediated endocytosis (*8*). Among well-known CPPs such as Tat,
penetratin, Arg_9, or transportan, proline-rich peptides have demonstrated good
cell uptake capacities (*9*).

From: *Methods in Molecular Biology, vol. 386: Peptide Characterization and Application Protocols*
Edited by: G. Fields © Humana Press Inc., Totowa, NJ

Several drawbacks for the physiological use of CPPs have recently been reviewed *(2)*, including lability to proteolytic cleavage. Protease-resistant compounds, such as loligomers *(10,11)*, peptides with their L-amino acids residues replaced for their nonnatural D-analogs *(12)*, as well as β-peptides *(13–16)* have therefore been evaluated as drug delivery agents. Our contribution in this area has been the description of new classes of peptidomimetics: γ-peptides derived from the γ–amino-L-proline, with the peptide backbone formed through the γ-amino function *(17,18)*, and polyproline dendrimers *(19–23)*. Our group has also reported other proline-rich peptides *(24–27)* with the ability to be taken up into eukaryotic cells. Despite the hydrophobic character of proline-rich peptides, they are highly water-soluble, an invaluable property for life science applications. This chapter describes the solid-phase syntheses of γ-amino-L-proline derived peptides and polyproline dendrimers, including their 5(6)-carboxyfluorescein (CF) labeling. Furthermore, protocols for the study of their cell-uptake capacities are described, including enzymatic stability, plate fluorimetry and flow cytometry analyses, cytotoxicity analysis, and confocal laser scanning microscopy.

1.1. γ-Amino-L-Proline-Derived Peptides

γ-Peptides are constructed using a common hexameric backbone of *cis*-γ-amino-L-proline. The skeleton is elongated through the γ-amino group of the proline by repetitive monomer couplings and the α-amino group of the proline is distinctly modified to obtain molecules that can imitate the side chains of natural amino acids **(Fig. 1)**.

Representative examples of these compounds include the following three γ-peptide families: N^α-acyl-γ-peptides, N^α-alkyl-γ-peptides, and N^α-guanidylated-γ-peptides. In each family, two different types of compounds can be synthesized: (a) homooligomers (identical side chains; **Fig. 2**, strategy 1) and (b) heterooligomers (different side chains; **Fig. 2**, strategy 2).

Orthogonal protecting groups are required for both α- and γ-amino functions. As a robust synthetic strategy for these CPPs is required, the

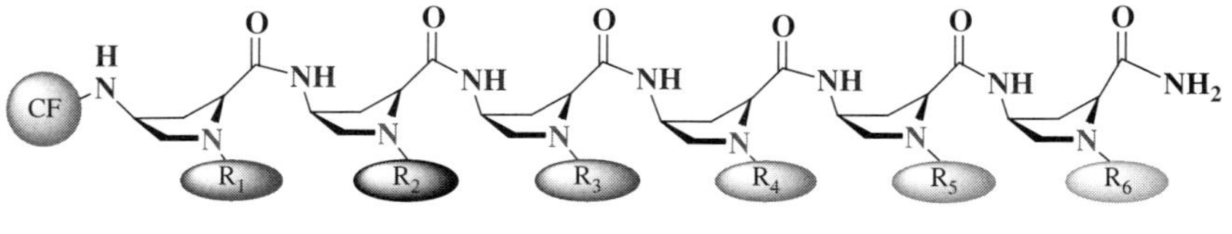

Fig. 1. Chemical structure of *cis*-γ-amino-L-proline oligomers.

7. SpectraMAX GeminiEM 96-well plate spectrafluorometer (Molecular Devices Corp.).
8. SoftMax Pro 4.3LS software (Molecular Devices Corp.).

2.2. Cells, Media and Reagents

1. M14 human melanoma cell line (obtained as a generous gift from Dr. Barbara M. Mueller, La Jolla Institute for Molecular Medicine, La Jolla, CA).
2. HS 895.T and HS 895.SK, human melanoma and normal skin fibroblast cell lines, respectively, (American Type Culture Collection, Manassas, VA; CRL-7637 & CRL 7636).
3. All reagents, including those used in tissue culture, from Fisher Scientific (Atlanta, GA) (unless otherwise stated).
4. RPMI 1640, supplemented with 10% fetal bovine serum (BioWhittaker, Inc. Walkersville, MD), and antibiotics: 0.1 mg/mL gentamicin sulfate, 50 U/mL penicillin, and 0.05 mg/mL streptomycin sulfate.
5. Adhesion media: RPMI-1640 supplemented only with 20 mM HEPES.
6. Trypsin-EDTA (pH 7.4): 137 mM NaCl, 5.4 mM KCl, 5.6 mM D-glucose, 7 mM NaHCO$_3$, 0.05% 1:250 trypsin (Difco™), 7 mM EDTA (tetrasodium salt).
7. Phosphate-buffered saline (PBS): 2.68 mM KCl, 1.38 mM K$_2$HPO$_4$, 137 mM NaCl, 9.58 mM NaH$_2$PO$_4$.
8. Nunc TC dish (100 X 15 SI).
9. ProBind™ 96-well plates (BD Biosciences, Bedford, MA).
10. Peptide-amphiphiles: C$_{16}$-(α1(IV)382-393)-NH$_2$ THP ligand for the $\alpha_2\beta_1$ integrin and C$_{16}$-(α1(IV)1263-1277)-NH$_2$ THP ligand for CD44/CSPG.
11. S.N.A.P.™ eukaryotic total RNA isolation kit and DNase I (Invitrogen Corp., Carlsbad, CA).
12. RT-PCR reagents, including SUPERSCRIPT II, Platinum® Taq DNA Polymerase, Platinum Quantitative PCR SuperMix-UDG and GAPDH-certified LUX™ primer set (Invitrogen Corporation, Carlsbad, CA).
13. 3% agarose gels poured and electrophoresed in 89 mM Tris Base, 89 mM H$_3$BO$_3$ and 8 mM EDTA (tetrasodium salt).
14. SYBR green I (BioWhittaker Molecular Applications, Rockland, ME).
15. S.N.A.P. gel purification kit (Invitrogen Corp.).
16. MMP capture antibodies (Chemicon International, Temecula, CA): MMP-1, monoclonal antibody (mAb) MAB1346; MMP-13, mAb MAB3321; and MMP-14, polyclonal antibody (pAb) AB815.
17. ELISA kits (R&D, Minneapolis, MN).
18. MMP blocking buffer: PBS containing 0.05% Tween™ 20 and 2 mg/mL bovine serum albumin (BSA).
19. Enzyme assay buffer: 50 mM Tricine, 50 mM NaCl, 10 mM CaCl$_2$, 0.05 % brij-35.

20. ProMMP-1 and proMMP-3 is expressed in *Escherichia coli* and folded from inclusion bodies as described previously *(110)*. ProMMP-1 is activated by reacting with 1 m*M* *p*-aminophenylmercuric acetate (APMA) and an equimolar amount of MMP-3 at 37°C for 6 h. After activation, MMP-3 is completely removed from MMP-1 by affinity chromatography using an anti-MMP-3 immunoglobulin (Ig)G Affi-Gel 10 column. ProMMP-3 is activated by reacting with 5 µg/mL chymotrypsin at 37°C for 2 h. Chymotrypsin is inactivated with 2 m*M* diisopropylfluorophosphate.

21. ProMMP-2 is purified from the culture medium of human uterine cervical fibroblasts *(111)* and activated by reacting with 1 m*M* APMA at 25°C for 2 h.

22. ProMMP-9 (Chemicon International) is activated with 1 m*M* APMA at 25°C for 2 h.

23. ProMMP-13 (R&D Systems, Minneapolis, MN) is activated with 1 m*M* APMA at 25°C for 2 h.

24. The amounts of active MMP-1, MMP-2, MMP-3, MMP-9, and MMP-13 are determined by titration with recombinant TIMP-1 *(112)* over a concentration range of 0.1–3 µg/mL.

25. Recombinant MMP-14 with the linker and *C*-terminal hemopexin-like domains deleted [residues 279–523; designated MMP-14($\Delta_{279-523}$)] is purchased from Chemicon International. MMP-14($\Delta_{279-523}$) is expressed in the active form with Tyr112 as the *N*-terminus. MMP-14($\Delta_{279-523}$), in contrast to MMP-14, does not undergo rapid autoproteolysis, and is used in the present studies due to the relatively small differences in MMP-14($\Delta_{279-523}$) and MMP-14 triple-helical peptidase activities noted previously *(113)*. The amount of active MMP-14 is determined by titration with TIMP-2 *(31,113)*.

26. Costar® 384-well plates (Corning Inc., Corning, NY).

27. Fluorogenic THP substrates (**Table 1**).

28. Precast Tris-HCl gels (Bio-Rad Laboratories, Inc., Hercules, CA).

29. Silver stain (Bio-Rad Laboratories, Inc.).

30. Microcon Spin Columns, 3000 MWCO (Millipore Corp., Bedford, MA).

31. 2X gel sample buffer: 0.125 *M* Tris-HCl, pH 6.8, 4% [w/v] sodium dodecyl sulfate (SDS), 20% (v/v) glycerol.

32. SDS-polyacrylamide gel electrophoresis (PAGE) (from Mini-PROTEAN II, Electrophoresis Cell Instruction Manual, Bio-Rad Laboratories) separating gel: 0.375 *M* Tris-HCl, 7.5% acrylamide, 1% SDS, 0.05% ammonium persulfate (APS), 0.05% TEMED and 1 mg/mL gelatin (type A, porcine skin, Sigma, St. Louis, MO); stacking gel: 0.125 *M* Tris-HCl, 4% acrylamide, 1% SDS, 0.05% APS, and 0.1% TEMED.

33. Collagenase buffer: 50 m*M* Tris-HCl, 200 m*M* NaCl, 5 m*M* CaCl₂, and 0.1% Brij-35, pH 7.6.

34. Coomassie brilliant blue (Acros, Morris Plains, NJ).

35. 1 m*M* phenylmethylsulfonylfluoride (PMSF) (Sigma) (dissolved in ethanol, diluted in collagenase buffer),1 m*M* *N*-ethylmaleimide (Sigma) (dissolved in water,

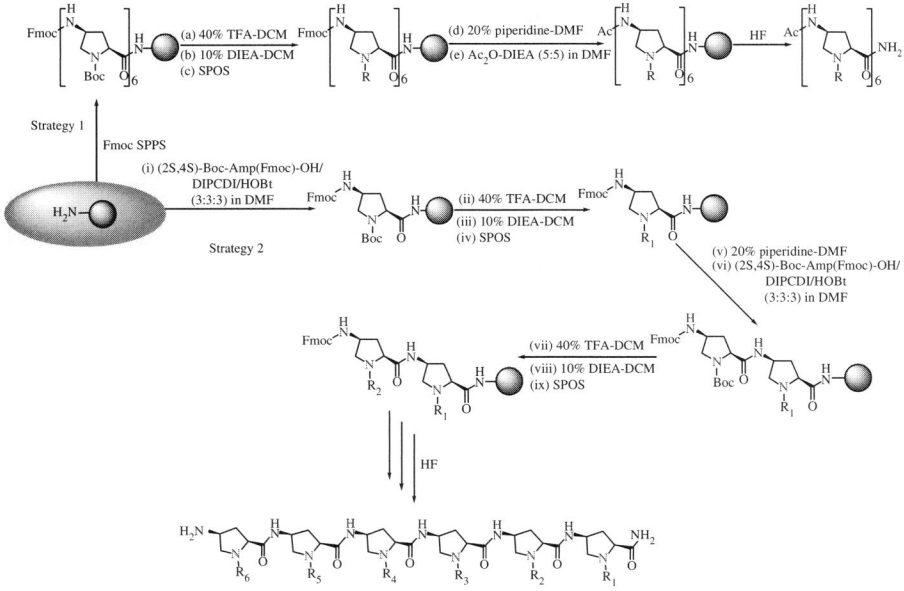

Fig. 2. Synthesis of N^{α}-alkyl-γ-hexapeptides and the N^{α}-acyl-γ-hexapeptides (synthesis of homo-oligomeric systems and of oligomers with different side chains for both cases). (Reprinted from **ref. 17**, with permission from the American Chemical Society.)

most obvious choice would seem to be the use of *tert*-butoxycarbonyl (Boc) and 9-fluorenylmethoxycarbonyl (Fmoc) groups, as the manipulation of these orthogonal protecting groups is well established and friendly based on the experience acquired over many years of continuous use *(28)*. γ-Peptide syntheses are carried out on a *p*-methylbenzhydrylamine resin (MBHA) using an Fmoc/Boc hybrid strategy, in which Fmoc is the temporary protecting group for the γ-amino functionality of each monomer, and Boc is the semipermanent protecting group for the α-amino group through which the side chain has been introduced. The amino acid used is *cis*-γ-Fmoc-amino-α-Boc-L-proline [Boc-Amp(Fmoc)-OH]. Acylations are carried out using the corresponding carboxylic acid and N, N'-diisopropylcarbodiimide (DIPCDI) in the presence of N-hydroxybenzotriazole (HOBt). Alkylations at the α-amino group of the proline employ reductive amination strategy with the corresponding aldehyde with NaBH$_3$CN. The N^{α}-amino guanidylation is carried out using N, N'-di-Boc-N''-trifluoromethanesulfonyl guanidine *(29)* and Et$_3$N. Before cleavage of the peptide from the resin, the Fmoc is removed and the resulting terminal amine is acetylated or functionalized with a fluorescent probe.

1.2. Synthesis of Poly(proline) Dendrimers

1.2.1. Synthesis of Repetitive Branched Poly(proline) Peptides (19)

Although monomers such as diaminoacetic acid (Daaa), L-diaminopropionic acid (Dapa), and L-diaminobutyric acid (Daba) can be used as branching units, 4-amino-L-proline (Amp) and imidazolidine-2-carboxylic acid (Imd) are preferable as they maintain structural coherence with the rest of the sequence (**Fig. 3**).

The preparation of the branched polyproline peptide Fmoc-Pro$_5$-L-Amp(Fmoc-Pro$_5$)-OH is illustrated in **Fig. 4**. Boc-L-Amp(Fmoc)-OH is incorporated onto the Merrifield resin using DIPCDI and catalytic amounts of 4-dimethylaminopiridine (DMAP) in dichloromethane (DCM). After removal of the Boc group, Boc-Pro-OH is coupled. The Fmoc group is then removed and a second Boc-Pro-OH is coupled, followed by the simultaneous elongation of the chains using Boc-Pro-OH for all residues with the exception of Fmoc-Pro-OH for terminal position. *In situ* neutralization should be used for the incorporation of the third residue to avoid DKP formation. Cleavage and purification by RP-HPLC affords the expected polyproline peptide in good yield and with purities up to 95%.

1.2.2. Assembly of Poly(proline) Dendrimers (20)

Several polyfunctional compounds such as spermidine or Imd can be used as a core.

Synthesis of a first-generation dendrimer using a spermidine core on a Wang carbonate resin is shown in **Fig. 5**. Chloranil and ninhydrin tests are used to ensure quantitative couplings.

The use of the orthogonal Dde and Mmt groups allows sequential incorporation of the two branching units, which can be the same or different. Final Fmoc removal followed by cleavage using trifluoroacetic acid (TFA) gives crude

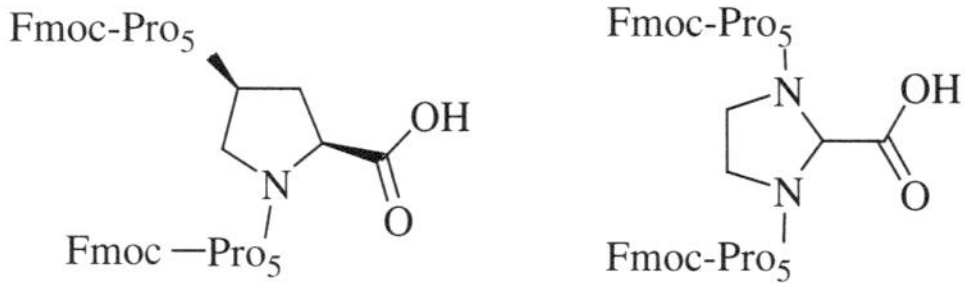

Fmoc-Pro$_5$ Fmoc-Pro$_5$

Fmoc—Pro$_5$ Fmoc-Pro$_5$

Fig. 3. Repetitive branched polyproline peptides synthesized using different branching units.

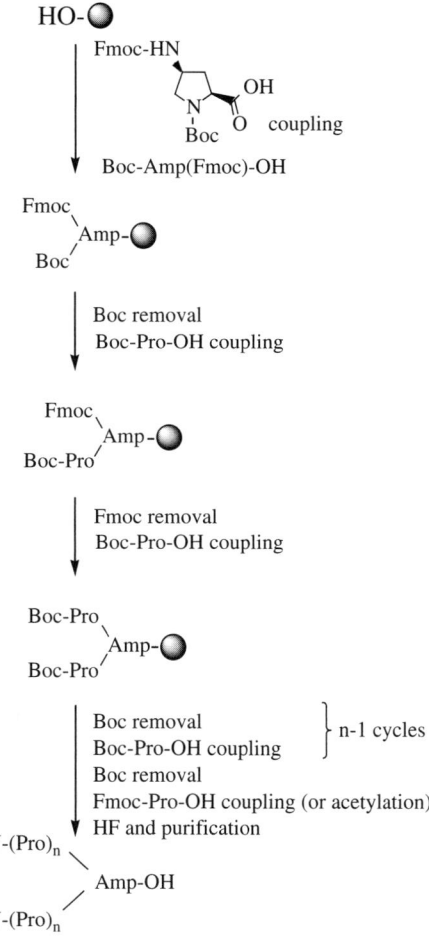

Fig. 4. Synthesis of repetitive branched poly(proline) peptides. (Reprinted from **ref. 20**, with permission from the American Chemical Society.)

dendrimers in very good yield (>65%). Purification by reversed-phase (RP)-high-performance liquid chromatography (HPLC) gives dendrimers with >95% purity. If needed, fluorescein labeling is performed on the solid-phase after the Fmoc removal using benzotriazol-1-yl-*N*-oxytris(pyrrolidino)phosphonium hexafluorophosphate (PyBOP) and *N,N*-diisopropylethylamine (DIEA) as coupling reagents.

The preparation of a second-generation of polyproline dendrimers *(21)* starting from either Amp or Imd is shown in **Fig. 6**. Extension of the dendrimer

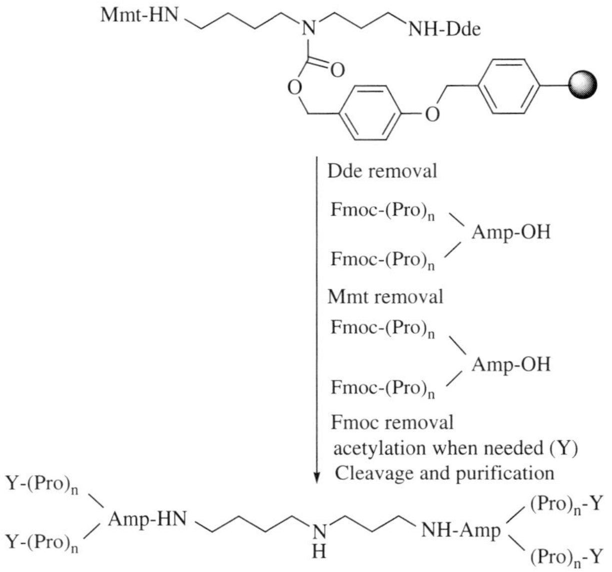

Fig. 5. Solid-phase synthesis of dendrimers using a spermidine core. (Reprinted from **ref. 20**, with permission from the American Chemical Society.)

to the second-generation is not trivial as a result of the difficulty of incorporating peptide fragments onto the secondary amine of the proline residue. However, fragment coupling can be facilitated by the incorporation of a small and flexible amino acid such as Gly at the N-terminal position. Thus, the Fmoc-Gly-Pro_5-L-Imd(Fmoc-Gly-Pro_5)-OH fragment is synthesized on Merrifield resin as described above (*see* **Fig. 4**), with the exceptions that an extra residue, such as Fmoc-Gly-OH, is introduced at each N-terminal position, and that imida-zolidine is the branching unit. The fragment (>95% by analytical RP-HPLC) is coupled on Rink-resin using Fmoc chemistry and the final compounds are cleaved from the resin with TFA, affording 88–90% pure products as determined by analytical HPLC.

Following this approach, the polyproline dendrimer could be further expanded via repetitive condensations of the building block, ultimately giving rise to a fourth generation *(22)*.

1.2.3. Labeling of Dendrimers

The fluorescent label can be added at either the N- or the C-terminus. In both cases, CF, which is stable to both fluorhydric acid (HF) and TFA, is coupled using DIPCDI in the presence of HOBt. The N-terminal label is intro-

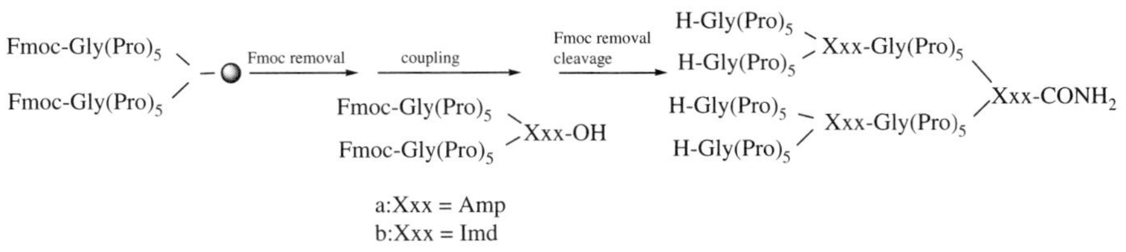

Fmoc-Gly(Pro)$_5$
Fmoc-Gly(Pro)$_5$ — ⊙ — Fmoc removal → coupling →

Fmoc-Gly(Pro)$_5$
Fmoc-Gly(Pro)$_5$ ＞Xxx-OH

Fmoc removal cleavage →

H-Gly(Pro)$_5$
H-Gly(Pro)$_5$ ＞Xxx-Gly(Pro)$_5$

H-Gly(Pro)$_5$
H-Gly(Pro)$_5$ ＞Xxx-Gly(Pro)$_5$ ＞Xxx-CONH$_2$

a:Xxx = Amp
b:Xxx = Imd

Fig. 6. Synthesis of second-generation polyproline dendrimers. (Reprinted from **ref. 21**, with permission from Wiley Inter Science.)

duced just before cleavage, whereas C-terminal labeling requires introduction of an orthogonally protected Lys residue, such as N^α(Fmoc)-N^ε(Mtt)-Lys-OH (Fmoc/tBu strategy), at the beginning of the synthesis. In the latter case, the 4-methyltrityl (Mtt) protecting group is ultimately removed with 3% TFA, at which point the CF is introduced.

1.2.4. Functionalization of Dendrimers

Polyproline dendrimer surfaces can be functionalized before fluorescent labeling and cleavage from the resin. For example, the free surface amine can be transformed into a free guanidine, a substituted guanidine, an Arg-rich surface, and a spermidine-rich surface **(Fig. 7)** *(23)*. The reaction of the free amino of the dendrimer with (1) 1,3-di-Boc-2-methylisothiourea and DIEA provides the dendrimer with protected guanidines; (2) iminium/uronium salts such as HBPyU in the presence of DIEA gives substituted guanidines; (3) repetitive couplings of Fmoc-Arg(Pmc)-OH and DIPCDI and HOBt affords the arginine rich (Arg$_4$) sequence; or (4) *N, N*-bis(*N'*-Fmoc-3-aminopropyl)-glycine potassium hemisulfate and DIPCDI and HOBt gives spermidines. Once the dendrimer surface has been modified, the synthesis continues, and the dendrimer can be labeled with CF if needed before cleavage from the resin.

Furthermore, if a Lys is incorporated at the C-terminal, the ε-amino function can also be labeled with CF. If polyethylene glycol (PEG) chains are required, for example, they can be introduced at the beginning of the synthesis by coupling to the Lys Fmoc-PEG$_{3400}$-OSu in the presence of DIEA.

2. Materials

2.1. Solid-Phase Synthesis

1. MBHA, Merrifield (hydroxymethyl resin), PAM (hydroxymethyl-phenoxy resin), and Wang carbonate resins (Calbiochem-Novabiochem AG).
2. Protected amino acids (Neosystem, Strasbourg, France).

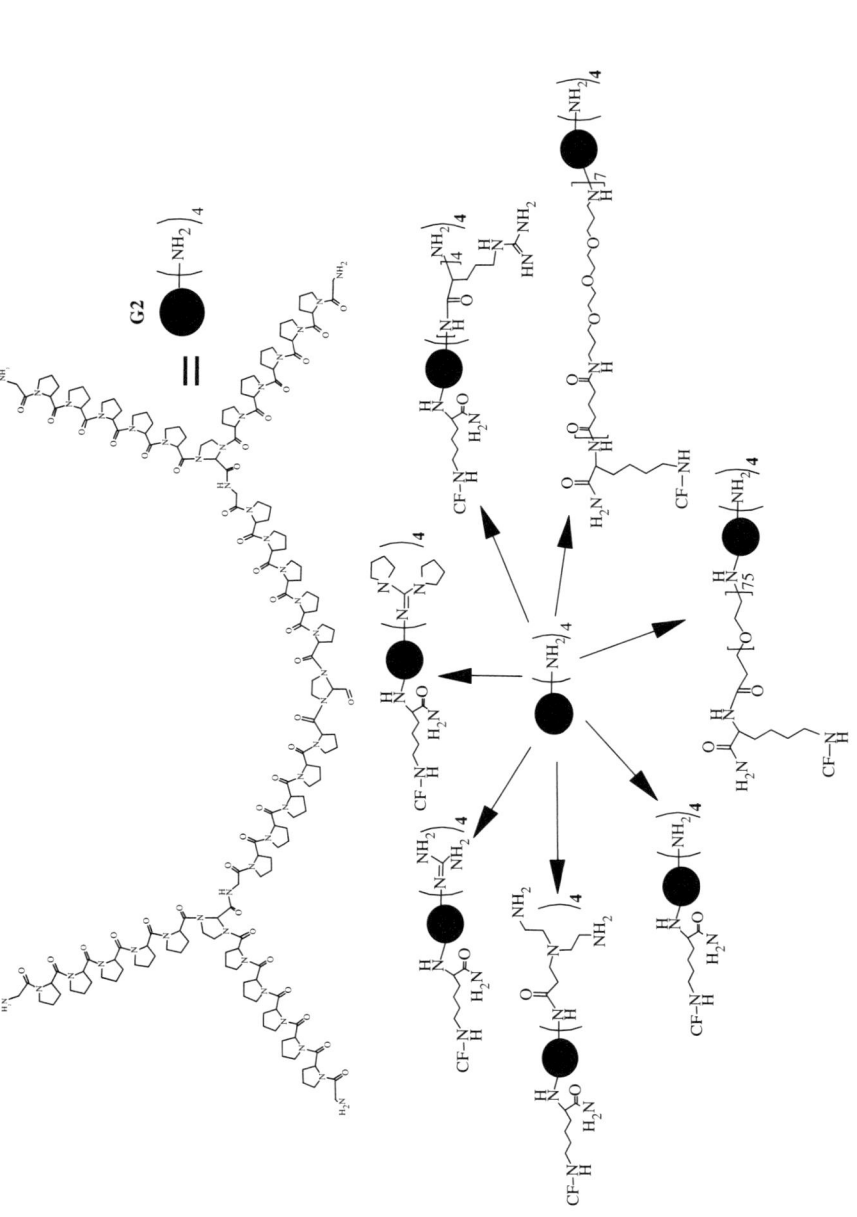

Fig. 7. Second-generation polyproline dendrimer structures tested as drug delivery systems. (Reprinted from **ref. 23**, with permission from Wiley InterScience.)

3. Solvents (e.g., CH$_3$CN, DCM, *N*, *N*-dimethylformamide [DMF], and *t*-butylmethyl ether [MTBE]) and other reagents (e.g., CF, AcOH, DIEA, DIPCDI, 7-aza-benzotriazol-1-yl-*N*-oxy-tri(pyrrolidino)-phosphonium hexafluorophosphate [PyAOP], PyBOP, HOBt, linker AM, linker AB, piperidine, sodium cyanoborohydride, Et$_3$N, and TFA), commercially available from different sources (e.g., Sigma-Aldrich, Milwaukee, WI; Across, NJ; and Fluka Chemika, Buchs, Switzerland).

4. HF (Air Products and Chemicals, Inc., Allentown, Canada), and related equipment (Peptide Institute Inc., Minoh, Osaka, Japan).

5. Analytical RP-HPLC performed using Waters (Milford, MA) chromatography systems with reversed-phase Symmetry C$_{18}$ (150 × 4.6 mm) 5 μm columns with ultraviolet (UV) detection at $\lambda = 220$ nm. Semipreparative RP-HPLC is performed on a Waters (Milford, MA) chromatography system using Symmetry C$_8$ (3 × 10 cm, 5 μm) columns. Compounds are detected by UV absorption at $\lambda = 220$ nm.

6. Mass spectra are recorded on a matrix-assisted laser desorption/ionization (MALDI) Voyager DE RP time-of-flight (TOF) spectrometer (Applied Biosystems, Framingham). 3,5-dihydroxybenzoic acid (DHB) is used as a matrix and is purchased from Aldrich.

2.2. Enzymatic Stability, Cell Culture, and Lysis

1. For the experiments carried out with the γ-peptides, cell lines are routinely grown at 37°C in a humidified atmosphere with 5% CO$_2$. HeLa and COS-1 cells are maintained in Dulbecco's modified Eagle's medium (DMEM) (1000 mg/mL glucose for HeLa and 4500 mg/mL for COS-1) culture medium containing 10% fetal calf serum (FCS), 2 m*M* glutamine, 50 μ/mL penicillin and 0.05 g/mL streptomycin. For the experiments carried out with the dendrimers, HeLa S-3 (human epithelial) cells in MEM (Eagle) supplemented with 10% (v/v) FBS; Jaws II (Dendritic Cells) in Alpha minimum essential medium supplemented with 4 m*M* L-glutamine, 1 m*M* sodium pyruvate and 5 ng/mL of granulocyte/macrophage colony-stimulating factor (GM-CSF) and 20 FBS; COS-7 (fibroblasts *Cercopithecus aethiops* [African green monkey]) in DMEM supplemented with 10% (v/v) FBS.

2. Trypsin from bovine pancreas E.C. 3.4.21.4 (Roche, Basel) and human serum (Sigma-Aldrich, Milwaukee, WI) are aliquoted and stored at −80°C.

3. Samples are sterilized with 0.22 μm filters (Millex-GV, PVDF, Durapore, Millipore). Concentrated samples are prepared in sterile PBS, filtered, and stored at 4°C for no more than 1 mo.

4. Culture flasks, plastic dishes, glass coverslips, eight-well Lab-Teck™ chambered coverglass, and 96-well plates (Nalge Nunc International, Naprville, IL).

5. Other reagents (e.g., HBBS buffer, DABCO, HEPES, RNAsa, Mowiol, MTT) are commercially available from different sources (e.g., Biological Industries Ltd., Kibbutz Beit Haemek, Israel; Calbiochem, San Diego, CA; and Sigma-Aldrich, Milwaukee, WI).

6. Texas Red–Dextran (TR-DX; 3 mg/mL, molecular weight [MW] = 10.000, Molecular Probes).
7. Lysis buffers are prepared in the laboratory (0.1% Triton X-100 in 50 mM Tris, pH 8.5).
8. Plate fluorimetry experiments are carried out in a FL600 Microplate Fluorescence Reader (Bio-Tek) using $\lambda_{excitation} = 485 \pm 20$ nm and a $\lambda_{emission} = 530 \pm 25$ nm.
9. Flow cytometry experiments are carried out in an Epics XL flow cytometer (Coulter).
10. Formazan absorbance in the MTT assay is measured at $\lambda = 570$ nm in a Universal Microplate Reader Spectrophotometer Elx800 (Bio-Tek).
11. Confocal laser scanning microscopy is performed with an Olympus Fluoview 500 confocal microscope using a × 60/1.4 NA plan-apochromatic objective and in a Zeiss confocal microscope (magnification; ×63 W) using LSM 510 software.

3. Methods

3.1. General Solid-Phase Synthesis Procedures

Peptides and dendrimers are synthesized using a combined Fmoc/Boc solid phase strategy. Each synthesis is performed manually in a polypropylene syringe fitted with a porous polyethylene disc (*see* **Note 1**). Solvents and soluble reagents are removed by suction. Washings between deprotection, coupling and subsequent deprotection steps are carried out with DMF (five times for 1 min) and DCM (five times for 1 min) using 10 mL solvent/g resin each time.

3.1.1. Coupling of the First Amino Acid onto an MBHA Resin

1. The resin is first pretreated with DCM (five times for 1 min) and then with TFA–DCM (4:6, once for 1 min + once for 30 min).
2. The resin is then washed with DCM (five times for 1 min), treated with DIEA–DCM (5:95, three times for 3 min) and washed with DCM (five times for 1 min).
3. The amino acid (5 eq) is then introduced with DIPCDI (5 eq) and HOBt (5 eq) in the solvent (DMF for Fmoc-amino acids and DCM for Boc-amino acids) for 2 h, followed by washes with DMF (five times for 1 min) and DCM (five times for 1 min).
4. The amino acid-resin complex is then acetylated to cap any remaining free amino groups (*see* **Subheading 3.1.9.**).

3.1.2. Reduction of Initial MBHA Resin Loading

1. A reduced resin loading, which may be required for dendrimer synthesis, is attained by incorporation of Fmoc-Gly-OH (0.5 eq) onto the MBHA resin using DIPCDI (0.5 eq) in the presence of HOBt (0.5 eq).

2. After the coupling, the resin is washed with DMF (four times for 1 min) and DCM (four times for 1 min), and any free remaining amino groups are capped by acetylation (*see* **Subheading 3.1.9.**).
3. After the capping, the resin is washed with DMF (four times for 1 min) and DCM (four times for 1 min) and an aliquot is treated with piperidine (*see* **Subheading 3.1.4.**).
4. The loading of the resin is monitored by measuring the concentration of the liberated dibenzofulvene via UV spectroscopy. Once the desired loading of the resin is attained, the remaining amino groups on the resin are blocked by acetylation (*see* **Subheading 3.1.9.**).

3.1.3. Coupling of the First Amino Acid to the Hydroxymethyl Resin

1. The resin is first pretreated with DCM (five times for 1 min) and then with TFA–DCM (4:6, once for 1 min + once for 30 min), washed with DCM (five times for 1 min), DIEA–DCM (5:95, three times for 3 min), DCM (five times for 1 min) and DMF (five times for 1 min).
2. The resin is subsequently treated with the amino acid (3 eq), DIPCDI (3 eq), and DMAP (0.3 eq) in DMF (twice for 1 h), followed by washes with DMF (five times for 1 min) and DCM (five times for 1 min).
3. The amino acid-resin complex is then treated with an acetylating agent to cap any remaining hydroxyl groups (*see* **Subheading 3.1.9.**).

3.1.4. Fmoc Removal and Fmoc-Amino Acid Coupling

1. The resin is washed with DCM (five times for 1 min), followed by washes with DMF (five times for 1 min), and the Fmoc group is removed using piperidine–DMF (1:4, once for 1 min + twice for 15 min), followed by washes with DMF (five times for 1 min) and DCM (five times for 1 min).
2. For those cases in which Fmoc removal is incomplete or difficult, the standard protocol is replaced by treatment with 1,8-diazabicyclo[4.3.0]undec-7-ene (DBU)/piperidine/toluene/DMF (5:5:20:70).
3. Fmoc-amino acids (5 eq) are coupled to the resin with DIPCDI (5 eq) in the presence of HOBt (5 eq) in DMF for 2 h.
4. The resin is then washed with DMF (five times for 1 min) and DCM (five times for 1 min).

3.1.5. Boc Removal and Boc-Amino Acid Coupling

1. The resin is washed with DCM (five times for 1 min), and the Boc protecting group is removed with TFA–DCM (4:6, once for 1 min + once for 30 min).
2. The resin is then washed with DCM (five times for 1 min), neutralized with DIEA–DCM (5:95, three times for 3 min) and washed with DCM (five times for 1 min).
3. Boc-amino acids (Boc-Pro-OH) are coupled to the resin using Boc-Pro-OH (5 eq) and DCC (5 eq) in DCM for 2 h.

4. The resin is then washed with DCM (five times for 1 min) and DMF (five times for 1 min).

3.1.6. Removal of Mtt

1. The resin is washed with DCM (five times for 1 min), and the Mtt group is removed by treatment with TFA–TES–DCM (3:1:96) (twice for 10 min + once for 30 min).
2. The resin is washed with DCM (five times for 1 min), neutralized with DIEA–DCM (1:19) (twice for 1 min), and washed with DCM (five times for 1 min) and DMF (five times for 1 min).

3.1.7. Removal of Dde

1. The resin is washed with DCM (four times for 1 min) and DMF (four times for 1 min) and the Dde group is removed by treatment with hydrazine–DMF (2:98) (three times for 4 min).
2. The resin is washed with DMF (four times for 1 min) and DCM (four times for 1 min).

3.1.8. Removal of Mmt

1. The resin is washed with DCM (four times for 1 min), and the Mmt group is removed by treatment with TFA–TIS–DCM (1:5:94) (four times for 2 min).
2. The resin is washed with DCM (four times for 1 min), neutralized with DIEA-DCM (1:19) (twice for 1 min), and washed with DCM (five times for 1 min) and DMF (five times for 1 min).

3.1.9. Acetylative Capping of Free Amine or Hydroxyl Groups

1. The resin is treated with acetic anhydride (50 eq) and DIEA (50 eq) in DMF for 30 min to cap amino groups and with AcOH (5 eq), DIPCDI (5 eq), and DMAP (0.5 eq) in DCM for 20 min to cap hydroxyl groups.
2. The resin is washed with DCM (five times for 1 min) after acetylation.

3.1.10. CF Labeling

1. CF (10 eq) is coupled to the amino group using DIPCDI (10 eq) in the presence of HOBt (10 eq) in DMF for 2 h.
2. Before cleavage from the resin, the peptide is treated with 20% piperidine in DMF (twice for 30 min) to remove any overincorporated CF (*see* **Note 2**).

3.1.11. Acidolytic Cleavage with TFA

1. The resin is washed with DCM (four times for 1 min), dried under vacuum, and treated with different cocktails of TFA ([a] TFA–H_2O [95:5], [b] TFA–H_2O–TIS [95:2.5:2.5], and [c] TFA–thioanisole–H_2O–phenol–EDT [82.5:5:5:5:2.5]) for 1 h at 25°C.

2. The TFA solution containing the peptide is then filtered, and the crude peptide is precipitated with cold anhydrous MTBE and subsequently centrifuged.
3. The MTBE is decanted off, and the peptide is dissolved in AcOH and then lyophilized.

3.1.12. Acidolytic Cleavage with HF

1. The peptidyl resin is washed with MeOH (three times for 1 min), and dried under vacuum.
2. The resin is then treated with anhydrous HF in the presence of 10% anisole for 1 h at 0°C.
3. The HF is removed by evaporation, the crude peptide is precipitated with cold anhydrous MTBE, and the resulting mixture is filtered through a syringe fitted with a porous polypropylene disk.
4. The peptide is eluted with 10% AcOH and then lyophilized.

3.2. Solid-Phase γ-Peptide Syntheses

The γ-peptides are synthesized on an MBHA resin using an Fmoc/Boc combined solid-phase strategy (*see* **Note 1**). Coupling of Boc-L-Amp(Fmoc)-OH (5 eq) is carried out with DIPCDI (5 eq) and HOBt (5 eq) in DMF for 2 h at 25°C. After coupling, the resin is washed with DMF (five times for 1 min) and DCM (five times for 1 min). Couplings are monitored by the Kaiser test (*see* **Note 3**).

When introducing a short PEG chain, two units of Fmoc-8-amino-3,6-dioxaoctanoic acid (5 eq) are first coupled to the resin using the same coupling reagents mentioned above. Peptide synthesis then proceeds as described for the other compounds.

3.2.1. Synthesis of Fmoc-[Amp(Boc)]₆-MBHA Resin

Fmoc-[Amp(Boc)]₆-MBHA resin is synthesized by repetitive couplings of Boc-Amp(Fmoc)-OH (5 eq) according to the protocol described under **Subheading 3.1.4.** and using piperidine for deprotection.

3.2.2. Synthesis of Homo Nᵅ-Acyl-γ-Hexapeptides Via Acylation

1. Starting from Fmoc-[Amp(Boc)]₆-MBHA resin (*see* **Subheading 3.2.1.**), the N^α-Boc protecting groups are removed with TFA (*see* **Subheading 3.1.5.**).
2. The α-amino groups are then acylated using the appropriate carboxylic acid (RCOOH [30 eq, 5 eq for each amine]), DIPCDI (30 eq), and HOBt (30 eq) in DMF for 2 h at 25°C.
3. The resin is then washed with DMF (five times for 1 min) and DCM (five times for 1 min).
4. Reaction progress is determined using the chloranil test (*see* **Note 4**).

3.2.3. Synthesis of Homo N^{α}-Alkyl-γ-Hexapeptides Via Reductive Amination

1. The synthesis starts from Fmoc-[Amp(Boc)]$_6$-MBHA resin (*see* **Subheading 3.2.1.**). The α-amines are first deprotected (*see* **Subheading 3.1.5.**), then alkylated via reductive amination. Thus the resin is treated with the appropriate aldehyde (RCHO [30 eq, 5 eq for each amine]) in 1% AcOH in DMF-MeOH (1:1) for 30 min, at which point NaBH$_3$CN (30 eq) in 1% AcOH in DMF-MeOH is added and the mixture is allowed to react for 2 h (*see* **Note 5**).
2. The resin is washed with DMF-MeOH (1:1) (five times for 1 min) and DCM (five times for 1 min).
3. Alkylations are monitored by the chloranil test (*see* **Note 4**).

3.2.4. Synthesis of Hetero N^{α}-Acyl-γ-Hexapeptides

The hetero N^{α}-acyl-γ-hexapeptides are synthesized stepwise, whereby each monomer is functionalized after its coupling. Thus, once the monomer is introduced by coupling of the corresponding protected monomer Boc-L-Amp(Fmoc)-OH following the protocol described under **Subheading 3.1.4.**, the N^{α}-Boc protecting group is removed (*see* **Subheading 3.1.5.**), and the acylation is carried out as described under **Subheading 3.2.2.** (using only 5 eq reagent excess). After N^{γ}-Fmoc deprotection (*see* **Subheading 3.1.4.**), the reaction sequence is repeated.

3.2.5. Synthesis of Hetero N^{α}-Alkyl-γ-Hexapeptides

The hetero N^{α}-alkyl-γ-hexapeptides are synthesized stepwise, whereby each monomer is functionalized after its coupling, in the same manner as under **Subheading 3.2.4.**, with the exception that N^{α}-acylation is substituted with alkylation as described under **Subheading 3.2.3.** (using only 5 eq reagent excess).

3.2.6. Synthesis of N^{α}-Guanidilate-γ-Hexapeptide

1. Starting from Fmoc-[Amp(Boc)]$_6$-MBHA resin (*see* **Subheading 3.2.1.**), the N^{α}-amino groups are first deprotected following the protocol described unde **Subheading 3.1.5.**, then guanidylated using N, N'-di-Boc-N''-trifluoromethanesulfonyl guanidine (30 eq, 5 eq for each amine) and Et$_3$N (30 eq, 5 eq for each amine) in DCM for 4 h at room temperature.
2. The resin is then washed with DCM (five times for 1 min).
3. The guanidylation is monitored by the chloranil test (*see* **Note 4**).

3.2.7. Functionalization of the N$^\gamma$-Terminal by Acetylation or CF Labeling

Once the γ-peptide backbone is synthesized, the N$^\gamma$-terminal Fmoc group is removed with piperidine (*see* **Subheading 3.1.4.**) and the α-amino group is acetylated (*see* **Subheading 3.1.9.**). In the event that fluorescent labeling is required instead of acylation, CF is coupled to the N$^\gamma$-terminal (*see* **Subheading 3.1.10.**).

3.3. Solid-Phase Synthesis of Dendrimers

3.3.1. Solid-Phase Synthesis of Branching Units

3.3.1.1. Synthesis of (Fmoc-Pro$_N$)$_2$-L-Amp-OH ($n = 5, 14$)

1. The first amino acid [Boc-L-Amp(Fmoc)-OH] is coupled to the Merrifield resin as described under **Subheading 3.1.3.**.
2. The reaction is monitored via the toluenesulfonyl chloride (TsCl)-4-(*p*-nitrobenzyl)pyridine (PNBP) test for hydroxyl groups (*see* **Note 6**).
3. The Boc group is removed and the Boc-Pro-OH is coupled following the protocol described under **Subheading 3.1.5.**.
4. The Fmoc group is then removed (*see* **Subheading 3.1.4.**) and the peptide is elongated by iterative coupling of Boc-Pro-OH units using Boc chemistry (*see* **Subheading 3.1.5.**; **Note 7**). In order to avoid DKP formation, the third proline is introduced using *in situ* neutralization (*see* **Note 8**) by treatment with Boc-Pro-OH (5 eq), BOP (5 eq), or PyBOP (5 eq), DIEA (15 eq), and HOBt (5 eq) in DMF for 2 h.
5. The resin is then washed with DMF (five times for 1 min) and DCM (five times for 1 min).
6. The last amino acid for both chains is introduced as an Fmoc-Pro-OH (*see* **Subheading 3.1.4.**).
7. Couplings are simultaneously monitored with the Kaiser and De Clerk tests (*see* **Note 9**).
8. Peptide cleavage is carried out with anhydrous HF in the presence of anisole (*see* **Subheading 3.1.12.**).

3.3.1.2. Synthesis of (Fmoc-Gly-Pro$_5$)$_2$-Imd-OH

The building block is synthesized as described in **Subheading 3.3.1.1.** on a hydroxymethyl resin, with the exceptions that an extra residue, such as Fmoc-Gly-OH, is introduced at each N-terminal position, and that Imd is used as a branching unit. The crude peptide is purified by semi-preparative RP-HPLC using a linear gradient of H$_2$O–ACN to obtain 95–99% pure peptide.

3.3.2. Solid-Phase Synthesis of First-Generation (G1) Polyproline Dendrimers using a Spermidine Core, [(CF-Pro₁₄)(H-Pro₁₄)-Amp]-Spermidine-[(H-Pro₁₄)₂-Amp]

1. The spermidine core, orthogonally protected with Dde and Mmt groups, is anchored through its central amino group to a carbonate Wang resin.
2. The Dde group of the 1*N*-Dde-8*N*-Mmt-spermidine-4-yl-carbonyl Wang resin is then removed (*see* **Subheading 3.1.7.**), followed by coupling of the fragment (Fmoc-Pro₁₄)₂-L-Amp-OH (1.5 eq) as described under **Subheading 3.3.1.1.**) using PyAOP (1.5 eq) and DIEA (1.5 eq, double addition) in DMF (three times for 8 h) at 25°C.
3. After washings with DMF (twice for 1 min) and DCM (twice for 1 min), the Mmt group is removed (*see* **Subheading 3.1.8.**) and a second (Fmoc-Pro₁₄)₂-L-Amp-OH is coupled using the same procedure described above.
4. The Fmoc groups are then removed with piperidine (*see* **Subheading 3.1.4.**) and the resin is treated with fluoresceine (2 eq), PyBOP (2 eq), and DIEA (4 eq) in DMF (once for 2 h), followed by washes with DMF (five times for 1 min) and DCM (five times for 1 min).
5. The fluoresceinated G1 dendrimer is then cleaved from the resin with TFA (*see* **Subheading 3.1.11.**) to provide a 65–75% pure dendrimer, which is further purified (*see* **Subheading 3.4.**).

3.3.3. Solid-Phase Synthesis of Second-Generation (G2) CF-Labeled Polyproline Dendrimers

3.3.3.1. SYNTHESIS OF N^α(FMOC)-N^ε(MTT)-LYSINE-AM-GLY-MBHA RESIN

1. The initial 0.7 mmol/g loading of the MBHA resin is reduced to 0.18 mmol/g as described under **Subheading 3.1.2.**
2. The handle Fmoc-AM-OH (10 eq) is then coupled to the resin with DIPCDI (10 eq) and HOBt (10 eq) in DMF for 16 h at 25°C.
3. Fmoc deprotection (*see* **Subheading 3.1.4.**) followed by coupling of N^α(Fmoc)-N^ε(Mtt)-Lys-OH (5 eq) using DIPCDI (5 eq) and HOBt (5 eq) gives the N^α(Fmoc)-N^ε(Mtt)-Lys-AM-Gly-MBHA resin.

3.3.3.2. SYNTHESIS OF ([FMOC-GLY-PRO₅]₂-IMD)₂-(GLY-PRO₅)₂-IMD-N^ε(MTT)-LYSINE-AM-GLY-MBHA RESIN

1. The Fmoc group of N^α(Fmoc)-N^ε(Mtt)-Lys-AM-Gly-MBHA resin (synthesized under **Subheading 3.3.3.1.**) is removed as described under **Subheading 3.1.4.** and (Fmoc-Gly-Pro₅)₂-Imd-OH (1.5 eq) (*see* **Subheading 3.3.1.2.**) is incorporated using DIPCDI (1.5 eq) in the presence of HOBt (1.5 eq) in DMF for 3 h at 25°C.
2. An additional fragment of (Fmoc-Gly-Pro₅)₂-Imd-OH (3 eq, 1.5 eq for each amino group) is then incorporated using the same procedure.
3. The reaction is monitored with the De Clercq and ninhydrin tests (*see* **Note 9**).

3.3.3.3. Solid-Phase Modification and CF Labeling of Dendrimer Surfaces

1. Starting from ([Fmoc-Gly-Pro$_5$]$_2$-Imd)$_2$-(Gly-Pro$_5$)$_2$-Imd-Lys(Mtt)-AM-Gly-MBHA resin (*see* **Subheading 3.3.3.2.**), Fmoc groups are removed and the surface is modified according to needs (*see* **Subheading 3.3.4.**).
2. The Mtt group of ([X-Gly-Pro$_5$]$_2$-Imd)$_2$-(Gly-Pro$_5$)$_2$-Imd-Lys(Mtt)-AM-MBHA resin (X = group introduced onto the surface) is then removed (*see* **Subheading 3.1.6.**) and CF is coupled to the N^ε-amino group of the lysine following the protocol described under **Subheading 3.1.10.**
3. The dendrimers are cleaved from the resin with TFA-H$_2$O-TIS (*see* **Subheading 3.1.11.**) to obtain ([X-Gly-Pro$_5$]$_2$-Imd)$_2$-(Gly-Pro$_5$)$_2$-Imd-Lys[N^ε(CF)]-NH$_2$.

3.3.4. Surface Modification of Second-Generation Polyproline Dendrimers

3.3.4.1. Synthesis of Dendrimers Containing Free Surface Amines

This dendrimer is synthesized as described under **Subheading 3.3.3.3.**, without surface modification. The surface amino groups are deprotected before cleavage of the product from the resin. In the event that CF labeling is required, it is performed after amine deprotection, followed by the surface amino group deprotection and cleavage of the product from the resin.

3.3.4.2. Modification of Dendrimer Surfaces with Free Guanidines

1. The free surface-amine dendrimer is treated with DIEA (4 eq).
2. After 1 min of preactivation, the starting material is reacted with 1,3-di-Boc-2-methylisothiourea (24 eq) and DIEA (12 eq) in anhydrous DMF (three times for 2 h, then once for 16 h) at 25°C.
3. After washing with DMF (four times for 1 min) and DCM (four times for 1 min), the synthesis continues as described under **Subheading 3.3.3.**

3.3.4.3. Modification of Dendrimer Surfaces with Substituted Guanidines

1. The dendrimer that contains free surface amines is treated with HBPyU (12 eq) in the presence of DIEA (4.4 eq) in DMF (twice for 2 h, then an additional 16 h) at 25°C.
2. The resin is washed with DMF (four times for 1 min) and DCM (four times for 1 min), and the synthesis continues as described under **Subheading 3.3.3.**

3.3.4.4. MODIFICATION OF DENDRIMER SURFACES WITH ARGININE-RICH
SEQUENCES

A sequence of Arg_4 is incorporated stepwise to the free amine surface of the
dendrimer by mediated couplings of Fmoc-Arg(Pmc)-OH (24 eq) using DIPCDI
(24 eq) and HOBt (24 eq) in DMF for 2 h at 25°C, followed by washes with
DMF (twice for 1 min) and DCM (twice for 1 min). Once the Arg_4 sequence
has been introduced, synthesis continues as described under **Subheading 3.3.3.**

3.3.4.5. MODIFICATION OF DENDRIMER SURFACES WITH SPERMIDINE

The free amine surface dendrimer is treated with N, N-bis(N'-Fmoc-3-
aminopropyl)-glycine potassium hemisulfate (24 eq, 6 eq per amino group),
liberated from a $KHSO_4$ solution by organic extraction, in the presence of
DIPCDI (24 eq) and HOBt (24 eq) in DMF (twice for 1 h) at 25°C. The resin
is washed with DMF (twice for 1 min) and DCM (twice for 1 min) and the
synthesis continues as described under **Subheading 3.3.3.**

3.3.5. Second-Generation Polyproline Dendrimers Modified with PEG Sequences

3.3.5.1. MODIFICATION OF FREE SURFACE-AMINE DENDRIMERS
WITH MONODISPERSE PEG

Starting from N^α(Fmoc)-N^ε(Mtt)-lysine-AM-Gly-MBHA resin (prepared
under **Subheading 3.3.3.1.**), a monodisperse polyamide pseudo-PEG_{2219}
(prepared from glycolic anhydride and PEG_{220}) is attached stepwise using the
methodology described by Rose et al. *(30)*. The synthesis then proceeds as
described under **Subheading 3.3.3.**, keeping the surface amino groups free.

3.3.5.2. SYNTHESIS OF FREE AMINE—PEG POLYDISPERSE

After removal of the Fmoc group of N^α(Fmoc)-N^ε(Mtt)-lysine-AM-Gly-
MBHA peptidyl resin (*see* **Subheading 3.3.3.1.**), Fmoc-PEG_{3400}-OSu (3 eq)
is coupled with DIEA (3 eq) in DMF (once for 2 h). After washing with
DMF (twice for 1 min) and DCM (twice for 1 min), the dendrimer synthesis
continues as described under **Subheading 3.3.3.**, keeping the amino groups of
the dendrimer surface free.

3.4. Purification and Characterization of Peptides and Dendrimers

Crude peptides and dendrimers are purified by preparative HPLC using
different linear gradients (depending on each product) of CH_3CN (containing
1% of TFA) and H_2O (containing 1% of TFA). Semipreparative and preparative

RP-HPLC employs a Waters 600 Controller Chromatography system with a reversed-phase Symmetry® C8 (30×100 mm, 5μm) column at flow rate of 25 mL/min and compounds are detected by a Waters 2487 Dual Absorbance Detector at $\lambda = 220$ and 443 nm. The purity of each fraction is verified by analytical HPLC and a range of purities is observed (from 95 to 99%). Dendrimers are also analyzed by SEC. Analytical size-exclusion chromatography (SEC) is carried out using an isocratic gradient of 5% NaH_2PO_4 with 3% CH_3CN at a flow rate of 0.5 mL/min and detection at $\lambda = 220$ nm. Ultrahydrogel™ 250 and 500 are used consecutively for all samples in order to achieve good separation. Products are also characterized by MALDI-TOF and/or electrospray ionization (ESI).

3.5. Enzymatic Stability Studies

The stability of the peptides and dendrimers to trypsin or human serum is measured over time in degradation experiments using HPLC. Experiments are carried out using 1 mL of solution in Eppendorf and heating the solution at 37°C in a water bath.

3.5.1. Trypsin Degradation Protocol

1. A master solution of trypsin in 1:1 glycerol-H_2O (6.25 mg/mL) is prepared, aliquoted and stored at -80°C. A single aliquot is used for each experiment.
2. Peptides are incubated at 37°C with the enzyme in 100 mM Tris-HCl buffer, pH 8.0. The final trypsin–peptide ratio is normally 1:100 (*see* **Note 10**), using a final peptide concentration of 125 μM.
3. An aliquot (50 μL) is periodically removed (from 5 min to 48 h), treated with 150 μL of 1 N HCl, and the solution is cooled in an ice bath.
4. An analytical sample of aliquot is run on HPLC (50 μL injection), and the intensity of the peptide or dendrimer peak is noted for each chromatogram. A constant peak intensity over the duration of the experiment indicates that the peptide is stable to trypsin.

3.5.2. Serum Degradation Protocol

1. The compound is diluted in HBBS buffer and the solution is incubated at 37°C. The peptide–serum ratio is 9:1 and compounds are used at a final concentration of 125 μM (added to the serum dissolved in the HBBS buffer).
2. The test compound solution is divided into batches, and each batch is heated at 37°C for a different length of time (ranging from 2 to 120 h).
3. After each batch has been heated for the appropriate time, an aliquot (50 μL) is removed, poured into 200 μL of MeOH (*see* **Note 11**) to precipitate the proteins, and cooled in an ice bath.

4. After 30 min, the sample is centrifuged and the supernatant is analyzed by HPLC (50 μL injection). The intensity of the peptide or dendrimer peak is noted for each chromatogram. A constant peak intensity over the duration of the experiment indicates that the peptide is stable to human serum.

3.6. Treatment of Cell Cultures With CF-Labeled Peptides or Dendrimers

For all experiments, exponentially growing cells are detached from the culture flasks using a trypsin-0.25% ethylenediamine tetraacetic acid (EDTA) solution and the cell suspension is seeded at different concentrations onto plastic dishes, glass coverslips, eight-well Lab-Teck chambered coverglass, or 96-well plates, depending on the experiment. Experiments are carried out 24 h later, once the confluence has reached approx 60–70%. The carboxyfluoresceinated compounds are dissolved in phosphate-buffered saline (PBS) and sterilized with 0.22 μm filters. The peptides and CF stock solutions are diluted with the cell culture medium. Nonadherent cells are washed away and attached cells are incubated with the peptides in DMEM at 37°C in CO_2 atmosphere or in 25 mM HEPES-buffered DMEM at 4°C.

3.6.1. Plate Fluorimetry Measurements of Cellular Uptake of Peptides or Dendrimers

1. COS-1 and HeLa cells are seeded onto 96-microwell plates at a concentration of 24×10^4 and 5×10^4 cells/cm^2, respectively, for 24 h.
2. The culture medium is discarded and replaced by fresh, serum-free medium containing CF-peptide/dendrimer at different concentrations (from 0.01 μM to 60 μM), using TAT or penetratin as positive controls and CF as a negative control.
3. The cells are then incubated for different periods of time (from 2 to 24 h) at 4°C and at 37°C. In the experiments at 4°C, HEPES is used at a final concentration of 25 mM.
4. At selected intervals of incubation, the medium is discarded and the cells are washed three times with PBS containing 1.1 mM CaCl$_2$ and 1.3 mM MgCl$_2$ (*see* **Note 12**).
5. The cells are lysed by adding 200 μL of lysis buffer to each well (with 0.1% Triton X-100 in 50 mM pH 8.5 Tris buffer for the γ-peptides and with MPER® mammalian protein extraction reagent for the dendrimers).
6. After 20 min, the internalized CF-peptide in the supernatant is analyzed by measuring its fluorescence intensity using a plate fluorimeter with $\lambda_{\text{excitation}} = 485 \pm 20$ nm and $\lambda_{\text{emission}} = 530 \pm 25$ nm.
7. Each experiment is performed in triplicate, and the fluorescence values obtained are corrected by subtracting the fluorescence value of the blank.

3.6.2. Flow Cytometry Measurements of Cellular Uptake of Peptides or Dendrimers

1. COS-1 and HeLa cells are seeded onto 35-mm plates at a concentration of 21.4×10^3 cells/cm^2 for 24 h.
2. The culture medium is discarded and replaced by fresh, serum-free medium containing CF-peptide/dendrimer at different concentrations (from $0.01 \mu M$ to $60 \mu M$), using TAT or penetratin peptides as a positive controls and CF as a negative control.
3. The cells are then incubated for different periods of time (from 2 to 24 h) at 4°C and at 37°C. In the experiments at 4°C, HEPES is used at a final concentration of 25 mM.
4. At selected intervals of incubation, the medium is discarded and the cells are washed three times with PBS containing 1.1 mM CaCl$_2$ and 1.3 mM MgCl$_2$ (*see* **Note 12**).
5. The cells are then detached with trypsin-EDTA 0.25% and are centrifuged for 4 min at 180 g at 4°C.
6. Finally, the supernatant is discarded by aspiration and the cells are resuspended in serum-free medium containing 0.1 mM of propidium iodide (PI) and maintained in an ice bath.
7. The cells are analyzed immediately in the flow cytometer.
8. In order to remove fluorescence from residual CF or CF-peptides bound to the plasma membrane, the pH of the medium/PI solution is brought to 6.0 by the addition of 1 N HCl (*see* **Note 13**), and the cells are re-analyzed in the cytometer.
9. Cells stained with PI are excluded from further analysis. Each sample is tested in triplicate for each condition and the results from independent experiments are normalized by first subtracting the autofluorescence control value from the fluoresence value obtained for each sample and then dividing by the fluorescence value obtained from the CF control under the same experimental conditions. Fluorescence differences between cells before and after addition of HCl are mainly attributed to the cell surface attached peptide.

3.6.3. MTT Cytotoxicity Assay (31)

1. COS-1 and HeLa cells are seeded onto 96-microwell plates at a concentration of 24×10^4 and 5×10^4 cells/cm^2, respectively, for 24 h. In order to avoid saturation in cell growth after 24 h of peptide incubation, COS-1 and HeLa cells are seeded onto 96-microwell plates at a concentration of 7×10^3 cells/cm^2 for 24 h.
2. The culture medium is discarded and replaced by fresh, serum-free medium containing CF-peptide/dendrimer at different concentrations (from $0.01 \mu M$ to $25 \mu M$), using TAT peptide as a positive control and CF as a negative control.
3. The cells are then incubated for different periods of time (from 2 to 24 h) at 37°C under 5% CO$_2$ atmosphere.

4. MTT dissolved in PBS is added 2 h before the end of the incubation. A small volume of the master (ca. 2–5 μL) solution is then directly added to each well until the desired final concentration (0.5 mg/mL) is obtained.
5. After 2 h of incubation with MTT, the medium is discarded by aspiration and 200 μL of isopropanol are added to dissolve formazan, a dark-blue-colored crystalline substance observed in the wells.
6. Absorbance at $\lambda = 570$ nm is measured in a spectrophotometer 30 min after the addition of isopropanol.
7. Cell viability is expressed as a percent representing the ratio of peptide-treated cells to untreated cells, which are used as a control. Each experiment is performed in triplicate.

3.6.4. MTS Cytotoxicity Assay

1. HeLa cells are seeded onto 96-microwell plates at a concentration of 7×10^3 cells/cm^2 and 2×10^4 cells/cm^2, respectively, for 20 h.
2. The culture medium is discarded and replaced by fresh, serum-free medium containing CF-dendrimer at different concentrations (0.4 to 60 μM). Penetratin is used as a positive control and CF as a negative control.
3. The cells are then incubated for 2 h at 37°C under 5% CO_2 atmosphere.
4. The medium is discarded, the cells are then washed with PBS and incubated for additional 2 h with 0.1 mL of fresh medium containing MTS at final concentration of 0.5 mg/mL.
5. The optical density of each well is measured at $\lambda = 490$ nm in a microplate reader. Cell viability is expressed as a percent ratio of cells treated with dendrimer to untreated cells, which are used as a control.

3.6.5. Confocal Laser Scanning Microscopy

1. COS-1 and HeLa cells are seeded onto glass coverslips at 21.4×10^3 cells/cm^2.
2. After 24 h, the culture medium is discarded and replaced by new medium containing different CF-peptides concentrations (from 0.01 μM to 60 μM).
3. The cells are then incubated for different periods of time (from 2 to 24 h) at 37°C under 5% CO_2 atmosphere. For the experiments carried out at 4°C, HEPES buffer is used at final concentration of 25 mM.
4. At selected time intervals of incubation, medium is discarded and cells are washed three times with PBS containing 1.1 mM CaCl$_2$ and 1.3 mM MgCl$_2$ (*see* **Note 12**).
5. Cells are fixed in 3% paraformaldehyde–2% sucrose in 0.1 M phosphate buffer (PB) for 15 min, followed by three washes with PBS.
6. PI (1 μg/mL) staining is performed at room temperature for 15 min in the presence of RNAsa in PBS (1 mg/mL). After that, coverslips are washed three times with PBS containing 1.1 mM CaCl$_2$ and 1.3 mM MgCl$_2$.
7. Cells are then mounted in Mowiol with 2.5% DABCO. The coverslips are carefully inverted into a drop of Mowiol on a microscope slide. Samples are allowed to dry

at room temperature (in dark conditions) and maintained in the dark at 4°C until confocal laser analysis for up to a month.

8. The slides are viewed under phase contrast microscopy in order to locate the cells and with confocal laser scanning microscopy. CF is excited with the 488 nm line of an argon laser and its emission is detected in a range of $\lambda = 515$–530 nm. The microscope settings are maintained identical for each peptide and dose. PI and TR-DX are excited at $\lambda = 543$ nm and detected with a 560 nm long pass filter. In order to avoid crosstalk, the two-fluorescence scanning is performed in a sequential mode.

3.6.6. Confocal Laser Live-Cell Imaging

1. COS-1 and HeLa cells are seeded onto glass bottom Lab-Tek chambers for live-cell imaging $(21.4 \times 10^3$ cells/cm^2).
2. After 24 h, the culture medium is discarded and replaced by new medium containing different CF-peptides concentrations (from $0.01 \mu M$ to $60 \mu M$). For endocytosis experiments, TR-DX (3 mg/mL) is incubated together with CF-peptide.
3. The cells are then incubated for different periods of time (from 15 min to 2 h) at 37°C under 5% CO$_2$ atmosphere.
4. After CF-peptide incubation, the cells are washed three times with PBS containing 1.1 mM CaCl$_2$, 1.3 mM MgCl$_2$, and 25 mM HEPES.
5. The samples are maintained in HBBS buffer containing 25 mM HEPES and images are acquired within the 30 min.
6. Confocal laser scanning microscopy is performed with an Olympus Fluoview 500 confocal microscope equipped with a ×60/1.4 NA plan-apochromatic objective. The CF fluorescence is excited with the 488 nm line of an argon laser and its emission is detected in the range of $\lambda = 515$–530 nm. The microscope settings are maintained for each peptide and dose. PI and TR-DX are excited at $\lambda = 543$ nm and detected with a 560 nm long-pass filter. In order to avoid crosstalk, the two-fluorescence scanning is performed in a sequential mode.

4. Notes

1. Each manual solid-phase synthesis is carried out in a polypropylene syringe equipped with a porous polypropylene disk at the bottom. The volume of the syringe used depends on the initial quantity of dry resin. Typically, resin is added to the syringe, and then the solvent used in the following reaction is added to create a slurry. The resin is washed with the solvent (3 mL of solvent per 1 mL of swollen resin). The mixture is stirred using a Teflon rod for a given time, and after finishing the treatment, the solvent is removed by suction. At the onset of each reaction, the bottom part of the syringe is capped using a septum, at which point the solvents and reagents are added. After manual stirring using a Teflon rod for 3 min, the mixture is allowed to react for a given time with agitation by a shaker.

2. These washes are required in order to remove overincorporated CF. When these washes are not carried out, compounds containing two and three extra units of CF are observed in the crude product. The washes are carried out until the washing solution is uncolored.

3. The γ-amino functions of these building blocks anchored on the resin give a clear positive Kaiser test *(32)*.

4. Although the De Clercq *(33)* test seems more sensitive than the chloranil *(34)* (i.e., the De Clercq gives slightly positive results for cases in which the chloranil reads negative), the chloranil test is easier to handle.

5. The use of methanol in the reduction with NaBH₃CN is essential for solubility reasons. Reactions carried out using only DMF do not work as well as when using a mixture of DMF-MeOH. Furthermore, longer reaction times are needed to obtain quantitative yields.

6. Couplings to the hydroxymethyl resin are monitored with the TosCl-PNBP test for hydroxyl groups *(35)*. Freshly prepared test solution gives a clear positive (brown) in all the cases and a clear negative test (uncolored resin). Solutions that are not fresh give slightly positive results for all the tests, including negative controls.

7. Incorporation of Boc-Pro-OH onto the Pro-Amp(Pro)-resin bound fragment is carried out by *in situ* neutralization with DIEA, using BOP or PyBOP as coupling reagents in the presence of HOBt in DMF to avoid diketopiperazine formation.

8. In order to achieve greater control of the building block synthesis, the introduction of the second proline residue in the orthogonally protected branching unit is carried out in two steps: in the first, a Boc-Pro-OH residue is coupled to the secondary α-amino position of the branching unit; in the second, a Boc-Pro-OH residue is introduced at the primary amino position of the Amp.

9. The simultaneous use of the De Clercq and Kaiser tests during the fragment couplings enables more accurate assessment of coupling results than when either test is used alone. For cases in which recoupling is required, a mixture of anhydrous 1-methyl-2-pyrrolidone (NMP)–DMF (1:1) has been found to facilitate the reaction. Secondary amine couplings are monitored with the Chloranil and De Clercq tests.

10. Other enzyme/peptide ratios were tested. γ-Peptides are not trypsin sensitive and a concentrated solution of the enzyme can be used.

11. Different solvents were tested in order to obtain good protein precipitations and, consequently, clean HPLC chromatograms. The best results were obtained using MeOH and only the peak corresponding to the peptide is observed.

12. These washes are carried out using PBS at the same temperature that the experiment is carried out. The use of different temperatures promoted premature cell detachment.

13. At pH 6.0, extracellular CF fluorescence is quenched without altering cell mechanisms, whereas lower pHs produce rapid cell death. The cells are immediately analyzed after addition of HCl.

References

1. Tréhin, R., and Merkle, H. P. (2004) Chances and pitfalls of cell penetrating peptides for cellular drug delivery. *Eur. J. Phar. Biophar.* **58,** 209–223.
2. Zorko, M., and Langel, Ü. (2005) Cell-penetrating peptides: mechanism and kinetics of cargo delivery. *Adv. Drug Delivery Rev.* **57,** 529–545.
3. Davidson, B. L., and Breakefield, X. O. (2004) Neurological diseases: viral vecors for gene delivery to the nervous system. *Nat. Rev. Neurosci.* **4,** 353–364.
4. Connor, J., and Huang, L. (1985) Efficient cytoplasmatic delivery of a fluorescent dye by pH-sensitive immunoliposomes. *J. Cell. Biol.* **101,** 582–589.
5. Foldvari, M., Mezei, C., and Mezei, M. (1991) Intracellular delivery of drugs by liposomes containing P0 glycoprotein from peripheral nerve myelin into human M21 melanoma cells. *J. Pharm. Sci.* **80,** 1020–1028.
6. Gentile, F. T., Doherty, E. J., Rein, D. H., Shoichet, M. S., Winn, S. R. (1995) Polymer science for macroencapsulation of cells for central nervous system transplantation. *Reactive Polymers* **25,** 207–227.
7. Chakrabarti, R., Wylie, D. E., and Schuster S. M. (1989) Transfer of monoclonal antibodies into mammalian cells by electroporation. *J. Biol. Chem.* **264,** 15, 494–15,500.
8. Leamon, C. P., and Low, P. S. (1991) Delivery of macromolecules into living cells: a method that exploits folate receptor endocytosis. *Proc. Natl. Acad. Sci. USA* **88,** 5572–5576.
9. Sadler, K., Eom, K. D., Yang, J-L., Dimitrova, Y., and Tam, J. P. (2002) Translocating proline-rich peptides from the antimicrobial peptide Bactenecin 7. *Biochemistry* **41,** 14,150–14,157.
10. Singh, D., Kiarash, R., Kawamura, K., LaCasse, E. C., and Gariépy, J. (1998) Penetration and intracellular routing of nucleus-directed peptide-based shuttles (loligomers) in eukaryotic cells. *Biochemistry* **37,** 5798–5809.
11. Brokx, R. D., Bisland, S. K., and Gariépy, J. (2002) Designing peptide-based scaffolds as drug delivery vehicles. *J. Controlled Release* **78,** 115–123.
12. Elmquist, E., Lindgren, M., Bartfai, T., and Langel, Ü. (2001) VE-cadherin-derived cell penetrating peptide, pVEC, with carrier functions. *Exp. Cell Res.* **269,** 237–244.
13. Umezawa, N., Gelman, M. A., Haigis, M. C., Raines, R. T., and Gellman, S. H. (2002) Translocation of a beta-peptide across cell membranes. *J. Am. Chem. Soc.* **124,** 368–369.
14. Rueping, M., Mahajan, Y., Sauer, M., and Seebach, D. (2002) Cellular uptake studies with beta-peptides. *ChemBioChem* **3,** 257–259.
15. Potocky, T. B., Menon, A. K., and Gellman, S. H. (2003) Cytoplasmic and nuclear delivery of a TAT-derived peptide and a beta-peptide after endocytic uptake into HeLa cells. *J. Biol. Chem.* **278,** 50,188–50,194.
16. Garcia-Echeverria, C., and Ruetz, S. (2003) Beta-Homolysine oligomers: a new class of Trojan carriers. *Bioorg. Med. Chem. Lett.* **13,** 247–251.

17. Farrera-Sinfreu, J., Zaccaro, L., Vidal, D., et al. (2004) A new class of foldamers based on *cis*-γ-amino-L-proline. *J. Am. Chem. Soc.* **126,** 6048–6057.
18. Farrera-Sinfreu, J., Giralt, E., Castel, S., Albericio, F., and Royo, M. (2005) Cell-penetrating *cis*-γ-amino-L-proline-derived peptides. *J. Am. Chem. Soc.* **127,** 9459–9468.
19. Crespo, L., Sanclimens, G., Royo, M., Giralt, E., and Albericio, F. (2002) Branched poly(proline) peptides: an efficient new approach to the synthesis of repetitive branched peptides. *Eur. J. Org. Chem.* **11,** 1756–1762.
20. Crespo, L., Sanclimens, G., Montaner, B., et al. (2002) Peptide dendrimers based on polyproline helices. *J. Am. Chem. Soc.* **124,** 8876–8883.
21. Sanclimens, G., Crespo, L., Giralt, E., Royo, M., and Albericio, F. (2004) Solid-phase synthesis of second-generation polyproline dendrimers. *Biopolymers (Pept. Sci.)* **76,** 283–297.
22. Sanclimens, G., Crespo, L., Giralt, E., Albericio, F., and Royo, M., (2005) Preparation of de novo globular proteins based on proline dendrimers. *J. Org. Chem.* **70,** 6274–6281..
23. Sanclimens, G., Shen, H., Giralt, E., Albericio, F., Saltzman, M. W., and Royo, M. (2005) Synthesis and screening of a small library of proline based biodendrimers for use as delivery agents. *Biopolymers* **80,** 800–814.
24. Fernández-Carneado, J., Kogan, M. J., Castel, S., Pujals, S., and Giralt, E. (2004) Potential peptide carriers: amphipathic proline-rich peptides derived from the N-terminal domain of γ-zein. *Angew. Chem. Int. Ed.* **43,** 1811–1814.
25. Fernández-Carneado, J., Kogan, M. J., Pujals, S., and Giralt, E. (2004) Amphipathic peptides and drug delivery. *Biopolymers (Pept. Sci.)* **76,** 196–203.
26. Foerg, C., Ziegler, V., Fernández-Carneado, J., et al. (2005) Decoding the entry of two novel cell-penetrating peptides in HeLa cells: lipid raft-mediated endocytosis and endosomal escape. *Biochemistry* **44,** 72–81.
27. Fernández-Carneado, J., Kogan, M. J., Van Mau, N., et al. (2005) Fatty acyl moieties: improving Pro-rich peptide uptake inside HeLa cells. *J. Pept. Res.* **65,** 580–590.
28. Lloyd-Williams, P., Albericio, F., and Giralt, E. (1997) *Chemical Approaches to the Synthesis of Peptides and Proteins.* CRC, Boca Raton, FL.
29. Feichtinger, K., Zapf, C., Sings, H. L., and Goodman, M. (1998) Diprotected triflylguanidines: a new class of guanidinylation reagents. *J. Org. Chem.* **63,** 3804–3805.
30. Rose, K., and Vizzavona, J. (1999) Stepwise solid-phase synthesis of polyamides as linkers. *J. Am. Chem. Soc.* **121,** 7034–7038.
31. Liu, Y., Peterson, D. A., Kimura, H., and Schubert, D. (1997) Mechanism of cellular 3-(4,5-dimethylthiazol-2-yl)-2,5-diphenyltetrazolium bromide (MTT) reduction. *J. Neurochem.* **69,** 581–593.
32. Kaiser, E., Colescott, R. L., Bossinger, C. D., and Cook, P. I. (1970) Color test for detection of free terminal amino groups in solid-phase synthesis of peptides. *Anal. Biochem.* **34,** 594–598.

33. Madder, A., Farcy, N., Hosten, N. G. C., et al. (1999) A novel sensitive colorimetric assay for visual detection of solid-phase bound amines. *Eur. J. Org. Chem.* 2787–2791.
34. Christensen, T. (1979) A qualitative test for monitoring coupling completeness in solid-phase peptide synthesis using chloranil. *Acta Chem. Scan.* **33,** 760–766.
35. Kuisle, O., Lolo, M., Quiñoá, E., and Riguera, R. (1999) Monitoring the solid-phase synthesis of depsides and depsipeptides. A color test for hydroxyl groups linked to a resin. *Tetrahedron* **55,** 14,807–14,812.

10

Peptide-Mediated Targeting of Liposomes to Tumor Cells

Evonne M. Rezler, David R. Khan, Raymond Tu, Matthew Tirrell, and Gregg B. Fields

Summary

One of the biggest obstacles for efficient drug delivery is specific cellular targeting. Liposomes have long been used for drug delivery, but do not possess targeting capabilities. This limitation may be circumvented by surface coating of colloidal delivery systems with peptides, proteins, carbohydrates, vitamins, or antibodies that target cell surface receptors or other biomolecules. Each of these coatings has significant drawbacks. One idealized system for drug delivery combines stabilized "protein module" ligands with a colloidal delivery vehicle. Prior studies have shown that peptide-amphiphiles, whereby both a peptide "head group" and a lipid-like "tail" are present in the same molecule, can be used to engineer collagen-like triple-helical or α-helical miniproteins. The tails serve to stabilize the head group structural elements. These peptide-amphiphiles can be designed to bind to specific cell surface receptors with high affinity. Structural stabilization of the integrated targeting ligand in the peptide-amphiphile system equates to prolonged in vivo stability through resistance to proteolytic degradation. Liposomes have been prepared incorporating a melanoma targeting peptide-amphiphile ligand, and shown to be stable with retention of peptide-amphiphile triple-helical structure. Encapsulated fluorescent dyes are selectively delivered to cells. In this chapter we describe the methods and techniques employed in the preparation and characterization of peptide-amphiphiles and peptide-amphiphile-targeted large and small unilamellar vesicles (LUVs and SUVs). Fluorescence microscopy is subsequently utilized to examine the targeting capabilities of peptide-amphiphile LUVs, which should allow for improved drug selectivity towards melanoma vs normal cells based on differences in the relative abundance of the targeted cell surface receptors.

Key Words: Liposome; drug delivery; peptide-amphiphile; melanoma; targeted drug delivery; fluorescence microscopy.

From: *Methods in Molecular Biology, vol. 386: Peptide Characterization and Application Protocols*
Edited by: G. Fields © Humana Press Inc., Totowa, NJ

1. Introduction

Cancer treatment by chemotherapy is typically accompanied by deleterious side effects, attributed to the toxic action of chemotherapeutics on proliferating cells from nontumor tissues. The general challenges for improved drug delivery are (a) design of components that permit the delivery vehicle to bypass multiple anatomic and cellular barriers to reach the target site, (b) synthesis of ligands that are selective for particular cell types, and (c) development of mechanisms for transfer of encapsulated drug into the target *(1,2)*. Improved selectivity of chemotherapeutics would allow for lower dose administration and, presumably, decreased side effects. Encapsulation of therapeutic agents inside colloidal delivery systems, which include liposomes, nanoparticles, micelles, and polymeric microparticles, has been a widely applied formulation strategy *(3)*. For example, the prospect of using liposomes (lipid vesicles) as biomimetic enclosures capable of carrying a therapeutic agent into tissues was first suggested in the 1960s. A drug is stored within the liposome, preventing degradation (**Fig. 1**). Upon coming into contact with a cell membrane, the liposome itself can reorganize the membrane, allowing for delivery of the drug. Because of their size, liposomes can extravasate through gaps in angiogenic blood vessels to reach tumor interstitial space *(2)*. One potential drawback is that the liposome itself is not targeted for a specific cell type. Thus, the inability of these systems to associate with specific cells before uptake by macrophages/reticuloendothelial system (RES) prevented their extensive application in drug delivery *(4)*. RES uptake of liposomes can be reduced by polyethylene glycol (PEG)-coating, allowing for longer circulation time *(5–9)*. Although not targeted, PEG-stabilized liposomes are in clinical use for doxorubicin delivery to Kaposi's sarcoma patients (DaunoXome; Gilead Sciences) and ovarian carcinoma patients (Doxil; ALZA) *(2,10,11)*.

The delivery potential of liposomes has been shown to increase by the attachment of liposomes to cell surfaces *(12)*. Thus, the pursuit of targeted liposomes that can be co-localized to cell surfaces and then deliver more of their cargo more efficiently has become a very promising area of research. Liposomes can be targeted to tumors by surface coating with peptides, proteins, carbohydrates, vitamins, or antibodies that are ligands for tumor cell surface receptors or other biomolecules (**Fig. 1B**). For example, fibrinogen, a ligand for the αIIbβ3 integrin, has been conjugated to nanoparticles and liposomes, which in turn deliver encapsulated radiopharmaceuticals to tumor blood vessels *(13)*. Peptides derived from P_o protein increased liposome binding to melanoma cells; levels of increased binding correlated to cellular intercellular adhesion molecular (ICAM)-1 expression *(14)*. A cyclic peptide inhibitor of matrix

A

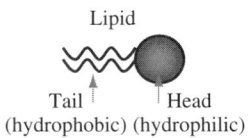

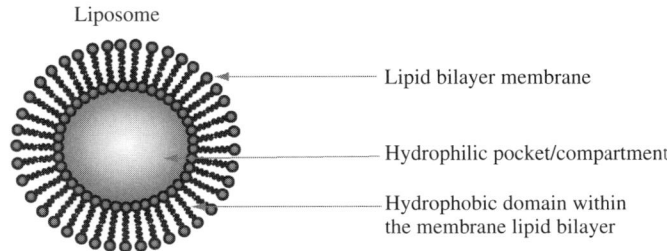

B

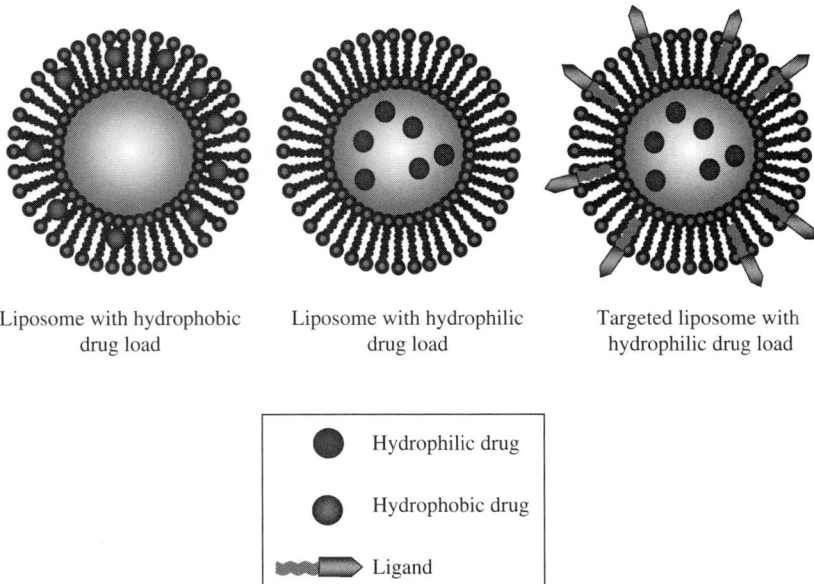

Fig. 1. Liposomes as delivery vehicles. (**A**) Lipid schematic (top) and liposome schematic (bottom) indicates hydrophilic compartment and hydrophobic inner-leaf of the lipid bilayer. (**B**) Drug or payload encapsulation into liposome hydrophilic compartment or incorporation into lipid bilayer.

metalloproteinase (MMP)-2 and -9 increased the tumor cell cytotoxicity of adriamycin-containing liposomes *(15)*. Hyaluronan-coated liposomes have been used to enhance the specificity of liposome-encapsulated doxorubicin for CD44-expressing tumor cells *(16,17)*. Folic acid has been used to target liposomes to folate receptor-overexpressing tumor cells *(18)*. Immunoliposomes targeted to human carcinoembryonic antigen (CEA) have been shown to accumulate in solid tumors *(8)*. Anti-HER2 immunoliposomes are internalized by HER2-overexpressing breast cancer cells *(19)*. A cationic lipid (3,5-dipentadecycloxybenzamidine hydrochloride) has been used to target cisplatin-containing liposomes to tumor cell surface chondroitin sulfate *(20)*. Cationic liposomes have also been shown to selectively deliver paclitaxel to tumor endothelial cells, preventing tumor growth and invasiveness in a humanized severe combined immunodeficient (SCID) mouse melanoma model *(21,22)*. Despite these numerous successes, there are significant drawbacks to peptide, protein, carbohydrate, vitamin, or antibody coating of liposomes. Peptides may be readily degraded prior to reaching desired targets, proteins are also subject to proteolysis and bind to multiple receptors, carbohydrates may be bound nonspecifically or by multiple receptors, vitamins are readily metabolized, and antibodies are limited by the small subsets of tumors that can be targeted and poor biodistribution into solid tumors *(23–25)*. In addition, immunoliposomes can be taken up even more rapidly by the RES than liposomes *(8)*. Liposomes may be encapsulated within larger vesicles, creating a structure known as a vesosome, which may prolong the circulation time and therefore the timescale of delivery of the drug within the interior liposomes *(26)*.

Another approach to improve drug selectivity is the construction of pro-drugs, from which the drug is released in the target tissue. This includes the concept of enzyme-activated targeting of liposomes *(27)*. The surface of liposomes may be modified to incorporate a sequence hydrolyzed by tumor-associated proteases. Upon proteolysis, the surface of the liposome becomes more positively charged, promoting liposome fusion (either with the tumor cell plasma membrane or endosomal membrane following endocytosis) and resulting in the delivery of a chemotherapeutic agent stored inside the liposome. Alternatively, hydrolysis of a protease sensitive protein or peptide sequence may sufficiently destabilize the liposome to allow drug release. For example, destabilization by trypsin has been achieved via incorporation of glycophorin into DOPE liposomes *(28)*, whereas MMP-9 "uncorked" liposomes possessing triple-helical models of the type I collagen α1(I)772–780 sequence *(29)*. Analogously, an MMP-sensitive sequence has been used for hydrogel drug delivery in vivo *(30)*. The problem with most pro-drug delivery systems is a lack of stability at other tissues and organs due to nonspecific activation. In addition, following proteolysis, the released drug may have significantly less activity than the unmodified drug *(31)*.

For both cell-surface-targeted and enzyme-activated delivery systems, specificity could be significantly improved if a peptide coating was introduced that was stable to the in vivo environment during delivery whilst maintaining a high degree of selectivity for a specific receptor or enzyme. As described above, there are several problems associated with simple peptide ligands serving as liposome or other drug delivery vehicle surface modifiers. First, the specificity and affinity of such ligands is usually not high. This problem can be obviated by induction of well-defined secondary and/or tertiary structures within peptide ligands, often accomplished via cyclization. Conformational restriction of Arg-Gly-Asp (RGD) sequences can provide receptor selectivity while enhancing affinity for a specific receptor *(32,33)*. Similarly, triple-helical ligand conformation may enhance the promotion of cell adhesion, spreading, and signaling activities by collagen-derived sequences *(34–36)*. Second, peptide ligands are extremely susceptible to proteolysis. The conformational restriction or stereochemical manipulation *(32,37)* of such sequences often reduces general proteolytic activity. Thus, one approach for improving peptide in vivo activity is to create biomolecules with distinct structural elements, i.e., that are conformationally constrained. The peptide-amphiphile or "PA" represents such a construct. The next two sections focus on PAs as targeting molecules, how this field has developed, procedures, and future challenges (such as proving targeting specificity, etc.).

PAs provide an approach by which structural elements such as collagen-like triple-helices, α-helices, or β-turns can be induced and/or stabilized. This approach utilizes non-covalent association of peptide segments. Pseudo-lipids (mono- and dialkyl chains) are covalently attached to peptides to create PAs that then associate via hydrophobic interactions to each other **(Fig. 2)** *(33,38–41)*. PAs are advantageous in that they present a multivalent ligand *(42)* that is chemically well defined, avoiding loss of activity that can occur during nonspecific coupling of peptides to lipids *(43)*. PA sequences, either individually or in combination *(44–47)*, may prove to be effective for enhancing biomaterial biocompatibility and function. In addition, PAs have been applied for combinatorial surface chemistry *(48)* and construction of nanofibers *(49)*. The amphiphilic character of PAs allows for the control of assembled structures by manipulating their molecular composition *(50)*. For example, the thermal stability of triple-helical and α-helical PA head groups can be modulated by the length of the lipophilic moiety *(35,38–40,45)*. Conversely, alterations in the

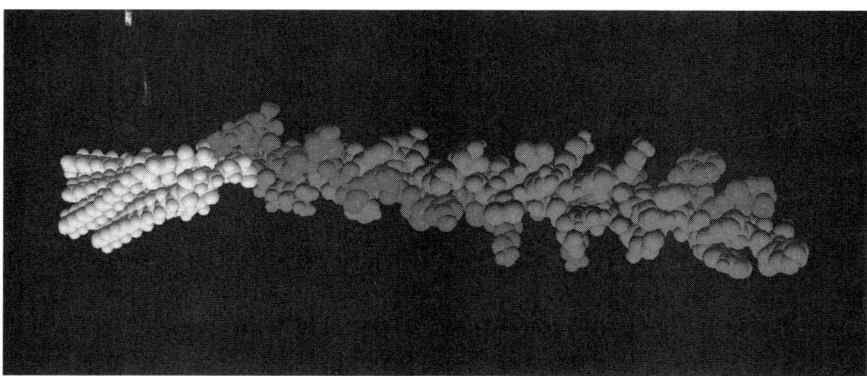

Fig. 2. Triple-helical PA composed of $(C_{16})_2$-Glu-C_2-(Gly-Pro-Hyp)$_4$-Gly-Val-Lys-Gly-Asp-Lys-Gly-Asn-Pro-Gly-Trp-Pro-Gly-Ala-Pro-NH$_2$. (Reproduced from **ref. 38**, by permission of the *Journal of the American Chemical Society*.)

pseudo-lipid tail composition effects PA aggregate structures *(40,50)*. Desirable peptide head group melting temperature (T_m) values can be achieved for in vivo use, as both triple-helical and α-helical PAs have been constructed with T_m values ranging from 30 to 70°C *(35,36,38–40,45,51–53)*. The range of PA thermal stabilities has allowed for their use in characterization of melanoma cell surface receptors for type IV collagen, as discussed in the next section. In addition, PAs have far greater stability to proteolysis than most peptides, as only certain MMPs are capable of efficiently hydrolyzing triple-helices *(54)*. Thus, the combination of thermal and proteolytic stability and cellular recognition indicate that PAs could potentially be applied to enhance the specificity of drug delivery vehicles such as liposomes.

The best characterized cell surface adhesion molecules are integrins, which are heterodimeric proteins composed of one α and one β subunit **(Fig. 3)**. The collagen binding integrins include α1β1, α2β1, α3β1, α10β1, and α11β1 *(55,56)*. Three of these collagen-binding integrins (α1β1, α2β1, and α3β1) are present on melanoma cells *(57,58)*. The α1β1 and α2β1 integrins are upregulated in metastatic melanoma, whereas the α3β1 integrin is upregulated in both primary and metastatic melanoma *(59–62)*.

In an effort to better understand the roles of individual integrins in the invasion process, triple-helical binding sites for the α1β1, α2β1, and α3β1 integrins have been identified within type IV collagen. The α1β1 integrin simultaneously binds Asp441 from two α1(IV) chains and Arg458 from the α2(IV) chain *(63,64)*. The Gly-Phe-Hyp-Gly-Glu-Arg motif, in triple-helical conformation, has

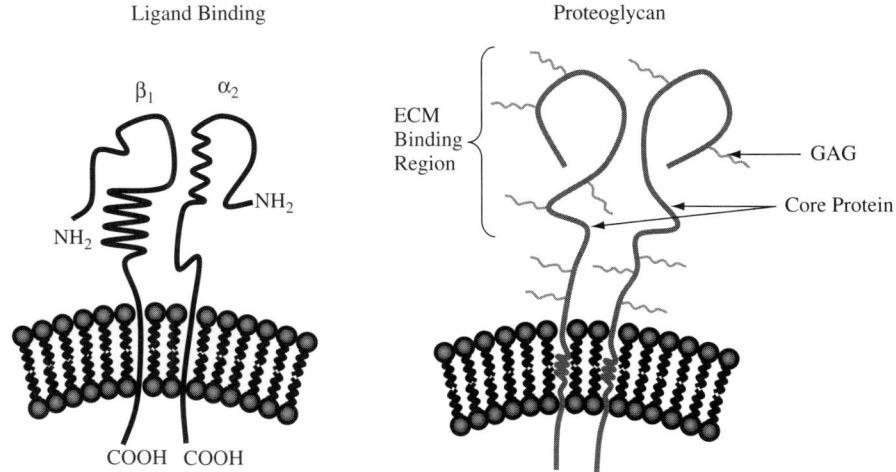

Fig. 3. Examples of cell surface adhesion molecules: (left) α2β1 integrin, (right) cell surface proteoglycan.

been shown to bind to the α2β1 integrin *(65–67)*. This motif is found within type IV collagen at α1(IV)385–390; a triple-helical model of α1(IV)382–393 binds to melanoma cells *(68,69)*. The melanoma cell α3β1 integrin binds to α1(IV)531–543 *(34,58,70)*. Binding of the α1β1 and α2β1 integrins is conformationally dependent, whereas binding of α3β1 is not *(55,70)*.

Melanoma cells have also been shown to possess at least two distinct cell surface chondroitin sulfate (CS) proteoglycans **(Fig. 3)**, CD44 and melanoma-associated proteoglycan/melanoma chondroitin sulfate proteoglycan (MPG/NG2/MCSP) *(71,72)*. CD44 (also known as CD44H and CD44s) is expressed by epidermal melanocytes in both an unmodified and CS-modified form *(73)*. Human CD44 is initially glycosylated via five potential N-linked and seven potential O-linked carbohydrate sites, resulting in a core protein of 85–95 kDa *(74–76)*. CS modification further increases the molecular mass to 180–200 kDa *(75)*.

A triple-helical ligand for CD44 was identified by a combination of methods, including (a) cell adhesion and spreading assays using triple-helical α1(IV)1263–1277 and an Asp^{1266}Abu variant, (b) inhibition of cell adhesion and spreading assays, and (c) triple-helical α1(IV)1263–1277 affinity chromatography with whole cell lysates and glycosaminoglycans *(77)*. Interaction of CD44 with this ligand was strongly dependent upon triple-helical conformation *(36,78)*.

How well the above peptide structures mimic the interactions between native proteins and cell receptors has been examined in detail, as binding affinities have been determined for triple-helical peptides and native collagens. For example, the $\alpha1\beta1$ integrin binds to $\alpha1(I)127–138$ and $\alpha1(I)496–507$ triple-helical peptides with $K_D = 7.0$ and 1.7 μM, respectively *(79)*. The $\alpha2$ I-domain binds to $\alpha1(I)127–138$ and $\alpha1(I)496–507$ triple-helical peptides with $K_D = 0.40$ and 1.1 μM, respectively *(79)*. The $\alpha1\beta1$ integrin binds to $[\alpha2(IV)454–465][\alpha1(IV)437–448][\alpha1(IV)437–448]$ triple-helical peptides with $K_D = 20 \mu M$ *(64)*. These results reflect similar affinities to those for integrin binding to collagen. For example, the recombinant $\alpha2$ A-domain was shown to bind to type I collagen with K_D approx 6–10 μM *(80)*. The $\alpha1$ A-domain has two classes of binding sites within type I collagen, with the higher-affinity site exhibiting a $K_D = 0.11 \mu M$ *(80)*. Thus, triple-helical peptides are accurate mimics of collagen activity, and could be used to selectively target liposomes towards collagen-binding cell surface receptors overexpressed in melanoma **(Fig. 4)**.

Preparations of the $(C_{16})_2$-Glu-C_2-(Gly-Pro-Hyp)$_4$-$[\alpha1(IV)1263–1277]$ PA in solution were found to form micelles *(50)*. Mixed liposomes have been prepared using the PA and dilauryl phosphatidylcholine (DLPC). Unilamellar vesicles composed of 0.15 mg/mL $(C_{16})_2$-Glu-C_2-(Gly-Pro-Hyp)$_4$-$[\alpha1(IV)1263–1277]$ and 12.5 mg/mL DLPC were extruded, and the stability of the liposomes was studied by cryo transmission electron microscopy (cryo-TEM). The morphology of liposomes could be directly visualized with cryo-TEM, as fluid specimens are rapidly fixed. Liposomes composed of just DLPC were not very stable, as the small size distribution of liposomes seen on day 1 had disappeared by day 42 **(Fig. 5)**. The liposomes had formed larger aggregates. In contrast, liposomes composed of $(C_{16})_2$-Glu-C_2-(Gly-Pro-Hyp)$_4$-$[\alpha1(IV)1263–1277]$ and DLPC appeared to be stable. The small size distribution seen on day 1 was retained by day 42 **(Fig. 5)**. Stable PA liposomes were also formed by a mixture of distearoyl-phosphatidyl choline (DSPC):distearoyl-phosphatidyl glycerol (DSPG):cholesterol:C_{16}-(Gly-Pro-Hyp)$_4$-$[\alpha1(IV)1263–1277]$-(Gly-Pro-Hyp)$_4$-NH$_2$ (4:1:5:0.1).

Subsequently, dimyristoyl phosphatidyl choline (DMPC) liposomes functionalized with the $(C_{16})_2$-Glu-C_2-(Gly-Pro-Hyp)$_4$-$[\alpha1(IV)1263–1277]$ PA were extruded through polycarbonate membranes to approx 110 nm, as verified by dynamic light scattering. A fluorescent amphiphilic dye molecule (Texas Red-DHPE) was also incorporated to image and quantify liposome binding to cellular surfaces. Triple-helical structure in the liposomal assemblies was examined by circular dichroism (CD) spectroscopy. The PA in liposomes possesses a super-secondary structure with a high triple-helical content **(Fig. 6)** *(81)*.

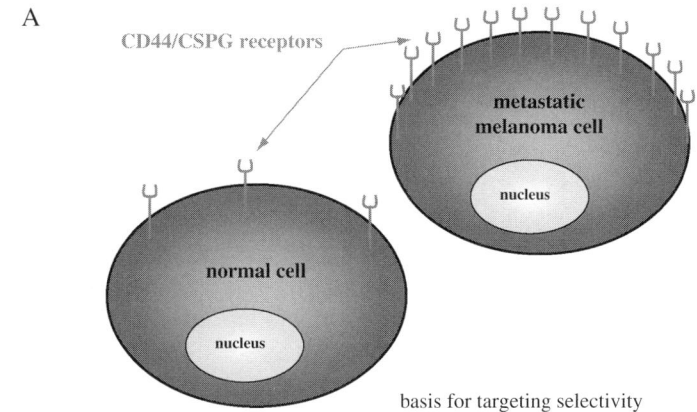

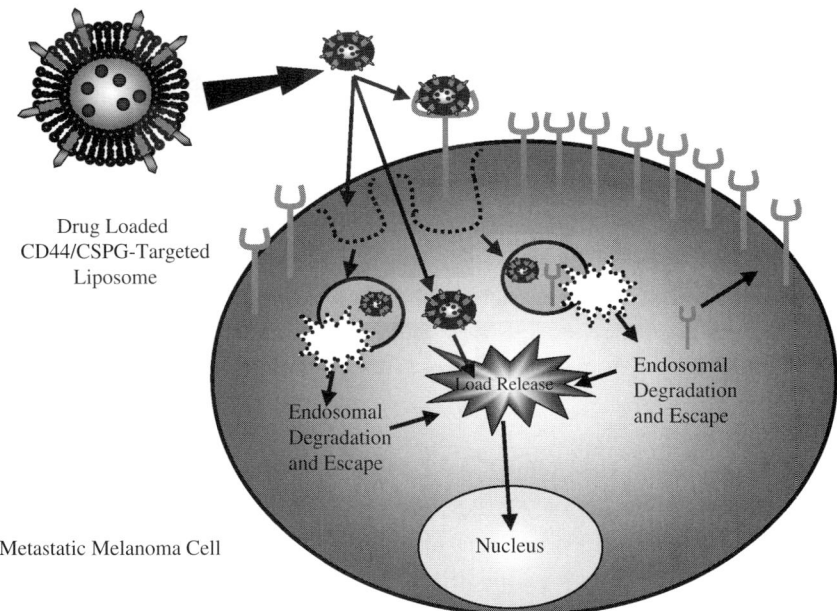

Fig. 4. Targeting liposomes for melanoma cells. (**A**) Targeting based on relative abundance of collagen-binding receptors. (**B**) Mechanisms for drug internalization following delivery by targeted liposomes.

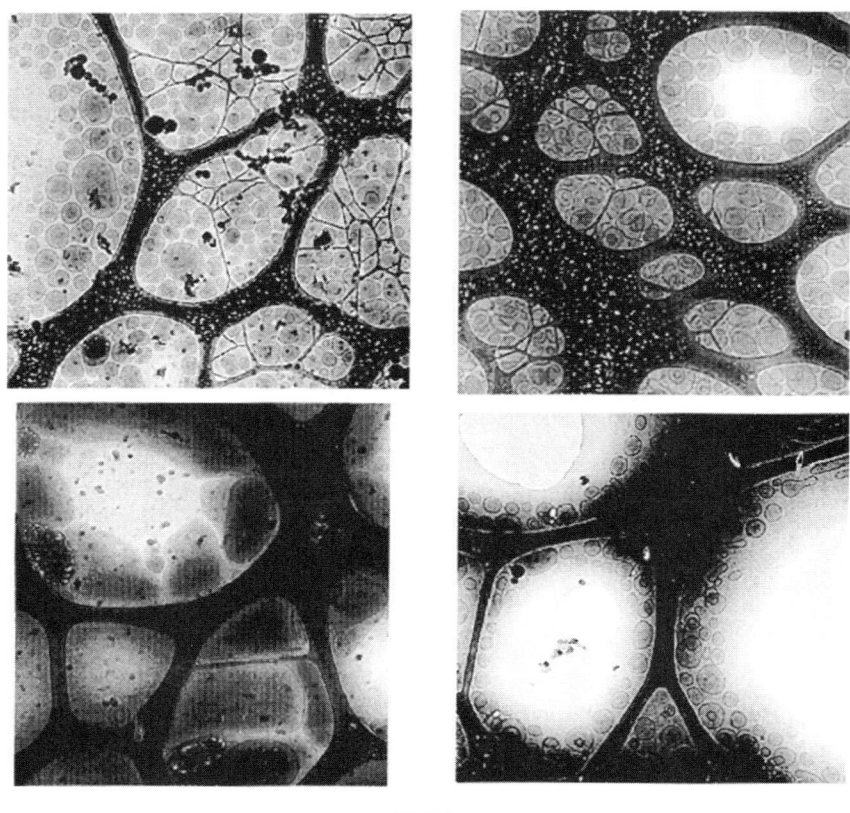

200 nm

Fig. 5. Analysis of liposome stability by cryo-transmission electron microscopy. Liposomes were composed of (left) DLPC or (right) a mixture of 0.15 mg/mL $(C_{16})_2$-Glu-C_2-(Gly-Pro-Hyp)$_4$-[α1(IV)1263–1277] and 12.5 mg/mL DLPC at 1 d (top panels) and 42 d (bottom panels).

Initial studies examined whether PAs are capable of promoting the binding of liposomes to cells *(81)*. DMPC liposomes containing a 5% mol fraction of the $(C_{16})_2$-Glu-C_2-(Gly-Pro-Hyp)$_4$-[α1(IV)1263–1277] PA were compared to DMPC liposomes alone using differential interference contrast (DIC) and fluorescence microscopy. The PA liposome has significantly enhanced cellular uptake compared to the liposome alone **(Fig. 7)**. The incorporation of the fluorescent dye could be quantified using fluorimetry **(Fig. 8)**.

The presence of the [α1(IV)1263–1277] sequence in DSPG:DSPC:cholesterol: C_{16}-(Gly-Pro-Hyp)$_4$-[α1(IV)1263–1277]-(Gly-Pro-Hyp)$_4$-NH$_2$ (4:1:5:0.1)

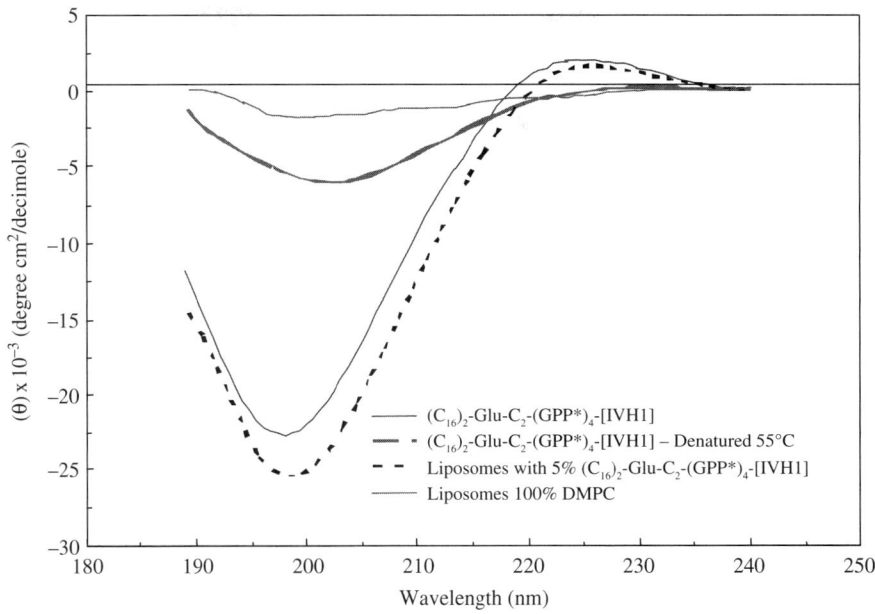

Fig. 6. Circular dichroism spectroscopic analysis of $(C_{16})_2$-Glu-C_2-(Gly-Pro-Hyp)$_4$-[α1(IV)1263–1277] or DMPC liposomes containing 5% $(C_{16})_2$-Glu-C_2-(Gly-Pro-Hyp)$_4$-[α1(IV)1263–1277]. Positive molar ellipticity at λ = 222–227 nm and strongly negative molar ellipticity at λ = 195–200 nm is characteristic of triple-helical structure. (Reproduced from **ref. 81**, by permission of American Pharmaceutical Review.)

liposomes was verified by fluorescence spectra of the Trp residue. CD44-positive cells were treated with rhodamine loaded DSPG:DSPC:cholesterol liposomes, with or without incorporated PA, and incubated at 37°C for 2 h. The liposome treated, unfixed cells were then stained with 4′, 6′-diamidino-2-phenylindole hydrochloride (DAPI) (nucleus dye) and fluorescein isothiocyanate (FITC) (nonspecific membrane dye) and visualized using fluorescence microscopy. Localization of DSPG:DSPC:cholesterol PA liposomes to cell membranes was observed (**Fig. 9**).

Subsequently, a variety of liposomes and PA liposomes (that include PEG lipids of various molecular weights) have been extruded. The presence of the hydrophilic polymer creates a steric barrier, which has been shown to dramatically increase the residence time of the liposomes in vivo (see earlier discussion). This "stealth" effect is desirable, but must be tuned such that the benefits of the barrier will not override the targeting effect of the PA. Melanoma cells were found to adhere and spread on mixtures of the PA with PEG lipids

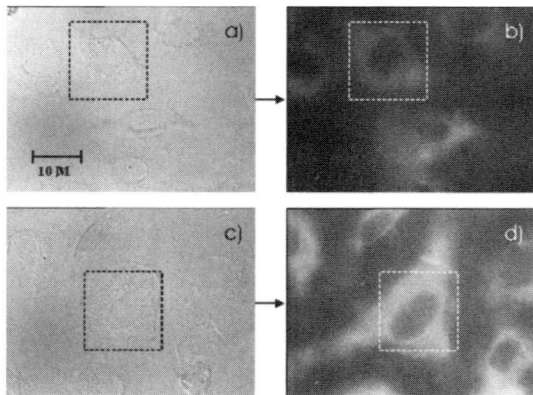

Fig. 7. Differential interference contrast **(A,C)** and fluorescence **(B,D)** microscopic images of cells complexed with fluorescently labeled liposomes (Texas Red DHPE, <1%), where liposomes are composed of pure DMPC **(A,B)** or 95% DMPC and 5% $(C_{16})_2$-Glu-C_2-(Gly-Pro-Hyp)$_4$-[α1(IV)1263–1277] **(C,D)**. The boxes highlight individual cells. (Reproduced from **ref. *81***, by permission of American Pharmaceutical Review.)

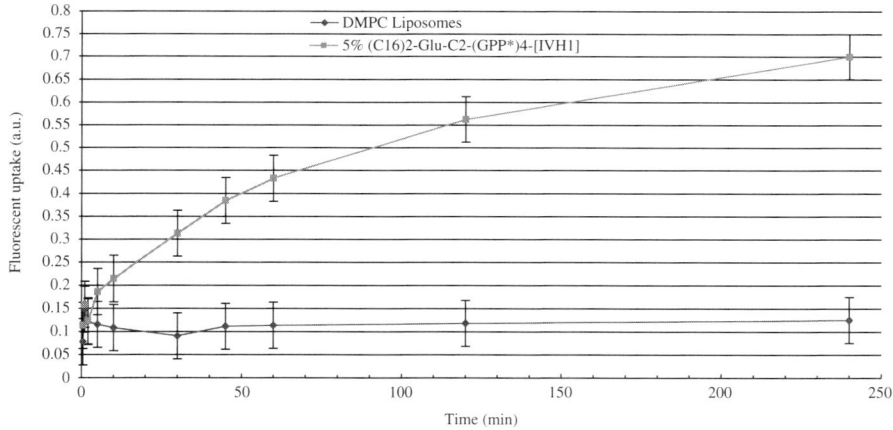

Fig. 8. Fluorescent dye (Texas Red DHPE, <1%) uptake by cells treated with pure DMPC liposomes or 95% DMPC and 5% $(C_{16})_2$-Glu-C_2-(Gly-Pro-Hyp)$_4$-[α1(IV)1263–1277] liposomes.

Fig. 9. DSPG:DSPC:cholesterol PA liposomes (red) co-localized with CD44-positive cells.

possessing PEG chains of 120 or 750 Da. In contrast, cells adhered but did not spread on the mixture containing a 2000 Da PEG, and did not adhere to pure PEG lipids nor the mixture containing a 5000 Da PEG *(82)*. A similar ligand "masking" effect was seen when PEG lipids were used in combination with the HIV-1 TAT peptide for liposomal delivery *(83)*. Neutron reflectivity analysis indicated that the $[\alpha1(IV)1263-1277]$ PA had a head group length of 8.8 nm. When $[\alpha1(IV)1263-1277]$ was mixed with lipids of similar head group length or shorter, melanoma cells adhered to the surface. However, cell spreading was only observed in mixtures where the lipid head group length was considerably shorter (>5 nm) than $[\alpha1(IV)1263-1277]$ (i.e., PEG 120 or 750 Da). These studies indicated that melanoma cells require a $[\alpha1(IV)1263-1277]$ surface in which virtually the entire ligand sequence is readily accessible for promotion of optimum cell spreading, but only a part of the sequence is required for binding to the receptor (CD44/CSPG). The targeting capability of the PA liposomes can indeed be masked by using PEG molecules of 2000 or 5000 Da **(Fig. 10)** *(81)*.

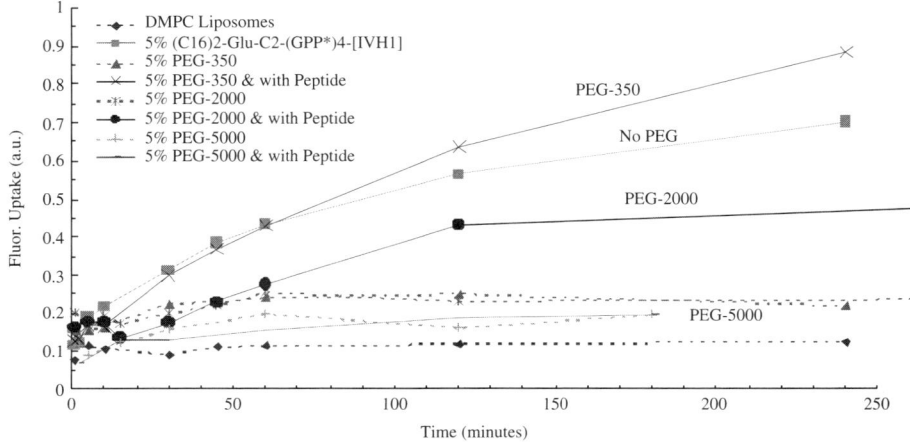

Fig. 10. Fluorescent dye (Texas Red DHPE, < 1%) uptake by cells treated with pure DMPC liposomes, DMPC/(C$_{16}$)$_2$-Glu-C$_2$-(Gly-Pro-Hyp)$_4$-[α1(IV)1263–1277] peptide-amphiphile liposomes, or DMPC/(C$_{16}$)$_2$-Glu-C$_2$-(Gly-Pro-Hyp)$_4$-[α1(IV)1263–1277]/polyethylene glycol peptide-amphiphile liposomes. (Reproduced from **ref. *81***, by permission of American Pharmaceutical Review.)

2. Materials

1. The appropriately protected amino acids, 1-hydroxybenzotriazole (HOBt), *N*-[(1*H*-benzotriazol-1-yl)(dimethylamino)methylene]-*N*-methylmethanaminium hexafluorophosphate *N*-oxide (HBTU), and Rink amide MBHA resin are all obtained from Novabiochem (La Jolla, CA), whereas *N, N*-dimethylformamide (DMF), *N*,*N*-diisopropylethylamine (DIEA), and *N*-methylpyrrolidine (NMP) are obtained from Fisher Scientific (Malvern, PA).
2. The monoalkyl chains hexanoic acid [CH$_3$-(CH$_2$)$_4$-CO$_2$H, designated C$_6$], octanoic acid [CH$_3$-(CH$_2$)$_6$-CO$_2$H, designated C$_8$], decanoic acid [CH$_3$-(CH$_2$)$_8$-CO$_2$H, designated C$_{10}$], dodecanoic acid [CH$_3$-(CH$_2$)$_{10}$-CO$_2$H, designated C$_{12}$], tetradecanoic acid [CH$_3$-(CH$_2$)$_{12}$-CO$_2$H, designated C$_{14}$], palmitic acid [CH$_3$-(CH$_2$)$_{14}$-CO$_2$H, designated C$_{16}$], and stearic acid [CH$_3$-(CH$_2$)$_{16}$-CO$_2$H, designated C$_{18}$] are purchased from Aldrich. The dialkyl ester (C$_{12}$)$_2$-Glu-C$_2$ and (C$_{16}$)$_2$-Glu-C$_2$ chains are prepared as described previously *(84)*. Briefly, synthesis first involves the acid-catalyzed condensation of Glu with the appropriate fatty acid alcohol to form the dialkyl ester of Glu. The free amino group of Glu is then treated with succinic anhydride and triethylamine to create a free carboxylic acid. The dialkyl ester tail may then be used directly for solid-phase synthesis or converted to a *p*-nitrophenyl ester with *p*-nitrophenol and dicyclohexylcarbodiimide and then coupled.

3. PAs (**Table 1**) are prepared utilizing well-established 9-fluorenylmethoxycarbonyl (Fmoc) solid-phase methodology on an ABI 433A Peptide Synthesizer (*see* **Note 1**) *(35,38–40,45,52,53,84)*.
4. The cleavage and side-chain deprotection of PA-resins employs H_2O–trifluoroacetic acid (TFA) (1:19) or ethanedithiol–thioanisole–phenol–water–TFA (2.5:5:5:5:82.5) as previously described *(85,86)*. All of these chemicals are available from ACROS Organics (Fairlawn, NJ) with the exception of phenol, which is obtained from Fisher Scientific (Malvern, PA).
5. The phospholipids used in the preparation of the vesicles are DSPC, dipalmitoylphosphatidylcholine (DPPC), DMPC, DLPC, DSPG, and dioleoylphosphatidylcholine (DOPC). All phospholipids and cholesterol are purchased from Avanti Polar Lipids (Birmingham, AL).
6. Other chemicals and solvents used in the synthesis of the vesicles, such as methanol, chloroform, and *tert*-butyl ether, are purchased from either Fisher Scientific (Pittsburg, PA), Sigma Chemicals Co. (St. Louis, MO), or Aldrich (Milwaukee, WI).
7. The fluorophores rhodamine 6G and fluorescein are obtained from ACROS Organics (Fairlawn, NJ) or Sigma Chemicals Co. (St. Louis, MO). Both the rhodamine and fluorescein stock solutions ($400\ \mu M$ in phosphate-buffered saline [PBS], 1X, pH 7.4) are stored in the dark at 4°C for no more then 3 mo.
8. MLVs are extruded through a Lipex Extruder (Northern Lipids Inc., Vancouver, British Columbia, Canada). Polycarbonate filters of various pore sizes (100–600 nm) employed in the extrusion process are obtained from SPI supplies (West Chester, PA).
9. The G-50 Sephadex medium grade resin (Amersham Biosciences) containing gravity column (20×1.5 cm, BioRad) pre-conditioned with PBS (1X, pH 7.4) is used to separate the unencapsulated fluorophores from the liposome-encapsulated fluorophores.
10. Latex spheres (Palo Alto, CA) of at least two different sizes, usually 100 and 200 nm for large and small unilamellar vesicles (LUVs and SUVs), are used in the dynamic light scattering assays for standardization and subsequent vesicle diameter determination.

Table 1
Triple-Helical Ligands Synthesized for PA Liposomes. $P^* = $ 4-Hydroxyproline

Receptor	Binding site location	Sequence
α2β1	α1(IV)382–393	C_n-(GPP*)$_4$-GAOGFOGERGEK-(GPP*)$_4$-NH$_2$
α3β1	α1(IV)531–543	C_n-(GPP*)$_4$-GEFYFDLRLKGDK-(GPP*)$_4$-NH$_2$
CD44	α1(IV)1263-1277 ([IV-H1])	C_n-(GPP*)$_4$-GVKGDKGNPGWPGAP-(GPP*)$_4$-NH$_2$

11. The matched Hs895T metastatic melanoma and Hs895Sk dermal fibroblast cell lines were established from the same 48-yr-old Caucasian female donor and are obtained from American Type Culture Collection (ATCC) (Manassas, VA).

12. Dulbecco's modified Eagle's medium (DMEM) medium is supplemented with 10% fetal bovine serum (FBS), 0.1 mg/mL gentamicin, 50 U/mL penicillin, and 0.05 mg/mL streptomycin. Trypsin-ethylenediamine tetraacetic acid (EDTA) solution is used for detaching adherent cells. Trypan Blue is used to access cell viability. The media and Trypan Blue are obtained from CellGro (Herndon, VA) and all reagents are purchased from Sigma (St. Louis, MO).

13. The DMEM adhesion media is prepared essentially by an identical procedure as described for the DMEM medium, with the exception of the presence of FBS and the antibiotics. Also, the DMEM used does not contain the phenol red indicator. The adhesion media is buffered to pH 7.4 using 400 mM HEPES.

14. Petri dishes are obtained from Fisher (Pittsburg, PA).

15. Olympus IX70 inverted Fluorescence Microscope equipped with a cool charge-coupled device (CCD) camera is used for the co-localization and fusion visualization studies of the fluorophore-loaded LUVs and cells. Either a 40× LUCPlanFl objective or a 100× oil immersion UPlanFl objective is used. Petri dishes with coverslip bottoms are employed from MatTek (Ashland, MA) when the 100× objective is used.

16. Only the cell nuclei are stained with DAPI. The cell membrane is not stained when using the 40× objective because phase-contrast and fluorescence images are overlaid using the Olympus DP Control Manager (DP70-BSW) software.

17. The 100× objective is used in conjunction with the membrane stain FITC (ACROS Organics, Fairlawn, NJ), a nonspecific membrane dye, as the 100 objective does not have the phase contrast capability. The three fluorescent images of the cells are then overlaid as described in the previous step.

18. The filters employed for the fluorescence images are: WG, wide green filter for rhodamine 6G; WIB, wide indigo blue filter for FITC; and WU, wide ultra violet filter for DAPI.

3. Methods

3.1. Synthesis and Purification of PAs

Procedures for the synthesis, purification, and analytical characterization of PAs are described in Chapter 5.

3.2. Structural Characterization of PAs

CD spectroscopy is used to evaluate the overall thermal stability of triple-helical PAs *(36,38,39,51–53,78,87,88)*. CD spectra are recorded over the range $\lambda = 190–250$ nm using a 1.0 or 10 mm path-length quartz cell and a sample concentration of 0.1 or 0.5 mg/mL. Thermal transition curves are obtained by recording the molar ellipticity ($[\Theta]$) at $\lambda = 225$ nm while the temperature is continuously increased in the range of 5–80°C at a rate of 0.2°C/min. For samples exhibiting sigmoidal melting curves, the inflection point in the transition region (first derivative) is defined as T_m. Alternatively, T_m is evaluated from the midpoint of the transition. Nuclear magnetic resonance (NMR) spectroscopy can also be used to study the relative thermal stability, alignment, and flexibility (backbone mobilities) of triple-helical PAs *(38,39,89)*.

3.3. Liposome Preparation

A good general guide for the preparation of multilamellar vesicles (MLVs), SUVs (50–100 nm in diameter), and LUVs (150–800 nm in diameter) is contained in **ref. 90**, and the reader is referred to the references cited within that publication.

3.3.1. LUV Preparation

The following methods describe the preparation of LUVs with and with out the incorporation of the PA. LUVs provide a number of advantages compared to other forms of liposomes, including high encapsulation of water-soluble drugs, economy of lipid, and reproducible drug release rates *(16,17,91)*. We include methods for the loading of the LUVs with fluorophores, which are used in the visualization of the vesicle targeting process to specific cell types, as well as for the monitoring of the stability of the different vesicle types with various lipid compositions. The methods of characterization are applicable to PA targeted and nontargeted LUVs as well as SUVs. Some of the primary physical properties affecting liposomal stability include length and structure of the diacyl chains, the charge on the head group of the phospholipids, the phase transition temperature (T_m), liposomal load, and the final liposome size.

The strategy for producing various liposome systems with different physical properties can be achieved by simply altering lipid composition. We have focused on the preparation of LUVs of various sizes between 100 and 600 nm to compare the effects of liposome size as well as lipid composition on the physical properties of the liposome. This includes the stability of LUVs over time, and the fusogenic properties of LUVs with cells when the LUVs are targeted with PAs or not. An important aspect of the MLV preparation process is that the lipid mixture must be heated for at least 30 min above a transition temperature of the most stable lipid in order to obtain formation of MLVs containing encapsulated aqueous phase in the hydrophilic core. The protocol that we employ to prepare LUVs of various sizes is as follows:

1. Prepare a solution containing two phospholipids and cholesterol, such as DSPG:DSPC:cholesterol. In a round bottom flask (250 mL), DSPG and DSPC lipids are combined with cholesterol in a fixed ratio of 1:4:5, respectively, and dissolved in an organic phase mixture of methanol, *tert*-butyl ether and chloroform (1:2:2.4) by either vortexing or stirring for 0.5 h at room temperature. The most stable lipid (DSPG in this case) is always the smaller component of the mixture.
2. The organic phase is then removed under reduced pressure by rotary evaporation, leaving a thin lipid film at the bottom of the flask which is dried over night *in vacuo*.
3. The aqueous phase, which is typically water with or without the rhodamine 6G or fluorescein fluorophore (400 μM), is then added to the lipid film and the resulting dispersion is extensively vortexed.
4. The dispersion is then stirred, for 30 min, at a temperature above the specific phase transition temperature of the most stable lipid in the system employed. Therefore, for the DSPC:DSPG:cholesterol system, where the transition phase temperature of DSPG is 57.5°C, the dispersion is maintained at 60°C.
5. The fluorophore loaded MLVs are then subjected to four freeze-thaw cycles and extruded 10 to 15 times through a double polycarbonate filter, at pressures typically at the lower end of the 250–700 psi range. Filters with a pore size ranging from 100 nm to 600 nm are employed depending on the desired liposome diameter.
6. Any unencapsulated fluorophore is then separated from the fluorophore-loaded liposomes by size-exclusion chromatography (SEC) using a G50 medium grade resin Sephadex column pre-conditioned with PBS (pH 7.4).

3.3.2. SUV Preparation

The following section describes the preparation of SUVs with and without the PA targeting ligand. SUVs can be advantageous due to enhanced circulation times and drug release *(92,93)*.

SUVs are prepared by using a high-pressure extruder. Lipids are dissolved in chloroform, either pure DLPC or DMPC (Avanti Polar Lipids) or mixtures of PAs and DLPC or DMPC. The solution is dried under N_2 for 6 h and the residual solvent removed using a vacuum evaporator. The dried amphiphile film is hydrated overnight in PBS buffer (pH 7.4), rhodamine solution, or 250 mM ammonium sulfate solution *(94)*, briefly vortexed, and then freeze-thawed five times. The freeze-thawed solution is extruded 10 times through a Lipex Biomembranes Extruder at 40°C using two nucleopore polycarbonate filters with 100-nm diameter pore sizes, resulting in a unilamellar dispersion with a uniform size distribution of less than 150 nm. The extruder works under high pressure, generally at around 100 psi, but in some cases up to 200 psi, which breaks up the multilamellae. Dynamic light scattering is used to confirm the formation of 100-nm diameter vesicles. The extruded solutions are kept at

5°C and used within a week. Examples of these SUVs incorporating the $(C_{16})_2$-Glu-C_2-(Gly-Pro-Hyp)$_4$-[α1(IV)1263–1277] PA have been described *(81)*.

3.4. LUV and SUV Characterization

The size, structure, and stability of liposomes can be evaluated by various different methods: cryo-TEM, fluorescence spectroscopy and microscopy, SEC, small-angle neutron scattering (SANS), dynamic light scattering, and CD spectroscopy. These procedures have been performed previously for studying PA aggregates *(40,45,50)* and charged-hydrophobic block copolymer micelles *(95–99)*. Below, each method is described.

1. In the cryo-TEM procedure, thin films of the liquid dispersion are layered on a holey carbon grid. The grid is placed in a Controlled Environment Vitrification System, and plunged into liquid ethane at a velocity of 2 m/s to form vitreous ice. The grid is stored under liquid N_2, and imaged on a cold stage holder. Cryo-TEM to determine size and stability of the vesicle samples is performed under a JEOL 1201 Transmission Electron Microscope at a focus of approx 4 µm *(50)*. The size and stability of LUVs and, more specifically, SUVs can be determined by this method.
2. Fluorescence microscopy of the fluorophore loaded vesicles is performed using an Olympus IX70 inverted microscope in conjunction with an Olympus Magnafire camera with Magnafire 1.0 software. Neutral density barrier filters that only allow either $\lambda_{emission} > 515$ nm or $\lambda_{emission} > 590$ nm and block all other wavelengths are used, thus helping to minimize photo bleaching of encapsulated fluorescein or rhodamine, respectively. Fluorescence microscopy is used to evaluate the size of the liposomes based on fluorescein or rhodamine encapsulation.
3. SANS is carried out on the 30-m beamline (NG-7) at the National Institute of Standards and Technology (NIST), Gaithersburg, MD. The wavelength of radiation, λ, is on average 6 Å, with a spread in wavelength of 10%. Data are collected with the detector at distances of 1.0, 4.0, and 13.7 m from the sample. Samples are held in 1.0-mm-thick quartz cells in a sample chamber that is temperature-controlled to 25.0 $\pm$0.1°C unless otherwise needed. A total of 10^6 or more counts are collected for each sample. To place the data on an absolute scale, corrections are made for detector efficiency, the presence of background radiation, and the scattering of the empty cell. The corrected data is then circularly averaged to obtain curves of the differential scattering cross section of the sample $dE(q)/d\Omega$, as a function of the scattering vector $q = 4\pi \sin \theta/\lambda$, where 2θ is the scattering angle. The size and shape of liposomes can be evaluated using SANS *(50)*.
4. The use of CD spectroscopy to analyze triple-helical structure in PA liposomes is performed as described previously in the "Structural Characterization of PAs" section *(50,81)*.
5. The size of PA liposomes is evaluated by SEC and dynamic light scattering *(40, 45,95,97)*. Dynamic light scattering analysis is carried out to specifically determine

the mean diameter of the liposomes from each batch prepared. Typically, we have found the following average diameters were obtained for the respective target LUV sizes: 114 ± 20 nm, for 100 nm liposomes; 206 ± 20 nm, for 200 nm liposomes; 343 ± 24 nm, for 400 nm liposomes; and 622 ± 106 nm, for 600 nm liposomes.

6. The liposome phospholipid content is determined by the Stewart (ammonium ferrothiocyanate) assay *(100,101)*.

7. PA incorporation into liposomes is evaluated by fluorescence spectroscopy, Fourier transform infrared (FTIR) spectroscopy, radiolabeling, and/or MS *(78,83)*.

3.5. Incorporation of PAs into the LUVs

The synthesis of targeted liposomes, which incorporate, for example, the α1(IV)1263–1277 PA that is specific for the CD44 receptor overexpressed by metastatic melanoma cells, is identical to that described in the above section with the exception that the α1(IV)1263–1277 PA is added to the lipid organic phase mixture in the very first step of the synthesis (which is also the case for SUVs). Typically, the amount of PA added is equivalent to 10% of the least abundant lipid in the lipid system. The melting temperature of the PA should be such that the triple-helical structure remains intact though out the entire preparative procedure.

3.6. PA-Targeted LUV Stability

As described above, several different techniques can be employed for the characterization of loaded vesicle stability. The method below is the method of choice for PA targeted, fluorophore-loaded LUVs, with the non-targeted, fluorophore-loaded LUVs used for comparison. This procedure has been found to be the most economical, simple, and predictive form of analysis for our vesicle systems.

The stability of the encapsulated fluorophore in PA LUV vs LUV systems is determined by monitoring release of the fluorophore from the vesicles. Liposome stability is usually dependent on the physical and chemical characteristics of the lipids composing the bilayer membranes. These include carbon chain or tail length and saturation and head group charge. For targeted liposomes, the incorporated targeting ligand, such as the PA, will also contribute to and/or affect the stability of the vesicles. Lastly, the hydrophobicity or hydrophilicity of the load (in this case, the fluorophore) will be a factor affecting the stability of the vesicle.

1. Vesicle solutions (200 μL, 3 mg/mL) are made up to 2 mL with either water or ethanol and then 200-μL aliquots are dispensed into 1.5 mL Eppendorf tubes and the samples are maintained at 37, 25, or 4°C over time, usually up to 1 mo.

2. At selected time points, 100-μL aliquots of the liposomes in water or ethanol are removed from the Eppendorfs, and dispensed into a 96-well quartz plate. The fluorescence intensity for each vesicle sample at each temperature is measured. The Spectra Max Gemini EM Fluorescent Plate Reader (Molecular Devices) is used to take readings at $\lambda_{excitation} = 525$ nm and $\lambda_{emission} = 555$ nm for rhodamine or $\lambda_{excitation} = 494$ nm and $\lambda_{emission} = 520$ nm for fluorescein. Complete release of fluorophore from the vesicles at each time point yields 100% dequenching, and is obtained from ethanol-treated liposome samples.

3. The percentage release of fluorophore from the vesicles is determined from the fluorescence intensity of each sample relative to 100% dequenching, which can then be expressed in terms of arbitrary fluorescent units (R_{fu}). Thus, the stability of rhodamine or fluorescein encapsulation is compared over time between liposome systems with different lipid and PA compositions. Typically, DSPC:DSPG:cholesterol LUVs in the size range of 160–200 nm were observed to be stable over a series of days, with only mild leakage being noted after 21 d.

3.7. PA Targeting of LUVs to Cells

Evidence for targeting specificity of PA liposomes for cells that overexpress the targeted receptor vs cells that do not should be established. Cells are treated with the fluorophore or drug-loaded targeted liposomes, and also the nontargeted loaded liposomes which serve as controls for non-specific binding with both cell types. The following materials and procedures are described for LUVs containing rhodamine as the fluorophore load.

3.8. Cell Culture

A general cell culture procedure is described below. Cells should be used before being passed eight times, at which point they should be discarded. This minimizes phenotypic drift and enables better reproducibility of results.

1. Cells are cultured with complete medium at 37°C in a humidified atmosphere of 5% CO_2 in air.
2. Cells are harvested from sub-confluent (<80%) cultures using a trypsin-EDTA solution and then re-suspended in fresh medium. Only cells with a >90% viability are used, as determined by Trypan Blue exclusion.
3. Cells are counted and seeded into appropriately sized plates. The cells are then allowed to adhere overnight in complete medium prior to use.

3.9. Fluorescence Microscopy

1. Cells are seeded at a density in the range of 1.0×10^5 to 1.0×10^6 and grown overnight in 60-mm Petri dishes or at a density in the range of 5.0×10^3 to 6.0×10^4

in 35-mm Petri dishes with coverslip bottoms. The DMEM medium is removed from the adherent cells. The cells are then washed twice with PBS (5 mL, pH 7.4).

2. An aliquot of DMEM adhesion media (1.5–2.0 mL, pH 7.4) is added to each plate.

3. The nucleus stain DAPI (0.5 mL, 0.1 mg/mL in water at pH 7.0) is added and the cells are incubated in the dark, at 37°C for 10 min. For studies using the 100× objective, the nonspecific membrane stain FITC (0.5 mL, 1 mg/mL, buffered with HEPES to pH 8.5) is added and the cells are incubated for a further 20 min at 37°C. For studies using the 40× objective the cells are not stained with FITC.

4. The supernatant solution is removed and the cells are washed with PBS (5 mL, pH 7.4), and 1–3 mL of DMEM adhesion media is added depending on the size of the Petri dish.

5. The rhodamine loaded PA or non-targeted liposomes in PBS, pH 7.4 (1 mL at phospholipid concentrations of either 50, 100, or 200 μM, prepared as previously described) are added to the cells (*see* **Note 3**).

6. The liposomes with cells are incubated for various periods of time and visualized using fluorescence microscopy.

7. For visualization studies employing the 40× objective, first locate the cells under visible light and take a phase-contrast image. The image of the cell nuclei is then taken using the ultraviolet WU filter. The image of the rhodamine in the liposomes as well as within the cells is taken using the green WG filter. The three images are overlaid.

8. For visualization studies employing the 100× (oil immersion) objective, first locate the cell nuclei using the WU filter, then take an image using the WIB filter that allows for membrane visualization. Lastly, the image of the rhodamine in the liposomes as well as within the cells is taken using the green WG filter. The three images are overlaid.

Acknowledgments

We gratefully acknowledge support of this work by the National Institutes of Health (CA 77402 and CA 98799 to GBF, EB 000289 to GBF/MT). In addition, this work was supported at UC Santa Barbara in part by the National Science Foundation under NSF awards NSF/MRSEC DMR-0080034, NSF/NIRT CTS-0103516, and the Army Research Office through the Institute for Collaborative Biotechnologies, and at Florida Atlantic University by the Center of Excellence in Biomedical and Marine Biotechnology (contribution #P200507).

4. Notes

1. As triple-helical PAs contain a *C*-terminal Hyp residue, they are synthesized as *C*-terminal amides [on Fmoc-4-(2′, 4′-dimethoxyphenyl-aminomethyl)phenoxy resin; Rink amide MBHA resin, Novabiochem] to prevent diketopiperazine (DKP) formation *(102)*.

2. As the synthesis of triple-helical peptides proceeds, removal of the Fmoc group becomes slower, particularly in the Gly-Pro-Hyp repeating sequence regions. It is recommended that deprotection of Gly, Pro, and Hyp proceed with extended treatment times and/or 2% DBU/2% piperidine in DMF.
3. At this stage, it is critical that the solution covers the cells completely.

References

1. Guo, X. and Szoka, F. C., Jr. (2003) Chemical approaches to triggerable lipid vesicles for drug and gene delivery. *Acc. Chem. Res.* **36**, 335–341.
2. Allen, T. M. and Cullis, P. R. (2004) Drug delivery systems: entering the mainstream. *Science* **303**, 1818–1822.
3. Martini, A. and Ciocca, C. (2003) Drug delivery systems for cancer drugs. *Expert Opin. Ther. Patents* **13**, 1801–1807.
4. Lasic, D. D. (1993) *Liposomes: From Physics to Applications.* Elsevier, Amsterdam.
5. Klibanov, A. L., Maruyama, K., Torchilin, V. P., and Huang, L. (1990) Amphipathic polyethyleneglycols effectively prolong the circulation time of liposomes. *FEBS Lett.* **268**, 235–237.
6. Allen, T. M., Hansen, C., Martin, F., Redemann, C., and Yau-Young, A. (1991) Liposomes containing synthetic lipid derivatives of poly(ethylene glycol) show prolonged circulation half-lives in vivo. *Biochim. Biophys. Acta* **1066**, 29–36.
7. Allen, T. M. and Hansen, C. (1991) Pharmacokinetics of stealth versus conventional liposomes: effect of dose. *Biochim. Biophys. Acta* **1068**, 133–141.
8. Maruyama, K., Ishida, O., Takizawa, T., and Moribe, K. (1999) Possibility of active targeting to tumor tissues with liposomes. *Adv. Drug Deliv. Rev.* **40**, 89–102.
9. Oku, N. (1999) Anticancer therapy using glucuronate modified long-circulating liposomes. *Adv. Drug Deliv. Rev.* **40**, 63–73.
10. Gabizon, A. A. (2001) Pegylated liposomal doxorubicin: metamorphosis of an old drug into a new form of chemotherapy. *Cancer Invest.* **19**, 424–436.
11. Jamil, J., Sheikh, S., and Ahmad, I. (2004) Liposomes: the next generation. *Mod. Drug Discovery* **7(1)**, 36–39.
12. Foldvari, M., Mezei, C., and Mezei, M. (1991) Intracellular delivery of drugs by liposomes containing Po glycoprotein from peripheral nerve myelin into human M21 melanoma cells. *J. Pharm. Sci.* **80**, 1020–1028.
13. Hallahan, D., Geng, L., Qu, S., et al. (2003) Integrin-mediated targeting of drug delivery to irradiated tumor blood vessels. *Cancer Cell* **3**, 63–74.
14. Jaafari, M. R. and Foldvari, M. (2002) Targeting of liposomes to melanoma cells with high levels of ICAM-1 expression through adhesive peptides from immunoglobulin domains. *J. Pharm. Sci.* **91**, 396–404.
15. Medina, O. P., Söderlund, T., Laakkonen, L. J., Tuominen, E. K. J., Koivunen, E., and Kinnunen, P. K. J. (2001) Binding of novel peptide inhibitors of type IV

collagenases to phospholipid membranes and use in liposome targeting to tumor cells in vitro. *Cancer Res.* **61**, 3978–3985.

16. Eliaz, R. E. and Szoka, J., F.C. (2001) Liposome-encapsulated doxorubicin targeted to CD44: a strategy to kill CD44-overexpressing tumor cells. *Cancer Res.* **61**, 2592–2601.

17. Eliaz, R. E., Nir, S., Marty, C., and Szoka, F., Jr. (2004) Determination and modeling of kinetics of cancer cell killing by doxorubicin and doxorubicin encapsulated in targeted liposomes. *Cancer Res.* **64**, 711–718.

18. Goren, D., Horowitz, A. T., Tzemach, D., Tarshish, M., Zalipsky, S., and Gabizon, A. (2000) Nuclear delivery of doxorubicin via folate-targeted liposomes with bypass of multidrug-resistance efflux pump. *Clin. Cancer Res.* **6**, 1949–1957.

19. Kirpotin, D., Park, J. W., Hong, K., et al. (1997) Sterically stabilized anti-HER2 immunoliposomes: design and targeting to human breast cancer cells in vitro. *Biochemistry* **36**, 66–75.

20. Lee, C. M., Tanaka, T., Murai, T., et al. (2002) Novel chondroitin sulfate-binding cationic liposomes loaded with cisplatin effectively suppress the local growth and liver metastasis of tumor cells in vivo. *Cancer Res.* **62**, 4282–4288.

21. Schmitt-Sody, M., Strieth, S., Krasnici, S., et al. (2003) Neovascular targeting therapy: Paclitaxel encapsulated in cationic liposomes improves antitumoral efficacy. *Clin. Cancer Res.* **9**, 2335–2341.

22. Kunstfeld, R., Wickenhauser, G., Michaelis, U., et al. (2003) Paclitaxel encapsulated in cationic liposomes diminishes tumor angiogenesis and melanoma growth in a "humanized" SCID mouse model. *J. Invest. Dermatol.* **120**, 476–482.

23. Dvorak, H. F., Nagy, J. A., and Dvorak, A. M. (1991) Structure of solid tumors and their vasculature: implications for therapy with monoclonal antibodies. *Cancer Cells* **3**, 77–85.

24. Shockley, T. R., Lin, K., Nagy, J. A., Tompkins, R. G., Dvorak, H. F., and Yarmush, M. L. (1991) Penetration of tumor tissue by antibodies and other immunoproteins. *Ann. NY Acad. Sci.* **618**, 367–382.

25. Jain, R. K. (1997) Delivery of molecular and cellular medicine to solid tumors. *Microcirculation* **4**, 1–23.

26. Kisak, E. T., Coldren, B., Evans, C. A., Boyer, C., and Zasadsinski, J. A. (2003) The vesosome—a multicompartment drug delivery vehicle. *Current Med. Chem.* **11**, 1241–1253.

27. Meers, P. (2001) Enzyme-activated targeting of liposomes. *Adv. Drug Deliv. Rev.* **53**, 265–272.

28. Hu, L., Ho, R. J. Y., and Huang, L. (1986) Trypsin induced destabilization of liposomes composed of dioleoylphosphatidylethanolamine and glycorphorin. *Biochem. Biophys. Res. Commun.* **141**, 973–978.

29. Sarkar, N. R., Rosendahl, T., Krueger, A. B., et al. (2005) "Uncorking" of liposomes by matrix metalloproteinase-9. *Chem. Commun.*, 999–1001.

30. Lutolf, M. P., Lauer-Fields, J. L., Schmoekel, H. G., et al. (2003) Synthetic matrix metalloproteinase-sensitive hydrogels for the conduction of tissue regeneration: engineering cell-invasion characteristics. *Proc. Natl. Acad. Sci. USA* **100**, 5413–5418.

31. Chau, Y., Tan, F. E., and Langer, R. (2004) Synthesis and characterization of dextran-peptide-methotrexate conjugates for tumor targeting via mediation by matrix metalloproteinase II and matrix metalloproteinase IX. *Bioconjugate Chem.* **15**, 931–941.

32. Pierschbacher, M. D. and Ruoslahti, E. (1987) Influence of stereochemistry of the sequence Arg-Gly-Asp-Xaa on binding specificity in cell adhesion. *J. Biol. Chem.* **262**, 17,297–17,298.

33. Pakalns, T., Haverstick, K. L., Fields, G. B., McCarthy, J. B., Mooradian, D. L., and Tirrell, M. (1999) Cellular recognition of synthetic peptide amphiphiles in self-assembled monolayer films. *Biomaterials* **20**, 2265–2279.

34. Lauer, J. L., Gendron, C. M., and Fields, G. B. (1998) Effect of ligand conformation on melanoma cell α3β1 integrin-mediated signal transduction events: implications for a collagen structural modulation mechanism of tumor cell invasion. *Biochemistry* **37**, 5279–5287.

35. Fields, G. B., Lauer, J. L., Dori, Y., Forns, P., Yu, Y.-C., and Tirrell, M. (1998) Proteinlike molecular architecture: biomaterial applications for inducing cellular receptor binding and signal transduction. *Biopolymers* **47**, 143–151.

36. Malkar, N. B., Lauer-Fields, J. L., Borgia, J. A., and Fields, G. B. (2002) Modulation of triple-helical stability and subsequent melanoma cellular responses by single-site substitution of fluoroproline derivatives. *Biochemistry* **41**, 6054–6064.

37. Li, C., McCarthy, J. B., Furcht, L. T., and Fields, G. B. (1997) An all-D amino acid peptide model of α1(IV)531-543 from type IV collagen binds the α3β1 integrin and mediates tumor cell adhesion, spreading, and motility. *Biochemistry* **36**, 15,404–15,410.

38. Yu, Y.-C., Berndt, P., Tirrell, M., and Fields, G. B. (1996) Self-assembling amphiphiles for construction of protein molecular architecture. *J. Am. Chem. Soc.* **118**, 12,515–12,520.

39. Yu, Y.-C., Tirrell, M., and Fields, G. B. (1998) Minimal lipidation stabilizes protein-like molecular architecture. *J. Am. Chem. Soc.* **120**, 9979–9987.

40. Forns, P., Lauer-Fields, J. L., Gao, S., and Fields, G. B. (2000) Induction of protein-like molecular architecture by monoalkyl hydrocarbon chains. *Biopolymers* **54**, 531–546.

41. Borgia, J. A. and Fields, G. B. (2000) Chemical synthesis of proteins. *Trends Biotech.* **18**, 243–251.

42. Mammen, M., Choi, S.-K., and Whitesides, G. M. (1998) Polyvalent interactions in biological systems: implications for design and use of multivalent ligands and inhibitors. *Angew. Chem. Int. Ed.* **37**, 2754–2794.

43. García, M., Alsina, M. A., Reig, F., and Haro, I. (2000) Liposomes as vehicles for the presentation of a synthetic peptide containing an epitope of hepatitis A virus. *Vaccine* **18**, 276–283.
44. Dillow, A. K., Ochsenhirt, S. E., McCarthy, J. B., Fields, G. B., and Tirrell, M. (2001) Adhesion of α5β1 receptors to biomimetic substrates constructed from peptide amphiphiles. *Biomaterials* **22**, 1493–1505.
45. Malkar, N. B., Lauer-Fields, J. L., Juska, D., and Fields, G. B. (2003) Characterization of peptide-amphiphiles possessing cellular activation sequences. *Biomacromolecules* **4**, 518–528.
46. Baronas-Lowell, D., Lauer-Fields, J. L., and Fields, G. B. (2004) Induction of endothelial cell activation by a triple-helical α2β1 integrin ligand derived from type I collagen α1(I)496-507. *J. Biol. Chem.* **279**, 952–962.
47. Kokkoli, E., Ochsenhirt, S. E., and Tirrell, M. (2004) Collective and single-molecule interactions of α5β1 integrins. *Langmuir* **20**, 2397–2404.
48. Huo, Q., Sui, G., Kele, P., and Leblanc, R. M. (2000) Combinatorial surface chemistry—is it possible? *Angew. Chem. Int. Ed.* **39**, 1854–1857.
49. Hartgerink, J. D., Beniash, E., and Stupp, S. I. (2002) Peptide-amphiphile nanofibers: a versatile scaffold for the preparation of self-assembling materials. *Proc. Natl. Acad. Sci. USA* **99**, 5133–5138.
50. Gore, T., Dori, Y., Talmon, Y., Tirrell, M., and Bianco-Peled, H. (2001) Self-assembly of model collagen peptide amphiphiles. *Langmuir* **17**, 5352–5360.
51. Lauer-Fields, J. L., Tuzinski, K. A., Shimokawa, K., Nagase, H., and Fields, G. B. (2000) Hydrolysis of triple-helical collagen peptide models by matrix metalloproteinases. *J. Biol. Chem.* **275**, 13,282–13,290.
52. Lauer-Fields, J. L., Nagase, H., and Fields, G. B. (2000) Use of Edman degradation sequence analysis and matrix-assisted laser desorption/ionization mass spectrometry in designing substrates for matrix metalloproteinases. *J. Chromatogr. A.* **890**, 117–125.
53. Lauer-Fields, J. L., Broder, T., Sritharan, T., Nagase, H., and Fields, G. B. (2001) Kinetic analysis of matrix metalloproteinase triple-helicase activity using fluorogenic substrates. *Biochemistry* **40**, 5795–5803.
54. Woessner, J. F. and Nagase, H. (2000) *Matrix Metalloproteinases and TIMPs.* Oxford University Press, Oxford.
55. Kühn, K. and Eble, J. (1994) The structural bases of integrin-ligand interactions. *Trends Cell Biol.* **4**, 256–261.
56. van der Flier, A. and Sonnenberg, A. (2001) Function and interactions of integrins. *Cell Tissue Res.* **305**, 285–298.
57. Kramer, R. H. and Marks, N. (1989) Identification of intracellular collagen receptor on human melanoma cells. *J. Biol. Chem.* **264**, 4684–4688.
58. Miles, A. J., Knutson, J. R., Skubitz, A. P. N., Furcht, L. T., McCarthy, J. B., and Fields, G. B. (1995) A peptide model of basement membrane collagen α1(IV)531-543 binds the α3β1 integrin. *J. Biol. Chem.* **270**, 29,047–29,050.

59. Klein, C. E., Dressel, D., Steinmayer, T., et al. (1991) Integrin α2β1 is upregulated in fibroblasts and highly aggressive melanoma cell in three-dimensional collagen lattices and mediates the reorganization of type I collagen fibrils. *J. Cell Biol.* **115**, 1427–1436.

60. Yoshinaga, I. G., Vink, J., Dekker, S. K., Mihm, M.C., Jr., and Byers, H. R. (1993) Role of α3β1 and α2β1 integrins in melanoma cell migration, *Melanoma Res.* **3**, 435–441.

61. Heino, J. (1996) Biology of tumor cell invasion: interplay of cell adhesion and matrix degradation. *Int. J. Cancer* **65**, 717–722.

62. Mizejewski, G. J. (1999) Role of integrins in cancer: survey of expression patterns. *Proc. Soc. Exp. Biol. Med.* **222**, 124–138.

63. Golbik, R., Eble, J. A., Ries, A., and Kühn, K. (2000) The spatial orientation of the essential amino acid residues arginine and aspartate within the α1β1 integrin recognition site of collagen IV has been resolved using fluorescence resonance energy transfer. *J. Mol. Biol.* **297**, 501–509.

64. Saccá, B., Sinner, E.-K., Kaiser, J., Lübken, C., Eble, J. A., and Moroder, L. (2002) Binding and docking of synthetic heterotrimeric collagen type IV peptides with α1β1 integrin. *ChemBioChem* **9**, 904–907.

65. Knight, C. G., Morton, L. F., Onley, D. J., et al. (1998) Identification in collagen type I of an integrin α2β1-binding site containing an essential GER sequence. *J. Biol. Chem.* **273**, 33,287–33,294.

66. Knight, C. G., Morton, L. F., Peachey, A. R., Tuckwell, D. S., Farndale, R. W., and Barnes, M. J. (2000) The collagen-binding A-domains of integrin α1β1 and α2β1 recognize the same specific amino acid sequence, GFOGER, in native (triple-helical) collagens. *J. Biol. Chem.* **275**, 35–40.

67. Emsley, J., Knight, C. G., Farndale, R. W., Barnes, M. J., and Liddington, R. C. (2000) Structural basis of collagen recognition by integrin α2β1. *Cell* **101**, 47–56.

68. Lauer-Fields, J. L., Sritharan, T., Stack, M. S., Nagase, H., and Fields, G. B. (2003) Selective hydrolysis of triple-helical substrates by matrix metalloproteinase-2 and -9. *J. Biol. Chem.* **278**, 18,140–18,145.

69. Baronas-Lowell, D., Lauer-Fields, J. L., Borgia, J. A., Sferrazza, G. F., Al-Ghoul, M., Minond, D., and Fields, G. B. (2004) Differential modulation of human melanoma cell metalloproteinase expression by α2β1 integrin and CD44 triple-helical ligands derived from type IV collagen. *J. Biol. Chem.* **279**, 43,503–43,513.

70. Miles, A. J., Skubitz, A. P. N., Furcht, L. T., and Fields, G. B. (1994) Promotion of cell adhesion by single-stranded and triple-helical peptide models of basement membrane collagen α1(IV)531-543: evidence for conformationally dependent and conformationally independent type IV collagen cell adhesion sites. *J. Biol. Chem.* **269**, 30,939–30,945.

71. Faassen, A. E., Drake, S. L., Iida, J., Knutson, J. R., and McCarthy, J. B. (1992) Mechanisms of normal cell adhesion to the extracellular matrix and alterations associated with tumor invasion and metastasis. *Adv. Pathol. Lab. Med* **5**, 229–259.

72. Iida, J., Meijne, A. M. L., Knutson, J. R., Furcht, L. T., and McCarthy, J. B. (1996) Cell surface chondroitin sulfate proteoglycans in tumor cell adhesion, motility and invasion. *Seminars Cancer Biol.* **7**, 155–162.

73. Herbold, K. W., Zhou, J., Haggerty, J. G., and Milstone, L. M. (1996) CD44 expression on epidermal melanocytes. *J. Invest. Dermatol.* **106**, 1230–1235.

74. Screaton, G. R., Bell, M. V., Jackson, D. G., Cornelis, F. B., Gerth, U., and Bell, J. I. (1992) Genomic structure of DNA encoding the lymphocyte homing receptor CD44 reveals at least 12 alternatively spliced exons. *Proc. Natl. Acad. Sci. USA* **89**, 12,160–12,164.

75. Naor, D., Slonov, R. V., and Ish-Shalom, D. (1997) CD44: structure, function, and association with the malignant process, in *Advances in Cancer Research* (Vande Woude, G. F. and Klein, G., eds.). Academic, Orlando, FL: pp. 241–319.

76. Lesley, J. and Hyman, R. (1998) CD44 structure and function. *Frontiers Biosci.* **3**, 616–630.

77. Lauer-Fields, J. L., Malkar, N. B., Richet, G., Drauz, K., and Fields, G. B. (2003) Melanoma cell CD44 interaction with the α1(IV)1263-1277 region from basement membrane collagen is modulated by ligand glycoslyation. *J. Biol. Chem.* **278**, 14,321–14,330.

78. Fields, C. G., Mickelson, D. J., Drake, S. L., McCarthy, J. B., and Fields, G. B. (1993) Melanoma cell adhesion and spreading activities of a synthetic 124-residue triple-helical "mini-collagen". *J. Biol. Chem.* **268**, 14,153–14,160.

79. Sweeney, S. M., DiLullo, G., Slater, S. J., et al. (2003) Angiogenesis in collagen I requires α2β1 ligation of a GFP*GER sequence, and possibly p38 MAPK and focal adhesion disassembly. *J. Biol. Chem.* **278**, 30,516–30,524.

80. Xu, Y., Gurusiddappa, S., Rich, R. L., et al. (2000) Multiple binding sites in collagen type I for the integrins α1β1 and α2β1., *J. Biol. Chem.* **275**, 38,981–38,989.

81. Tu, R., Mohanty, K., and Tirrell, M. (2004) Liposomal targeting through peptide-amphiphile functionalization. *Am. Pharm. Rev.* **7(2)**, 36–41.

82. Dori, Y., Bianco-Peled, H., Satija, S. K., Fields, G. B., McCarthy, J. B., and Tirrell, M. (2000) Ligand accessibility as a means to control cell response to bioactive bilayer membranes. *J. Biomed. Mater. Res.* **50**, 75–81.

83. Torchilin, V. P., Rammohan, R., Weissig, V., and Levchenko, T. S. (2001) TAT peptide on the surface of liposomes affords their efficient intracellular delivery even at low temperature and in the presence of metabolic inhibitors. *Proc. Natl. Acad. Sci. USA* **98**, 8786–8791.

84. Berndt, P., Fields, G. B., and Tirrell, M. (1995) Synthetic lipidation of peptides and amino acids: monolayer structure and properties. *J. Am. Chem. Soc.* **117**, 9515–9522.

85. King, D. S., Fields, C. G., and Fields, G. B. (1990) A cleavage method which minimizes side reactions following Fmoc solid phase peptide synthesis. *Int. J. Peptide Protein Res.* **36**, 255–266.

86. Fields, C. G. and Fields, G. B. (1993) Minimization of tryptophan alkylation following 9-fluorenylmethoxycarbonyl solid-phase peptide synthesis. *Tetrahedron Lett.* **34**, 6661–6664.

87. Fields, C. G., Lovdahl, C. M., Miles, A. J., Matthias-Hagen, V. L., and Fields, G. B. (1993) Solid-phase synthesis and stability of triple-helical peptides incorporating native collagen sequences. *Biopolymers* **33**, 1695–1707.

88. Grab, B., Miles, A. J., Furcht, L. T., and Fields, G. B. (1996) Promotion of fibroblast adhesion by triple-helical peptide models of type I collagen-derived sequences. *J. Biol. Chem.* **271**, 12,234–12,240.

89. Yu, Y.-C., Roontga, V., Daragan, V. A., Mayo, K. H., Tirrell, M., and Fields, G. B. (1999) Structure and dynamics of peptide-amphiphiles incorporating triple-helical proteinlike molecular architecture. *Biochemistry* **38**, 1659–1668.

90. Chapter 1 (2002), in *Liposome Methods and Protocols*, Methods in Molecular Biology Vol. 199 (Basu, S. B. and Basu, M., eds.). Humana, NJ.

91. Hope, M. J., Bally, M. B., Webb, G., and Cullis, P. R. (1985) Production of large unilamellar vesicles by a rapid extrusion procedure: characterization of size distribution, trapped volume and ability to maintain a membrane potential. *Biochim. Biophys. Acta* **812**, 55–65.

92. Drummond, D. C., Meyer, O., Hong, K., Kirpotin, D. B., and Papahadjopoulos, D. (1999) Optimizing liposomes for delivery of chemotherapeutic agents to solid tumors. *Pharmacol. Rev.* **51**, 691–744.

93. Nagayasu, A., Uchiyama, K., and Kiwada, H. (1999) The size of liposomes: a factor which affects their targeting efficiency to tumors and therapeutic activity of liposomal antitumor drugs. *Adv. Drug Deliv. Rev.* **40**, 75–87.

94. Charrois, G. J. R. and Allen, T. M. (2003) Rate of biodistribution of STEALTH liposomes to tumor and skin: influence of liposome diameter and implications for toxicity and therapeutic activity. *Biochim. Biophys. Acta* **1609**, 102–108.

95. Balsara, N. P., Stepanek, P., Lodge, T. P., and Tirrell, M. (1991) Dynamic light scattering from microstructured block copolymer solutions. *Macromolecules* **24**, 6227–6230.

96. Dan, N., and Tirrell, M. (1993) Self-assembly of block copolymers with a strongly charged and a hydrophobic block in a selective, polar solvent: micelles and adsorbed layers. *Macromolecules* **26**, 4310–4315.

97. Guenoun, P., Delsanti, M., Gazeau, D., et al. (1998) Structural properties of charged diblock copolymer solutions. *Eur. Phys. J. B* **1**, 77–86.

98. Guenoun, P., Davis, H. T., Doumaux, H. A., et al. (2000) Polyelectrolyte micelles: self-diffusion and electron microscopy studies, *Langmuir* **16**, 4436–4440.

99. Muller, F., Delsanti, M., Auvray, L., et al. (2000) Ordering of urchin-like charged copolymer micelles: electrostatic, packing and polyelectrolyte correlations. *Eur. Phys. J. E* **3**, 45–53.

100. Zuidam, N. J., de Vrueh, R., and Crommelin, D. J. A. (2003) Characterization of liposomes, in *Liposomes: A Practical Approach, Second Edition* (Torchilin, V. P. and Weissig, V., eds.). Oxford University Press, Oxford, UK: pp. 31–78.

101. Backer, M. V., Gaynutdinov, T. I., Patel, V., Jehning, B. T., Myshkin, E., and Backer, J. M. (2004) Adapter protein for site-specific conjugation of payloads for targeted drug delivery. *Bioconjugate Chem.* **15**, 1021–1029.

102. Fields, G. B., Lauer-Fields, J. L., Liu, R.-Q., and Barany, G. (2001) Principles and practice of solid-phase peptide synthesis, in *Synthetic Peptides: A User's Guide, 2nd Edition* (Grant, G. A., ed.). W.H. Freeman & Co., New York: pp. 93–219.

11

Peptide-Mediated Delivery of Nucleic Acids into Mammalian Cells

Sébastien Deshayes, Federica Simeoni, May C. Morris, Gilles Divita, and Frédéric Heitz

Summary

Control of gene expression using RNA interference (RNAi) technology constitutes a method of choice for investigating gene function in mammalian cells. However, like most oligonucleotide-based strategies, the major limitation of interfering RNA is their poor cellular uptake due to low permeability of the cell membrane to nucleic acids. Several strategies have been developed to improve delivery of oligonucleotides both in cultured cells and in vivo. So far, there is no universal method for their delivery, as they all present several limitations. Peptide-based strategies have been demonstrated to improve the cellular uptake of nucleic acids both in cultured cell and in vivo. This chapter describes a new peptide-based gene delivery system, MPG, which forms stable noncovalent complexes with oligonucleotides and promotes their delivery into a large panel of cell lines without the need for prior chemical covalent coupling. Protocols are described for both adherent and suspension cell lines.

Key Words: Peptide carrier; cell-penetrating peptide; cellular uptake; oligonucleotides; siRNA, gene silencing; gene and oligonucleotide delivery; NLS.

1. Introduction

The control of gene expression using RNA interference (RNAi) technology is essential both in fundamental and pharmaceutical research. However, the major limitation of cellular internalization of nucleic acids remains their poor cellular uptake associated with low permeability of the cell membrane to nucleic acids *(1,2)*. Several viral *(3–6)* and nonviral *(3,7)* strategies have been proposed to improve the delivery to both cultured cells and in vivo *(3)*. So far, although nucleic acid transfection can be achieved with classical laboratory cultured

From: *Methods in Molecular Biology, vol. 386: Peptide Characterization and Application Protocols*
Edited by: G. Fields © Humana Press Inc., Totowa, NJ

cell lines using lipid-based formulations, nucleic acid delivery remains a major challenge for many cell lines and there is still no reasonably efficient method for in vivo application *(3)*. The most efficient method for in vivo applications is the nonviral "hydrodynamic" tail-vein injection of mice with high doses of nucleic acid *(8–10)*.

Cell-penetrating peptides are powerful carriers for cellular uptake of a variety of macromolecules including proteins, peptides, and oligonucleotides *(11–14)*. Several peptide-based strategies have been developed to improve the delivery of oligonucleotides both in vitro and in vivo using either covalent or complex approaches *(15–17)*. We have described a new peptide-based gene delivery system, MPG, which forms stable noncovalent complexes with several nucleic acids (plasmid DNA, oligonucleotides) and promotes their delivery into a large panel of cell lines without the need for prior chemical covalent coupling *(18–20)*. We recently demonstrated that a variant containing a single mutation of this peptide carrier constitutes an excellent tool for the delivery of nucleic acids, including siRNA into different cell lines *(7)*. MPG is a peptide of 27 residues (GALFLGFLGAAGSTMGAWSQPKKKRKV) consisting of three domains with specific functions: a hydrophobic motif (GALFLGFLGAAGSTMGA) derived form the fusion sequence of the HIV protein gp41 required for efficient targeting to the cell membrane and internalization, a hydrophilic lysine-rich domain (PKKKRKV) derived from the Nuclear Localization Sequence (NLS) of simian virus (SV)-40 large T antigen, involved in the main interactions with nucleic acids and required to improve intracellular trafficking of the cargo, as well as a spacer domain (WSQ), which improves the flexibility and the integrity of both the hydrophobic and the hydrophilic domains *(20)*. Given that the mechanism through which MPG delivers nucleic acids and oligonucleotides does not involve the endosomal pathway, the stability of the macromolecules delivered is significantly improved and rapid dissociation of the MPG/cargo particle is favored as soon as it crosses the cell membrane. This peptide-based oligonucleotide delivery strategy presents several advantages including rapid delivery of nucleic acid into cells with very high efficiency, stability in physiological buffers, lack of toxicity, and lack of sensitivity to serum. MPG constitutes a powerful tool for basic research and several studies have demonstrated than this technology is extremely powerful for targeting specific genes or modifying gene function in vitro as well as in vivo *(7,20–22)*. The present chapter describes several protocols for the use of the noncovalent MPG technology for the delivery of nucleic acids into mammalian cells.

2. Materials

2.1. Cell Cultures

1. Phosphate-buffered saline (PBS) (cat no. 14190-169), Dulbecco's modified Eagle's medium (DMEM) (cat no. 41965-062), glutamine, and streptomycin/penicillin (cat no. 15140-130) are from Invitrogen Life Technologies (Carsbad, CA).
2. Fetal bovine serum (FBS) is from PERBIO (cat no. CH30160-03, Lot 3264EHJ).

2.2. Oligonucleotide and Transfection Reagents

1. Oligonucleotides are purchased from Sigma-Genosys. For transfection experiments double-stranded oligodeoxynucleotides (ODNs) are phosphorothioate-modified on the first base at the 5′ end and on the two last bases at the 3′ end.
2. Double-stranded phosphodiester oligonucleotides (18mer and 36mer), corresponding to the sequence of HIV natural primer binding site, are synthesized on an Applied Biosystems 380 B DNA synthesizer and purified by reversed-phase high-performance liquid chromatography (HPLC) *(22)*.
3. Double-stranded oligonucleotides (18/36mer) are formed by annealing an equimolar mixture of both single-stranded oligonucleotides in 20 mM Tris-HCl (pH 7.5) by heating for 15 min at 70°C, followed by cooling to room temperature over a period of 2 h in a water bath.
4. The siRNAs targeting GAPDH gene are obtained from Proligo and/or Ambion (Silencer™ kit).
5. Control small interfering (si)RNA transfection experiments are performed with the commonly used lipid-based Oligofectamine™ reagent (cat. no. 12252-011) from Invitrogen Life Technologies (Carlsbad, CA).

2.3. Peptide-Carrier MPG

MPG is a 27-residue-long peptide GALFLGFLGAAGSTMGAWSQP-KKKRKV, where the N-terminus is acetylated and the C-terminus bears a cysteamide group (NH-CH$_2$-CH$_2$-SH): (molecular weight: 2908 Da). The sequence variant of MPG contains a single mutation in the NLS sequence (K^{23}S), GALFLGFLGAAGSTMGAWSQPKSKRKV (molecular mass: 2867 Da) and is referred hereafter as MPG-Δ^{NLS} (**Note 1**).

Both N- and C-termini protections are essential for the stability of the peptide and the transfection mechanism *(7)*. MPG can be synthesized in house or obtained from commercial sources. Protocols for the synthesis and purification of MPG and derivatives have been described *(20,23,24)*. MPG and derivatives are stable at least 1 yr when stored at −20°C in a lyophilized form.

3. Methods

The methods described below outline (1) the formation of MPG/oligonucleotide complexes and handling, (2) protocol for the delivery of oligonucleotides into adherent cell lines, (3) protocol for their delivery into suspension cell lines, and (4) a control lipid-based method for transfection. The different procedures were modified based on prior reports *(7,20)*.

3.1. Preparation of MPG/Oligonucleotide Complexes

The procedure for formation of MPG/oligonucleotide complexes constitutes an important factor in the success of MPG technology and should be followed carefully (*see* **Notes 2** and **3**).

3.1.1. Preparation of the Solutions of MPG

1. Take the vial containing the peptide powder out of the freezer and equilibrate for 30 min at room temperature without opening the vial. This step is essential to limit hydration of the peptide powder. Resuspend the peptide at a concentration not higher than 1 mg/mL (concentration: 0.35 m*M*) in water.
2. Mix gently by tapping the tube or by vortexing at low speed for 20 s.
3. Repeated freeze/thaw cycles can induce peptide aggregation, therefore, it is recommended that one aliquot the MPG stock solution into tubes containing the amount one expects to use in a typical experiment prior to freezing. The MPG stock solution is stable for about 2 mo when stored at −20°C.

3.1.2. Formation of MPG/Oligonucleotide Complexes

1. Prepare a stock solution of single-stranded or annealed double-stranded duplex at 20 μ*M* in water. Dilute to 100 μL of water for each reaction. The oligonucleotides can be used at concentrations varying from 0.005 to 1 μ*M*, depending on the biological response expected. From our experience a concentration ranging from 20 to 50 n*M* of oligonucleotide is sufficient for a silencing response higher that 80%.
2. MPG solution must be diluted 1:10 in sterile H$_2$O (concentration: 35 μ*M*). At this stage, sonication of the peptide solution is recommended, to limit aggregation, for 5 min in a water bath sonicator. Alternatively, a probe sonicator can also be used; place the tube in water and sonicate for 1 min at an amplitude of 30%. Dilute the appropriate volume of MPG into 100 μL of sterile water for each reaction. For optimal transfection the MPG/oligonucleotide molar ratio is generally 10:1 to 40:1.
3. Add 100 μL of diluted oligonucleotide to 100 μL diluted MPG. Mix gently by tapping the tube. It is necessary to make the MPG/oligonucleotide complex in a concentrated solution, which will be added to the cells and then be diluted to the final transfection volume (600 μL for a 35-mm culture plate).

4. For multiple assays, a mix for six transfections can be used. Dilute the corresponding MPG volume into 600 μL of water and the oligonucleotide into 600 μL of PBS. Mix the two solutions, then proceed as described below. Do not exceed the volume required for six transfections, as this may cause aggregation.
5. Incubate at 37°C for 20 min to allow the complex to form, then proceed immediately to the transfection experiments. At this stage, the peptide/oligonucleotide complexes should not be stored.

3.2. Example of Application: An MPG-mediated siRNA Delivery

3.2.1. Delivery into Adherent Cell Lines

The protocol is described for HeLa and human fibroblast (HS-68) cell lines cultured in 35-mm culture plates, using an siRNA targeting the *GAPDH* gene. siRNA concentration dependence of the silencing response is analyzed by Western blot and the kinetics of siRNA-induced degradation of GAPDH mRNA are followed by Northern blotting (**Fig. 1**). The amount of siRNA and MPG, the transfection volume, and the number of cells should be adjusted accordingly to the size of the culture plate used (*see* **Notes 4** and **5**).

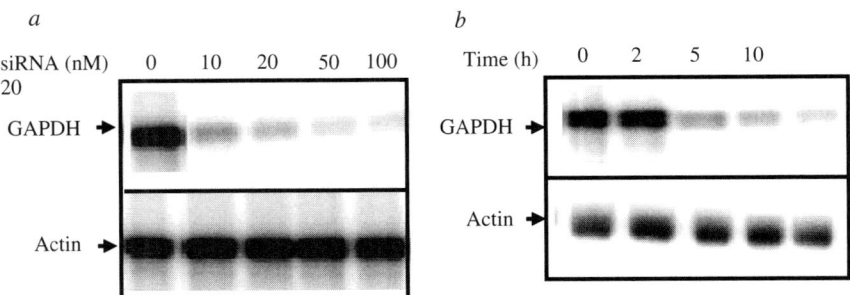

Fig. 1. MPG-mediated delivery of small interfering (si)RNA targeting *GAPDH* gene in Hela cells. (**A**) MPG-Δ^{NLS} -mediated delivery of siRNA targeting GAPDH gene: Western blot analysis. Different concentrations of siRNA (0, 10, 20, 50, 100 n*M*) were transfected with MPG-Δ^{NLS} at a molar ratio of 10:1 and the levels of GAPDH protein were analyzed by Western blotting 30 h posttransfection. Actin was used as a control to normalize protein loading. (**B**) MPG-Δ^{NLS} -mediated delivery of siRNA targeting GAPDH gene: Northern blot analysis. The kinetics of siRNA (50 n*M*)-induced degradation of GAPDH mRNA following transfection with MPG (molar ratio 10:1) were analyzed by Northern blotting (after 0, 2, 5, 10, and 20 h from left to right). Actin was used as a control to normalize mRNA levels in each sample. (Reproduced from **ref. 7**, with permission from Oxford University Press.)

1. In a six-well or a 35-mm tissue culture plate, seed 0.3×10^6 cells per well in 2 mL of complete growth medium. Incubate the cells at 37°C in a humidified atmosphere containing 5% CO_2 until the cells are 50–70% confluent. It is recommended that the cells be passed the day before treatment for a better response following transfection (*see* **Notes 3** and **5**).
2. Aspirate the medium from the cells to be transfected and wash the cells twice with PBS.
3. Overlay the cells with the 200 µL MPG-Δ^{NLS} /siRNA complex. Add 400 µL of serum-free medium to the overlay to achieve the final transduction volume of 600 µL for a 35-mm plate.
4. Incubate at 37°C in a humidified atmosphere containing 5% CO_2 for 30 min.
5. Add 1 mL of complete growth medium to the cells and adjust the level of serum to 10 or 20% depending upon the culture condition required. Do not remove the MPG/siRNA complex. Incubate at 37°C in a humidified atmosphere containing 5% CO_2 for 24 to 48 h, depending on the cellular response expected (*see* **Note 6**). The siRNA are fully released in the cells 1 and 2 h later, respectively.
6. Process the cells for observation or detection assays. GAPDH protein and mRNA levels were analyzed by Western and Northern blots, respectively. GAPDH protein is probed using an anti-GAPDH (Sigma). For Northern blotting, total RNAs are isolated from cells using TriReagent™ (Sigma, Saint Louis) according to the manufacturer's recommendations. RNAs are then purified by phenol extraction followed by ethanol precipitation. RNA samples (10 µg) are separated by electrophoresis in formaldehyde agarose gels (1.2%), transferred to nylon membranes (Hybond N+, Amersham Pharmacia Biotech), and hybridized with ^{32}P-labeled GAPDH and actin probes (the latter used to normalize RNA loading). Signals are detected by Phosphorimaging (Molecular Dynamics) and quantified using ImageQuant software. For morphology assays, the cells may be fixed or observed directly, using live imaging technology.

3.2.2. Delivery into Cells in Suspension

The protocol of MPG-Δ^{NLS} -mediated siRNA delivery is optimized on Hela and Jurkat T cells, using siRNA targeting GAPDH. However, efficient transfection may require optimization of MPG concentration, cell numbers, and exposure time of cells to the MPG/siRNA complex. A comparative study of different cell lines using a siRNA targeting GAPDH is reported in **Fig. 2**.

1. The same number of cells recommended for seeding adherent cells is recommended for suspension cells (confluency between 50 and 70%). Cells are cultured in standard medium in 35-mm dishes or six-well plates (*see* **Notes 3** and **5**).
2. The MPG/siRNA complexes are formed as described for adherent cells (**Subheading 3.2.1., steps 1–4**).

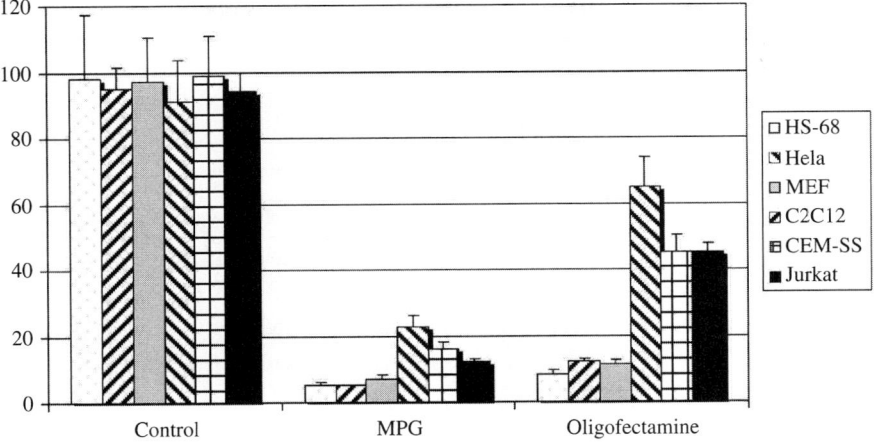

Fig. 2. MPG-Δ^{NLS}-mediated delivery of siRNA targeting GAPDH in adherent and suspension cell lines. MPG/siRNA targeting GAPDH gene complexes were formed at a molar ratio 10:1, then transfected on several adherent and suspension cell lines grown to 60% confluence. The levels of GAPDH protein were analyzed by Western blotting 30 h posttransfection. Control transfection experiments were performed using Oligofectamine™.

3. Collect the suspension cells by centrifugation at 400g for 5 min. Remove the supernatant and wash the cells twice with PBS.
4. Centrifuge at 400g for 5 min to pellet the cells. Remove the supernatant.
5. Solubilize the cell pellet in the MPG-Δ^{NLS}/siRNA complex solution (200 μL). Add serum-free medium to achieve a final transduction volume of 600 μL.
6. Incubate at 37°C in a humidified atmosphere containing 5% CO_2 for 30 min to 1 h depending on the cell lines.
7. Add complete growth medium to the cells and adjust serum levels according to culture requirements. Do not remove the MPG/siRNA complex. Continue to incubate at 37°C in a humidified atmosphere containing 5% CO_2 for 24 to 48 h depending upon the expected cellular response. As described for adherent cells, siRNA are fully released into cells 1 to 2 h later, respectively.
8. Process the cells for observation or detection assays (**Subheading 3.2.1., step 6**).

3.3. Lipid-Based Formulation for Oligonucleotide Delivery

Several lipid or cationic polymer-based formulations for oligonucleotide delivery are commercially available (reviewed in **ref. *3***). Control transfection experiments are performed for all the cell lines using the most commonly used Oligofectamine (Invitrogen). Transfection protocols for six-well plates are performed according to the manufacturer's guidelines using a concentration of oligonucleotide of 50 nM. In the case of a siRNA, assay for silencing is measured 48 h after transfection and conditions used lead to typical transfection efficiencies of about 80% for HeLa cell lines.

Acknowledgments

This work was supported in part by the Centre National de la Recherche Scientifique (CNRS) and by grants from the Agence Nationale de Recherche sur le SIDA (ANRS), the European Community (QLK2-CT-2001-01451), and the Association pour la Recherche sur le Cancer to MCM (ARC-4326) and to GD (ARC-5271). FS was supported by a grant from La Ligue de Recherche Contre le Cancer.

4. Notes

1. An important property of MPG-Δ^{NLS}-based siRNA delivery is that it allows for the relatively rapid introduction of oligonucleotides into cells. The lack of requirement of covalent coupling for formation of MPG/oligonucleotide particles favors rapid release of siRNA into the cytoplasm as soon as the cell membrane has been crossed *(26)*. Entry of MPG/siRNA complexes into the cell occurs in a short time (10 min), and release of macromolecules takes place within the first 2 h. The rapid release of siRNA into the cytoplasm favors rapid degradation of target mRNA. However, nuclear targeting can be achieved as a result of an MPG variant containing an integral NLS. This latter version of MPG was shown deliver siRNA efficiently, but exhibits slower degradation kinetics of mRNA. For long-term effects, the release of siRNA can be controlled by using this version of MPG leading to a more stable formulation *(7)*.

2. It is essential to perform complex formation between MPG and the siRNA in the absence of serum to limit degradation of siRNA and interactions with serum proteins. However, the transfection process itself is not affected by the presence of serum, which is a considerable advantage for most biological applications *(7)*.

3. Although this protocol was tested on several cells lines, conditions for efficient siRNA delivery should be optimized for every new cell line, including reagent concentration, cell number, and exposure time of cells to the MPG/siRNA complexes. A well characterized siRNA should always be used as a positive control of transfection.

4. A large variety of antisense DNA or siRNA have been successfully introduced into different cell lines using MPG-based strategy and have been shown to induce significant silencing activity *(7,20–22,25,26)*. Recently, MPG formulation was adapted to conditions required for in vivo applications and successfully used for intra-tumoral injection of siRNA *(21)*.

5. Low efficiency may be associated with several parameters:

a. *Cell confluency:* for adherent cells, the optimal confluence is of about 50–60%; higher confluence (90%) dramatically reduces the transduction efficiency.

b. *Formation of MPG/siRNA complexes:* conditions for the formation of MPG/siRNA complexes are critical and should be respected. Special attention should be paid to the recommended volumes, incubation times for the formation of the complexes, and time of exposure of these complexes to cells.

6. The advantages of MPG technology are directly associated with the mechanism through which this carrier promotes delivery of siRNA into cells. The independence of MPG-mediated siRNA transfection on the endosomal pathway significantly limits degradation and preserves the biological activity of internalized cargoes for prolonged time periods. Silencing effects were observed up to 7 d after transfection.

References

1. Luo, D. and Saltzman, M. W. (2000) Synthetic DNA delivery system. *Nat. Biotechnol.* **18**, 33–37.
2. Niidome, T. and Huang, L. (2002) Gene therapy progress and prospects: non viral vectors. *Gene Ther.* **10**, 991–998.
3. Rozema, D. B. and Lewis, D. L. (2003) siRNA delivery technologies for mammalian systems. *Target* **2**, 253–260.
4. Hommel, D. J., Sears, R. M., Georgescu, D., Simmons, D., and Dileone, R. J. (2003) Local gene knockdown in the brain using viral-mediated RNA inteference. *Nat. Med.* **9**, 1539–1543.
5. Brummelkamp, T. R., Bernards, R., and Agami, R. (2002) Stable suppression of tumorigenicity by virus-mediated RNA interference. *Cancer Cell.* **2**, 243–247.
6. Xia, H., Mao, Q., Paulson, H. L., and Davidson, B. L. (2002) siRNA-mediated gene silencing in vitro and in vivo. *Nat. Biotechnol.* **20**, 1006–1010.
7. Simeoni, F., Morris, M. C., Heitz, F., and Divita, G. (2003) Insight into the mechanism of the peptide-based gene delivery system MPG: implication for delivery of siRNA into mammalian cells. *Nucleic Acids Res.* **31**, 2717–2727.
8. McCaffrey. A. P., Meuse, L., Pham, T. T., Conklin. D. S., Hannon, G. J., and Kay, M. A. (2002) RNA interference in adult mice. *Nature* **418**, 38–39.
9. Song, E., Lee, S., Wang, J., et al. (2003) RNA interference targeting Fas protects mice from fulminant hepatitis. *Nat. Med.* **9**, 347–351.
10. Lewis, D. L., Hagstrom, J. E., Loomos, A. G., Wolff, J. A., and Herweijer, H. (2002) Efficient delivery of siRNA for inhibition of gene expression in postnatal mice. *Nat. Genet.* **32**, 107–108.
11. Gariepy, J. and Kawamura, K. (2000) Vectorial delivery of macromolecules into cells using peptide-based vehicles. *Trends Biotechnol.* **19**, 21–26.
12. Morris, M. C., Depollier, J., Mery, J., Heitz, F., and Divita, G. (2001) A peptide carrier for the delivery of biologically active proteins into mammalian cells. *Nat. Biotechnol.* **19**, 1173.

13. Wadia, J. S. and Dowdy, S. F. (2002) Protein transduction technology. *Curr. Opin. Biotechnol.* **13**, 52–56.
14. Langel, U. ed. (2002) *Cell Penetrating Peptides: Processes and Application.* CRC, Boca Raton, FL.
15. Morris, M. C., Chaloin, L., Heitz, F., and Divita, G. (2000) Translocating peptides and proteins and their use for gene delivery. *Curr. Opin. Biotechnol.* **11**, 461–466.
16. Järver, P. and Langel, U. (2004) The use of cell-penetrating peptides as a tool for gene regulation. *Drug Discovery Today* **9**, 395–402.
17. Gait, M. J. (2003) Peptide-mediated cellular delivery of antisense oligonucleotides and their analogues. *Cell. Mol. Life Sci.* 60, 1–10.
18. Morris, M. C., Vidal, P., Chaloin, L., Heitz, F., and Divita, G. (1997) A new peptide vector for efficient delivery of oligonucleotides into mammalian cells. *Nucleic Acids Res.* **25**, 2730–2736.
19. Vidal, P., Morris, M. C., Chaloin, L., Heitz, F., and Divita, G. (1997) New strategy for RNA vectorization in mammalian cells. Use of a peptide vector. *C R Acad Sci III* **320**, 279–287.
20. Morris, M. C., Chaloin, L., Mery, J., Heitz, F., and Divita, G. (1999) A novel potent strategy for gene delivery using a single peptide vector as a carrier. *Nucleic Acids Res.* **27**, 3510–3517.
21. Morris, M. C., Heitz, F., and Divita G., personal communication.
22. Marthinet, E., Divita, G., Bernaud, J., Rigal, D., and Baggetto, L. G. (2000) Modulation of the typical multidrug resistance phenotype by targeting the MED-1 region of human MDR1 promoter. *Gene Ther.* **14**, 1224–1233.
23. Mery, J., Granier, C., Juin, M., and Brugidou J. (1993) Disulfide linkage to polyacrylic resin for automated Fmoc peptide synthesis. Immunochemical applications of peptide resins and mercaptoamide peptides. *Int. J. Pept. Protein Res.* 42, 44–52.
24. Vidal, P., Chaloin, L., Mery, J., et al. (1996) Solid-phase synthesis and cellular localization of a C- and/or N-terminal labelled peptide. *J. Pept. Sci.* **2**, 125–133.
25. Morris, M. C., Chaloin, L., Choob, M., Archdeacon, J., Heitz, F., and Divita, G. (2004) Combination of a new generation of PNAs with a peptide-based carrier enables efficient targeting of cell cycle progression. *Gene Ther.* **11**, 757–764.
26. Deshayes, S., Plenat, T., Aldrian-Herrada, G., Divita, G., Le Grimellec, C., and Heitz, F. (2004) Primary amphipathic cell-penetrating peptides: structural requirements and interactions with model membranes. *Biochemistry* **43**, 7698–7706.

12

Broth Microdilution Antibacterial Assay of Peptides

Laszlo Otvos and Mare Cudic

Summary

Native peptides exhibit various biological activities from which the antimicrobial property is one of the most frequently studied. A convenient way of telling whether peptides influence the life cycle of bacteria is the broth microdilution assay. In this measure, growing bacteria are incubated with peptides and growth inhibition is detected with colorimetric methods. Highly charged and protease-sensitive peptides need special considerations in assay design and readout interpretation to reveal the true antimicrobial efficacy and potential utility as human or veterinary therapeutics. The broth microdilution assay is suitable for first assessment of antimicrobial resistance induction.

Key Words: Bacteria; electrolyte content; growth inhibition; microbiology media; resistance induction; serial dilution; serum stability; turbidity.

1. Introduction

Short and medium-sized native peptides exhibit a series of biological activities, and are potentially viable starters for drug design. In addition to their immunological activities as stand-alone fragments of proteins, major peptide therapeutic classes include peptide hormones *(1)*, toxins *(2)*, and antimicrobial peptides *(3)*. Indeed, many antimicrobial peptides reached various phases of clinical development *(4)*. The initial evaluation of the in vitro efficacy of natural or synthetic analogs always starts with a type of antimicrobial assay. While there are a plethora of suitable protocols to tell whether a certain molecule can influence the life cycle of bacteria, such as agar diffusion and turbidimetric, potentiometric, enzymatic, and radioimmune assays *(5)*, the most convenient measure for peptides is the broth microdilution growth inhibition assay *(6)*. This protocol seems to suit the needs of the readers of this book, who are likely

From: *Methods in Molecular Biology, vol. 386: Peptide Characterization and Application Protocols*
Edited by: G. Fields © Humana Press Inc., Totowa, NJ

to be peptide chemists and biochemists, familiar with 96-well plate solution assays, and interested in pharmacological properties such as half-inhibitory concentration (IC_{50}) or functional serum stability in addition to the minimal inhibitory concentration (MIC) that microbiologists are mainly concerned with.

Broth microdilution is convenient and widely used for susceptibility testing of several antibiotics on a large number of bacterial isolates in a short time (7). The laboratory method is transferable to commercially available automated systems, including MicroScan plates or the Vitek automated microbiology system (bio Merieux Vitek, Inc.), which are also based on the broth microdilution method. In this measure, bacteria are grown in plate wells and antimicrobial test compounds are added in varying concentrations. Growing bacteria cause turbidity in the wells that is cleared if the antimicrobials, in our case, peptides, inhibit bacterial growth. In broth microdilution, as with many susceptibility tests, problems might arise when determining the endpoints of growth. Turbidity is usually read spectrophotometrically, most frequently at $\lambda = 600\,nm$, rather than just visual inspection of the wells. The assay is a 3-day procedure, with growing bacteria at day 1, peptide addition at day 2, and reading the plates at day 3.

1.1. Working With Bacteria

The properties that make bacteria very attractive as compared to eukaryotic cells are that, in most cases, bacteria grow very well, and an enormous number of strains are available from the American Type Tissue Collection (ATCC) or private depositories for nominal costs. The two staple organisms that are used as targets for the first indication of peptide suitability for human or veterinary applications are the Gram-negative *Escherichia coli* and the Gram-positive *Staphylococcus aureus* (8). These strains propagate well in Muller-Hinton (MHB) or Luria-Bertani (LB) broth without any additive (9,10). Ready-to-use LB solution or MHB powder to be reconstituted are commercially available; MHB appears to be the medium of choice for liquid antimicrobial susceptibility assays. Other common enterobacteriaceae with therapeutic interest such as *Salmonella typhimurium*, *Klebsiella pneumoniae*, or *Pseudomonas aeruginosa* also grow well in MHB. Some fastidious strains require special handling; most notable are respiratory tract pathogens—*Streptococcus pneumoniae*, *Haemophilus influenzae*, or *Moraxella catarrhalis*—which fail to grow efficiently in nonsupplemented media, but propagate adequately if special liquid or gaseous additives are introduced (11,12). The commercial suppliers of the strains will provide detailed instructions for bacterial growth conditions.

Actually, bacteria frequently grow better than we really want. Before any bacterial work is initiated, peptide laboratories must equip themselves for biosafety level (BSL)-2 conditions, and the laboratory environment has to be approved for such work by the institutional laboratory safety committee. The risk assessment process for working with infectious agents in a laboratory is accomplished by placing the agent in any one of four BSLs as defined by the Centers for Disease Control and National Institutes of Health guidelines, *Biosafety in Microbiological and Biomedical Laboratories*. BSL-1 is appropriate for work with defined and characterized microorganisms that are not known to cause disease in healthy adult humans. BSL-2 is appropriate for a broad spectrum of indigenous moderate risk agents present in the community and associated with human disease of varying severity. BSL-3 is appropriate for work with indigenous or exotic agents with a potential for respiratory transmission, and which may cause serious and potential lethal infection. BSL-4 is appropriate for dangerous and exotic agents that pose a high individual risk of life-threatening disease that may be transmitted by aerosol route and for which there is no available vaccine or therapy.

There are various levels of containment in a laboratory. They include laboratory practice and technique, safety equipment, and facility design. Primary containment is meant for the protection of personnel and the immediate laboratory environment. This may include personal protective equipment and the use of Biosafety Cabinets and/or the use of other safety devices. Secondary containment is for the protection of the environment external to the laboratory. This is accomplished by a combination of facility design and operational practices.

1.2. Most Frequently Used Terms

For broth microdilution, the MIC is defined as the lowest concentration of an antimicrobial agent at which no bacterial growth is detected (*see* **Fig. 1**). A bit confusing is the microbiology use of the terms MIC_{50} and MIC_{90} that reflect the percentage of bacterial strains of a given, usually broad assay, where antibiotics show certain MIC values. The all-or-nothing view of antimicrobial therapies do not favor the evaluation of activity as an IC_{50} figure; yet for structure-activity relationship studies this way of expressing peptide activity is more convenient than the dry MIC terminology. IC_{50} is defined as an antibiotic concentration where the activity curve crosses the half-line between uninhibited bacterial growth and medium only at $\lambda = 600\,nm$ colorimetric readings (*see* **Fig. 1**).

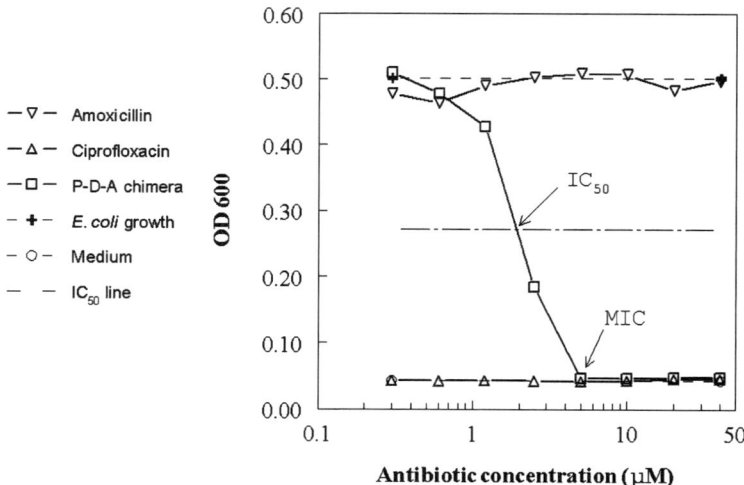

Fig. 1. Identification of the minimal inhibitory concentration and the half-inhibitory concentration during the broth microdilution antibacterial assay. Shown are bacterial growth inhibition curves of the peptidic antibiotic P-D-A chimera (description in **ref. 16**), amoxicillin (negative control), and ciprofloxacin (positive control) against the β-lactam and sulfonamide resistant *Escherichia coli* SEC102 strain. The assay was run in full-strength Muller-Hinton broth.

Another major difference between the microbiology and basic peptide antibiotic literature is the expression of active peptide concentration. For ensuing pharmacokinetics and dosing studies, microbiologists use microgram per milliliter concentration values. Peptide chemists prefer molarity terms (peptides usually kill bacteria in micromolar concentrations) not only because of a better grasp of the underlying molecular interactions, but also because microgram per milliliter figures would further favor small molecule conventional antibiotics by 5- to 10-fold over peptides that are significantly larger molecules. To satisfy both parties, we usually recommend providing both types of efficacy values.

1.3. Special Considerations for Peptide Antibiotics

For communicating the results with drug development companies, it is imperative that the assay is run according to recommendations by the National Committee for Clinical Laboratory Standards (NCCLS *[13]*). These recommendations require the use of high nutrient growth media in which bacteria propagate intensely. However, the jury is still out as to which environment mimics most faithfully the growth conditions present in mammalian organisms. For example, peptide antibiotics are inhibited by the salt content of test media *(14)*, and indeed, the renal tubules, target tissues of some antimicrobial peptides, have significantly lower salt content than other tissues or bodily fluids *(15)*. We often study the activity of peptide antibiotics in both full-strength MHB and in a one-quarter strength version (diluted with water 1:3). It must be added that bacterial growth is less intense in diluted medium, likely with a less healthy membrane development *(16)*. Nevertheless, conventional antibiotics, for which membrane penetration is not an issue, show identical MIC values irrespective of the electrolyte composition of the medium whereas peptide antibiotics exhibit a wide activity range (*see* **Table 1**).

Pharmacologists know too well that the major drawback to the therapeutic use of peptides is their sensitivity to serum degradation *(17)*. Indeed, some otherwise potent antimicrobial peptides lose activity in the presence of mammalian serum whereas structurally similar analogs retain high activity when serum is added to the growth medium (*see* **Fig. 2**). A useful exercise is to assay the antibacterial activity of peptide antibiotics on half of the plates without serum addition, and in the presence of 25% serum (replacing 25% of medium with serum) on the other half. As the colorimetric readings at $\lambda = 600$ nm (AUFS) are influenced

Table 1
Minimal Inhibitory Concentrations of the Peptidic Antibiotic Pip-pyrr-MeArg Dimer (description in ref. *14*) and the Fluoroquinolone Antibiotic Levofloxacin in the Broth Microdilution Assay Conducted Either in Full-Strength Muller-Hinton Broth or in a One-Quarter Strength Version

Bacterial strains and antibiotics	Minimal inhibitory concentration (μg/mL) in	
	undiluted MHB	one-quarter strength MHB
Escherichia coli ATCC25922		
Pip-pyrr-MeArg dimer	>128	8
Levofloxacin	0.06	0.03
Salmonella typhimurium ATCC14028		
Pip-pyrr-MeArg dimer	>128	16
Levofloxacin	0.06	0.03
Klebsiella pneumoniae ATCC13883		
Pip-pyrr-MeArg dimer	>128	8
Levofloxacin	0.06	0.06

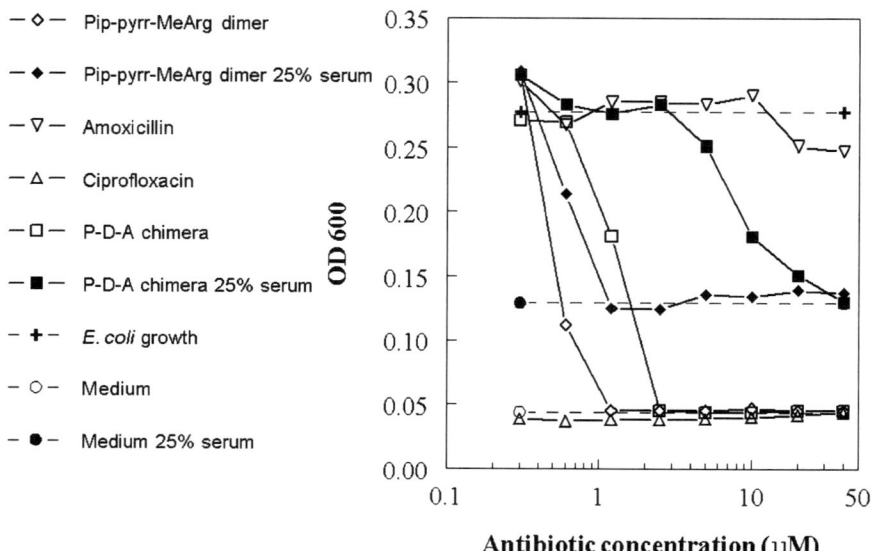

Fig. 2. Influence of serum on the activity of peptide antibiotics. The assay was run without or in the presence of 25% mouse serum. The bacterial strain was *Escherichia coli* SEC102, and the medium was one-quarter strength Muller-Hinton broth. The efficacy of neither the negative control amoxicillin nor the positive control ciprofloxacin was affected by serum addition.

by the presence of serum, both halves of the plates have to have their own uninhibited bacterial growth and medium only lanes (*see* **Fig. 2**). Do not be surprised if occasionally peptide activity improves in the presence of serum; this probably reflects peptide stability and the more in vivo-like bacterial growth conditions.

1.4. Resistance Induction

A well crafted recent review points out that bacteria can mutate their proteins easily, and antibacterial peptides with target proteins may be inferior to analogs acting on the cell membrane because of a relatively easier selection of resistant strains through mutations at the target level (*18*). To study this possibility, one can try to generate resistance against peptidic antimicrobials in vitro. Although a number of various methods are available for the evaluation of resistance induction, the most frequently used is the serial passage of bacteria that are subjected to sublethal doses of peptide antibiotics (*19*). For example, we co-cultured *E. coli* CFT073 once daily for 15 d with sublethal doses of a designed

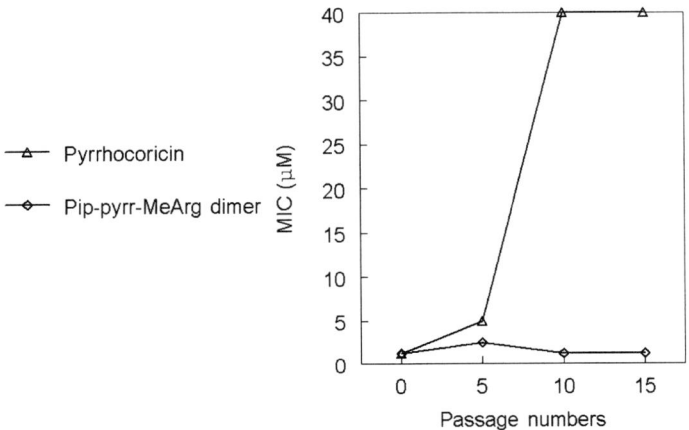

Fig. 3. Induction of microbial resistance by the peptide antibiotic Pip-pyrr-MeArg dimer (description in **ref.** *14*). The minimal inhibitory concentration against the clinical uropathogen *Escherichia coli* CFT073 was measured after 5, 10, and 15 passages in one-quarter strength Muller-Hinton broth.

antimicrobial peptide derivate, the Pip-pyrr-MeArg dimer, that is based on the native proline-rich peptide pyrrhocoricin *(15)*. The induction of resistant mutants, as indicated by the increase of the MIC values, was studied after every fifth passage not only against the co-culturing antibiotic, but a series of other antimicrobial compounds. As **Fig. 3** indicates, treatment of bacteria with the designer dimer failed to generate resistance to itself, but quickly turned the strains resistant to the native peptide. Although this exercise is usually useful to first assess the potential for resistance induction, it is not perfect. With this method, we failed to induce β-lactamase resistance to amoxicillin.

2. Materials

2.1. Equipment

1. UV/VIS Spectrophotometer (Beckman, model DU-50).
2. Shaking incubator at 37° C (Broekel Industries).
3. Automated Microplate Reader (Biotek Instruments, model #L311).
4. Vortex (Fisher).

2.2. Medium and Reagents

1. MHB medium (Difco 275730).
2. Muller-Hilton agar (Difco 225250).
3. Sterile mouse serum (Equitech-bio SM-0050).

4. Commercially available antibiotics as controls:

    a. Amoxicilin (Sigma A-8523).
    b. Ciprofloxacin (USBiological 5074).

5. *E. coli* (ATCC 25922).
6. *S. typhimurium* (ATCC 14028).
7. *K. pneumoniae* (ATCC 13883).
8. *S. aureus* (ATCC 27660).
9. Sterile snap cap tubes (Falcon 35–2059).
10. Sterile water (to dilute peptides with).
11. Sterile Eppendorf tubes, pipet tips, cryovials (Fisher).
12. Disposable plastic cuvettes (Fisher).
13. Flat bottom polypropylene 96 well plates (Falcon 35–3072) (*see* **Note 1**).

2.3. Preparation of Media and Agar for Bacterial Grow

1. MHB Medium: Dissolve 21 g in 1 L deionized water. Autoclave at 121°C for 20 min and store at room temperature.
2. MHB Agar Plates: Dissolve 38 g in 1 L deionized water. Autoclave at 121°C for 20 min, then allow agar to cool to 50°C, then pour into plates and let it to solidify (*see* **Note 2**).

2.4. Prepare Peptide/Antibiotic Solutions

1. Dissolve peptides to be tested such that the stock concentration is 400 μ*M* (peptide research approach) or 1.28 mg/mL (microbiology approach).
2. Make twofold serial dilutions from peptide stock starting with 400 μ*M* (400, 200, 100, 50, 25, 12.5, 6.25, and 3.12 μ*M*). Use sterile water and tubes for making the dilutions.
3. For controls use commercial antibiotics such as amoxicillin or ciprofloxacin.

3. Methods
3.1. Antibacterial Assay
3.1.1. Starting With Freeze-Dried Bacteria

1. Reconstitute lyophilized bacteria following the manufacturer's (for example ATCC) recommendation.
2. Transfer reconstituted cells to appropriate media, diluting 1:5 (original culture).
3. Pipet 200 μL of original culture to prewarmed agar plate and spread evenly using a sterile pipet tip or loop. Place plate into the incubator and let it grow overnight.

4. Dilute original culture 1:10 with media (400 μL bacterial suspension to 3.6 mL media) in a snap cap tube and place in shaking incubator to grow overnight (stock bacterial solution).
5. Use this stock solution to freeze cells in the morning for seed cell bank (*see* **Note 3**).

3.1.2. Starting With Frozen Bacteria

1. Remove one tube of frozen bacteria (seed tube) from the −80° C freezer and place it on dry ice immediately.
2. Take a stab from the top of the tube using a sterile pipet tip (you need to let the top of the tube thaw a little) and spread the cells on prewarmed agar plate by touching the plate gently. Place the plate into an incubator set to 37° C and let it grow overnight.
3. Place seed tube back to the −80° C freezer as soon as possible after finished, keeping it on dry ice in the meantime.

3.1.3. Growing Bacteria

1. Start growing cells in suspension by transferring a few isolated colonies from the agar plate to 4 mL media in a snap cap tube (starter culture).
2. Place tube into the shaking incubator at 37° C and let the cells grow for approx 2–3 h to medium opacity.
3. Transfer 1 mL starter culture to a sterile Eppendorf tube and centrifuge for 5 min at 4000 rpm. (Keep the rest of the starter culture at room temperature.)
4. Remove and discard the supernatant (*see* **Note 4**). Resuspend the cell pellet in 1 mL of fresh medium and transfer it to snap cap tubes containing 3 mL media (1:4 dilution).
5. Transfer 1 mL of this solution to a plastic cuvette and measure the optical density at 600 nm using medium only as baseline (*see* **Note 5**).
6. Keep the diluted starter culture on ice.

3.1.4. Assay Set-Up

1. Just before setting up the assay, further dilute the already diluted starter culture (from **Subheading 3.1.3.**) such that the final optical density at 600 nm will be 0.001. Use full or one-quarter strength media for the dilution depending upon the expected efficacy. Approximately 11 mL cell suspension will be needed for each plate assay.
2. Use flat-bottom 96-well plates for assay. Each well will hold 90 μL bacterial suspension and 10 μL diluted peptide. It is recommended that the assay be run in duplicates or triplicates (2–3 wells for every given peptide concentration) (*see* **Note 6**).
3. Use 100 μL media in the last lane of the plate for medium only control (*see* **Note 7**).
4. In the penultimate lane, use 90 μL bacterial suspension without any peptide + 10 μL water as uninhibited bacterial growth control.

5. Pipet 90 μL bacterial suspension to the rest of each well. Add 10 μL of serially diluted peptide to each well, starting with the highest concentration in the first row and getting more and more dilute further down the lane. Mark each peptide's position on a paper sheet copying the plate contents (see **Note 8**).
6. Cover the plate and place into an incubator set to 37°C and allow growth overnight.

3.1.5. Quantitation of the Assay

1. After approx 16 h of incubation measure the cell density in each well at $\lambda = 600$ nm using the plate reader.
2. Generate inhibition curves by plotting the colorimetric readings at $\lambda = 600$ nm on the Y axis and peptide concentration on the x-axis (as in **Figs. 1** and **2**).
3. Sterilize the plates with 10% bleach and discard them into red containers labeled as "Infectious Waste."

3.2. Resistance Induction

1. Determine the MIC (from **Subheading 3.1.5.**).
2. In a single well, incubate 90 μL bacterial suspension with 10 μL peptide antibiotics added at half or one-quarter of the MIC (one or two higher dilution) (see **Note 9**).
3. Incubate the plate at 37°C overnight.
4. Use this well as a starter culture and repeat **steps 2–3** for 10–15 consecutive days (see **Note 10**).
5. After the last subculturing cycle, use the well content as starter culture and repeat the entire broth microdilution assay (**Subheadings 3.1.4.** and **3.1.5.**).
6. Plot the MIC as a function of passage numbers (see **Fig. 3**).

4. Notes

1. Do not substitute polystyrene for polypropylene microtiter plates. Cationic peptides bind to the surface of polystyrene (especially "tissue-culture treated" polystyrene).
2. Spray and wipe the bench with 95% ethanol before making agar plates. Use flame to minimize contamination. If bubbles form while the agar is being poured into the plate, burst them by passing the flame on the agar quickly. Put the plates inside the bag upside down (to avoid drying out) and store at 4°C.
3. Cells are optimally stored in 15% glycerol at −80°C.
4. Whenever discard any bacterial material, it must be sterilized with 10% bleach solution and placed to infectious waste containers.
5. The desired cell density is 0.1–0.4 AUFS units. If the cells are not dense enough (optical density less than 0.1) place the starter culture back to the incubator and check again 30 min later (**Subheading 3.1.3., steps 3–5**). Discard the first set of diluted cells. Keep checking until the desired density is achieved.
6. The final peptide concentration in the first row (highest concentration) will be one-tenth of the values from **Subheading 2.4., step 2**.

7. Be careful not to cross-contaminate the rest of the lanes with the bacterial preparation.
8. For studying peptide activity in the presence of serum, add 75 μL bacterial suspension to the wells (**Subheading 3.1.4., step 3**) and 65 μL bacterial suspension (**Subheading 3.1.4., steps 4** and **5**), followed by 25 μL sterile mouse or human serum. Add 10 μL of peptide solution and continue.
9. Ideally, one would need only one dilution higher than the MIC, where some bacterial growth is still detected. However, the liquid broth microdilution assay is frequently reported to be inaccurate for one serial dilution. To make certain that bacteria still grow, the authors sometimes use peptides at 25% of the MIC.

References

1. Loffet, A. (2002) Peptides as drugs: is there a market? *J. Pept. Sci.* **8**, 1–7.
2. Jones, R. M., and Bulaj, G. (2000) Conotoxins—new vistas for peptide therapeutics. *Curr. Pharm. Des.* **6**, 1249–1285.
3. Cudic, M. and Otvos, L., Jr. (2002) Intracellular targets of antibacterial peptides. *Curr. Drug Targets* **3**, 101–106.
4. Zasloff, M. (2002) Antimicrobial peptides of multicellular organisms. *Nature* **415**, 389–395.
5. Sabath, L. D. (1976) The assay of antimicrobial compounds. *Hum. Pathol.* **7**, 287–185.
6. Amsterdam, D. (1996) Susceptibility testing of antimicrobials in liquid media, in: *Antibiotics in Laboratory Medicine* (Loman, V., ed.). Williams and Wilkins, Philadelphia, PA: pp. 52–111.
7. Rahman, M., Kuhn, I., Rahman, M., Olsson-Liljequist, B., and Mollby, R. (2004) Evaluation of a scanner-assisted colorimetric MIC method for susceptibility testing of gram-negative fermentative bacteria. *Appl. Environment. Microbiol.* **70**, 2398–2403.
8. Otvos, L., Jr. (2000) Antibacterial peptides isolated from insects. *J. Pept. Sci.* **6**, 497–511.
9. Stubbings, W. J., Bostock, J. M., Ingham, E., and Chopra, I. (2004) Assessment of a microplate method for determining the post-antibiotic effect in *Staphylococcus aureus* and *Escherichia coli*. *J. Antimicrob Chemother.* **54**, 139–143.
10. Rajagopal, S., Eis, N., and Nickerson, K. W. (2003) Eight gram-negative bacteria are 10,000 times more sensitive to cationic detergents than to anionic detergents. *Can. J. Microbiol.* **49**, 775–779.
11. Krisher, K. K. and Linscott, A. (1994) Comparison of three commercial MIC systems, E test, fastidious antimicrobial susceptibility panel, and FOX fastidious panel, for confirmation of penicillin and cephalosporin resistance in *Streptococcus pneumoniae*. *J. Clin. Microbiol.* **32**, 2242–2245.
12. Virella, G. (1996) *Microbiology and Infectious Diseases*. Williams and Wilkins, Philadelphia, PA.

13. National Committee for Clinical Laboratory Standards (1997) Methods for dilution antimicrobial susceptibility tests for bacteria that grow aerobically, 4th ed. Approved standard M7–A4. *National Committee for Clinical Laboratory Standards*, Wayne, PA.

14. Cudic, M., Condie, B. A., Weiner, D. J., et al. (2002) Development of novel antibacterial peptides that kill resistant clinical isolates. *Peptides* **23**, 271–283.

15. Cudic, M., Lockatell, C. V., Johnson, D. E., and Otvos, L., Jr. (2003) In vitro and in vivo activity of a designed antibacterial peptide analog against uropathogens. *Peptides* **24**, 807–820.

16. Otvos, L., Jr., Snyder, C., Condie, B., Bulet, P., and Wade, J. D. (2005) Chimeric antimicrobial peptides exhibit multiple modes of action. *Int. J. Pept. Res. Ther.* **11**, 29–42.

17. Powell, M. F., Stewart, T., Otvos, L., Jr., et al. (1993) Peptide stability in drug development. II. Effect of single amino acid substitution and glycosylation on peptide reactivity in human serum. *Pharm. Res.* **10**, 1268–1273.

18. Gennaro, R., Zanetti, M., Benincasa, M., Podda, E., and Miani, M. (2002) Pro-rich antimicrobial peptides from animals: structure, biological functions and mechanism of action. *Curr. Pharm. Des.* **8**, 763–778.

19. Ge, Y., MacDonald, D. L., Holroyd, K. J., Thornsberry, C., Wexler, H., and Zasloff, M. (1999) In vitro properties of pexiganan, and analog of magainin. *Antimicrob. Agents Chemother.* **43**, 782–788.

13

Depsipeptide Synthesis

Maciej Stawikowski and Predrag Cudic

Summary

Naturally occurring cyclic depsipeptides, peptides that contain one or more ester bonds in addition to the amide bonds, have emerged as an important source of pharmacologically active compounds or promising lead structures for the development of novel synthetically derived drugs. This class of natural products has been found in many organisms, such as fungi, bacteria, and marine organisms. It is very well known that cyclic depsipeptides and their derivatives exhibit a diverse spectrum of biological activities, including insecticidal, antiviral, antimicrobial, antitumor, tumor-promotive, anti-inflammatory, and immunosuppressive actions. However, they have shown the greatest therapeutic potential as anticancer and particularly antimicrobial agents. Difficulties associated with isolation and purification of larger quantities of this class of natural products and, particularly, unlimited access to their synthetic analogs significantly hampered cyclic depsipeptides exploitation as lead compounds for development of new drugs. As an alternative, total solution or solid-phase peptide synthesis of these important natural products and combinatorial chemistry approaches can be employed to elucidate structure–activity relationships and to find new potent compounds of this class. In this chapter, methods for formation of depsipeptide ester bonds, hydroxyl group protection, and solid-phase reaction monitoring are described.

Key Words: Depsipeptides; solution and solid-phase synthesis; ester bond formation; hydroxyl group protection; reaction monitoring.

1. Introduction

Natural products serve as an important source of pharmacologically active compounds or lead structures for the development of novel synthetically derived drugs *(1–3)*. This is particularly evident in the areas of cancer and infectious diseases, where in the period between 1981 and 2002 over 60% of the approved drugs and drug-candidates are of natural origin *(4)*. Among natural products,

From: *Methods in Molecular Biology, vol. 386: Peptide Characterization and Application Protocols*
Edited by: G. Fields © Humana Press Inc., Totowa, NJ

peptides are particularly interesting because of the key roles they play in physiological processes. According to recent literature data, there are more than 40 peptide drugs available on the market and more than 80 peptides in the clinical phase II and III trials (5–8). Although there are limitations for peptides as drugs per se (short half-life, rapid metabolism, and poor oral bioavailability), their potential high efficacy combined with minimal side effects made them to be widely considered as a lead compounds in drugs development. Nevertheless, pharmacokinetic properties of peptides can be improved by different types of modifications (9). Peptidomimetic modifications or cyclization of linear peptides are frequently used as an attractive method to provide more conformationally constrained and thus more stable and bioactive peptides (10–15). In addition to this, replacement of the amide groups that undergo proteolytic hydrolysis with ester groups may lead to longer-acting compounds not so prone to proteolysis (16–19). Considering all these modifications that can potentially improve peptide metabolic stability, naturally occurring cyclic depsipeptides that contain one or more ester bonds in addition to the amide bonds emerge as promising lead compounds for drug discovery.

Cyclic depsipeptides have been found in many natural organisms such as fungi, bacteria, and marine organisms (20,21). It is very well known that cyclic depsipeptides and their derivatives exhibit a diverse spectrum of biological activities including insecticidal, antiviral, antimicrobial, antitumor, tumor-promotive, anti-inflammatory, and immunosuppressive actions. However, they have shown the greatest therapeutic potential as anticancer and particularly antimicrobial agents. Depsipeptides such as didemnin B, dolastatin 10, kahalalide F, and FR901228 have entered clinical trials as potential anticancer agents. Among them, dolastatin 10 emerges as the most potent antineoplastic agent known (22). It inhibits microtubule assembly and induces apoptosis in numerous malignant cell lines. Dolastatin 10 is currently in phase-I and -II cancer clinical trials. In addition, occurrence of multidrug-resistant pathogens and urgent demands for new and more potent antimicrobials place also this class of natural products in the center of the attention for development of new antibacterial agents. Excellent examples of depsipeptide's clinical potentials as novel antimicrobial agents are naturally occurring cyclic lipodepsipeptides daptomycin 1 and ramoplanin 2 (Fig. 1). Very importantly, both of these lipodepsipeptides inhibit biosynthesis of Gram-positive bacterial cell wall by the mechanisms that differ from those characteristic for vancomycin, the most important drug in current use for the treatment of Gram-positive bacterial infections. Cyclic lipodepsipeptide daptomycin 1 (Cubicin®, Cubist Pharmaceuticals, Inc.) (23) was approved in September

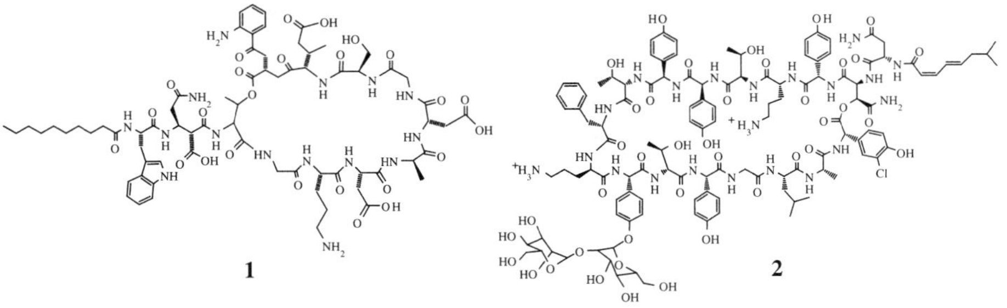

Fig. 1. Structures of lipodepsipeptides daptomycin 1 and ramoplanin 2.

2003 by the US Food and Drug Administration (FDA) for the treatment of complicated skin infections caused by Gram-positive bacteria, including methicillin-resistant *Staphylococcus aureus*. This cyclic lipodepsipeptide has a unique mechanism of action that involves disruptions of multiple aspects of bacterial membrane function. Daptomycin is the first antibiotic in a new structural class to be approved since the introduction of the oxazolidinone linezolid (Zyvox®, Pfizer) in 2000. Another cyclic lipodepsipeptide mentioned here, ramoplanin 2 (Oscient Pharmaceuticals), is among the agents in advanced stages of clinical development for eradication of vancomycin-resistant *Enterococcus faecium* and methicillin-resistant *S. aureus* and, at present, likely to proceed to licensing *(15,23)*. Ramoplanin disrupts bacterial cell wall biosynthesis by inhibiting glycosyltransferase- and transglycosylase-catalyzed peptidoglycan biosynthesis.

Difficulties associated with isolation and purification of larger quantities of naturally occurring depsipeptides and, particularly, unlimited access to their synthetic analogs significantly slower their exploitation as lead compounds for development of new drugs. As an alternative, total solution or solid-phase peptide synthesis of these important natural products and combinatorial chemistry approach can be employed to elucidate structure–activity relationship and to find new potent compounds of this class. Cyclic depsipeptide synthesis presents a challenging synthetic task because of depsipeptides' structural diversity and complexity, specifically in the macrocyclic domain, and their complex, mainly lipidic, side chains. Therefore, a general depsipeptide synthetic strategy can be outlined as follows:

1. Synthesis of unusual building blocks (amino acids, lipids, sugars etc.).
2. Incorporation of these building blocks into peptide chain by traditional peptide synthetic methodologies.

3. Cyclization in solution or on solid support *via* macrolactamization (amide bond formation), or macrolactonization (ester bond formation).

Although synthesis and incorporation of unusual building blocks into the peptide chain is not straightforward and very often poses a synthetic challenge, the key step in the synthesis of cyclic depsipeptides is the ring closure. The ring closure carries significant strategic importance and can dictate the level of success of the synthesis. For example, poor ring disconnection can lead to slow cyclization rates, thus facilitating side reactions such as dimerization, oligomerization, and/or epimerization of the C-terminal residue. Traditional methods to prepare cyclic peptides and, therefore, depsipeptides involve solid-phase synthesis of the partially protected linear precursor and cyclization in solution under high dilution conditions *(10,12)*. As an attractive alternative, cyclization could be performed while peptides still remain anchored to the polymeric support. The solid-phase method may be advantageous because of pseudo-dilution effect, a kinetic phenomenon that favors intramolecular reactions over intermolecular reactions *(24)*. Also, taking into consideration the limited stability of the ester bond and the possibility of racemization if basic conditions were to be used as well as the compatibility of the deprotection and cleavage conditions with the resin linkage, macrolactamization appears to be better choice for depsipeptide ring closure *(25–28)*. However, macrolactamization is not the only option, and examples of successful macrolactonization have been reported as well *(29,30)*. Cyclization strategies in peptide synthesis were subject of many recent reports *(10,12)*; therefore, they are not described in this chapter. Instead, methods for formation of depsipeptide ester bond, hydroxyl group protection, and methods for solid phase reaction monitoring are described.

2. Materials

1. All materials and reagents are commercially available and used as received.
2. Synthesis solvents, such as dichloromethane (DCM), ethyl acetate (EtOAc), *tert*-butanol (*t*-BuOH), 1-methyl-2-pyrrolidinone (NMP), tetrahydrofurane (THF), methanol (MeOH), toluene, were high-performance liquid chromatography (HPLC) or peptide-synthesis-grade and can be obtained from Sigma-Aldrich, Fisher, VWR, or other commercial sources.
3. Peptide coupling reagents such as diisopropylcarbodiimide (DIC), benzotriazole-1-yl-oxy-tris-pyrrolidino-phosphonium hexafluorophosphate (PyBOP), and bromo-tris-pyrrolidino-phosphonium hexafluorophosphate (PyBrop), may be obtained from Sigma-Aldrich, ChemImpex, Novabiochem.
4. Specific esterification reagents including diethyl azodicarobxylate (DEAD) and derivatives, triphenylphosphine (PPh$_3$) and 2,4,6-trichlorobenzoyl chloride, can be purchased from Sigma-Aldrich, Fisher, VWR or other commercial sources.

5. Hydroxyl protecting/deprotecting reagents such as trifluoromethanesulfonic acid *tert*-butyldimethylsilylester (TBDMS triflate), tetrabutylammonium fluoride (TBAF), dihydropyran, *p*-toluenesulfonic acid (*p*-TsOH), and hexafluoroacetone (HFA) can be purchased from Sigma-Aldrich, Fisher, Fluka or other suppliers.

6. Reagents for monitoring of the presence of free hydroxyl groups such as 4-(p-nitrobenzyl)pyridine, *p*-toluenesulfonyl chloride (*p*-TsCl), 4-methyl morpholine (NMM) and 2,4,6-trichloro-[1,3,5]-triazine (TCT), and dyes such as Alizarin R, fluorescein or fuchsin, can be purchased from Sigma, Fluka, Merck or other suppliers.

7. Resins for solid-support synthesis can be obtained from Rapp-Polymere (Germany), Novabiochem (USA), ChemImpex, Aapptec (USA) and other suppliers.

3. Methods

Since R. B. Merrifield's pioneering work in early 1960s, solid phase synthesis became a routine tool for the preparation of peptides and other natural oligomers, namely nucleotides and oligosaccharides. As a result of a high coupling efficiency and suppression of enantiomerization, aminium- (uranium) and phosphonium-salt based coupling reagents have become the preferred peptide synthetic tools. However, their application in depsipeptide ester bond formation was shown to be quite inefficient *(34)*. Because no general methodology has been established for the synthesis of depsipeptides, this chapter describes approaches most commonly reported in the literature for the synthesis of this important class of natural products. These include solid phase and solution procedures.

3.1. Carbodiimide/4-Dimethylaminopyridine Coupling Method

Carbodiimide reagents have been widely used in peptide synthesis because of their moderate activity and low cost *(31)*. They are used as a coupling reagents and esterification reagents during loading first amino acid on resin. The most commonly used carbodiimide reagent is 1,3-diisopropylcarbodiimide (DIC, DIPCI). By using a 2:1 molar ratio of amino acid to DIC, the symmetrical anhydride is formed which in turn reacts with free hydroxyl group and the ester bond is formed (**Fig. 2**). The reaction is catalyzed by the presence of 4-dimethylaminopyridine (DMAP), which increases the nucleophilicity of the hydroxyl group *(32)*.

When carbodiimide is used in 1:1 molar ratio with amino acid, the reaction proceeds *via* O-acylisourea mechanism and the corresponding ester bond is formed (**Fig. 3**).

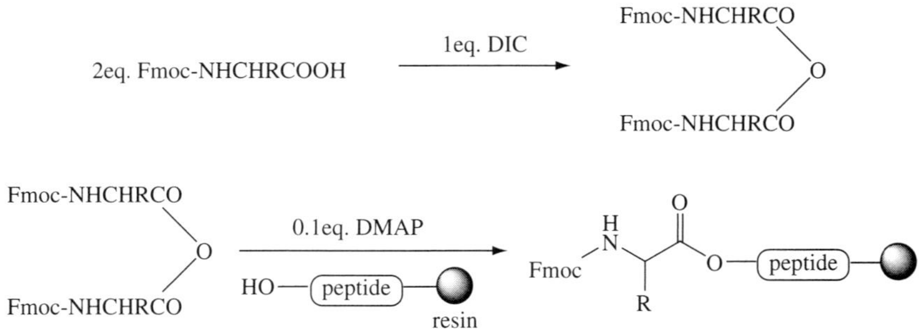

Fig. 2. Symmetrical anhydride method of ester bond formation.

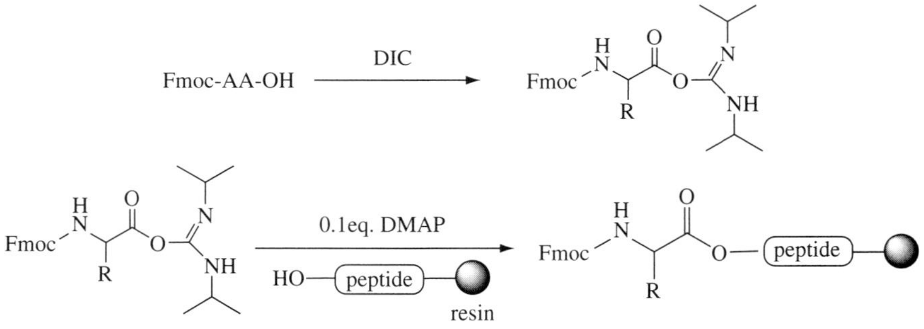

Fig. 3. Mechanism of ester bond formation via *O*-acylisourea.

The following general solution and solid phase synthetic protocols can be used for depsipeptide bond formation using DIC/DMAP coupling methodology *(33,34)*.

3.1.1. Depsipeptide Ester Bond Formation Using the DIC/DMAP Method on Solid Support

This protocol was adopted from **refs. *33*** and ***34***.

1. Place the resin in dry reaction vessel.
2. Wash the resin three times with THF and DCM and then swell in THF.
3. Dissolve DIC (5 eq), DMAP (0.1 eq relative to resin loading) and protected carboxylic acid derivative (protected amino acid derivative) in THF (1 mL/100 mg of the resin) (*see* **Note 1**). Add this solution to the resin.
4. Allow the resin to agitate at room temperature for 2 h.
5. Wash the resin (3 min each) with 3 × 10 mL of THF; 3 × 10 mL of acetone; 3 × 10 mL of DCM.

6. Transfer small amount of resin to the test tube and perform test for the presence of free hydroxyl groups (*see* **Subheadings 3.6.1.** and **3.6.2.**) or monitor the coupling by matrix-assisted laser desorption/ionization (MALDI)-time-of-flight (TOF) analysis.
7. If the resin gives the positive test, repeat **steps 2–5** with fresh reagents.

3.1.2. Depsipeptide Ester Bond Formation Using the DIC/DMAP Method in Solution

This protocol was adopted from **refs. *36–38***.

1. Dissolve protected amino acid (1 eq) and alcohol derivatives (1 eq) in DCM (10 mL/equivalent of amino acid).
2. Cool the reaction mixture to 0°C under an atmosphere of dry N_2.
3. Add DIC (1 eq) and DMAP (0.1 eq).
4. Stir the mixture at room temperature for 16 h.
5. Filter the reaction mixture to remove the precipitated diisopropylurea.
6. Concentrate the filtrate and purify by column chromatography.

3.2. Ester Bond Formation Using the Boc-Amino Acid N-Hydroxysuccinimide Ester Method

R. Katakai et al. *(39)* reported recently solution phase synthesis of Boc-tri- and terta-depsipeptides using Boc-amino acid *N*-hydroxysuccinimide ester (Boc-AA-ONSu) **(Fig. 4)**. The method first requires synthesis of didepsipeptide-free acids through the formation of an ester bond between the carboxyl group of an amino acid and the hydroxyl group of a free hydroxyl acid. This is achieved by the reaction of Boc-AA-ONSu with hydroxyl acid pyridinium salt in the presence of a catalytic amount of DMAP. In the second step, depsipeptide chain elongation was obtained by the reaction of free C-terminal carboxylic group with Boc-AA-ONSu. The method also allows the use of less polar organic solvents, which can be advantageous for the synthesis of highly hydrophobic depsipeptide sequences.

This protocol was adopted from **ref. *39***.

1. Dissolve Boc-protected hydroxy amino acid (1.2 eq) in THF (1.6 mL/mmol of amino acid) and pyridine (1.2 eq).
2. Add to the solution Boc-AA-ONSu (1 eq) and DMAP (0.1 eq).
3. Stir the reaction mixture at room temperature for 20 h.
4. Dilute the mixture with ethyl acetate, and wash with 1 *M* HCl and water.
5. Extract the mixture three times with saturated $NaHCO_3$ aqueous solution.
6. Combine aqueous layers, acidify with 1 *M* HCl and extract three times with ethyl acetate.
7. Wash organic layer with water and saturated NaCl aqueous solution and dry over Na_2SO_4.

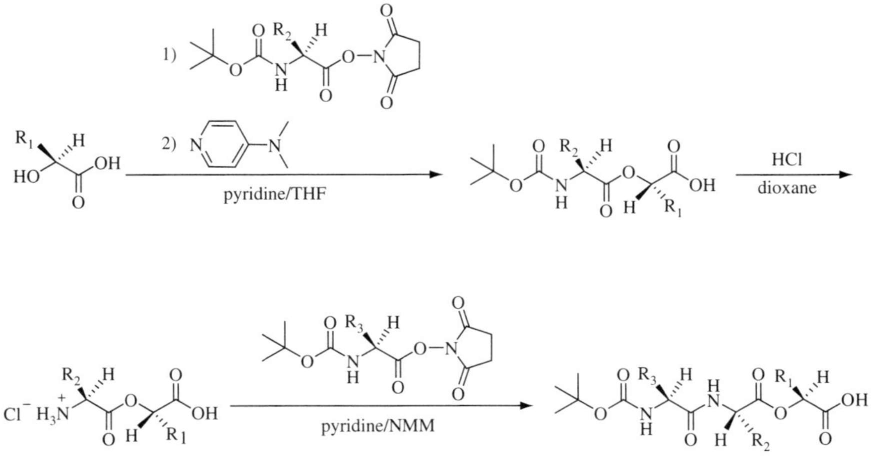

Fig. 4. Synthesis of Boc-depsipeptides using Boc-amino acid *N*-hydroxysuccinimide ester.

8. Concentrate the filtrate and crystallize the product by addition of hexane.
9. Purify the product by recrystalization from diethyl ether/hexane (*see* **Note 2**).

3.3. Mitsunobu Esterification

The Mitsunobu reaction *(40)* is widely used in synthetic organic chemistry because of reaction mildness and effectiveness, and provides an alternative method for esterification in which an alcohol, not a carboxylic component, is activated (**Fig. 5**). The esterification reaction via Mitsunobu mechanism is carried out in the presence of the redox system such as DEAD/PPh$_3$ and proceeds with complete inversion of configuration of the alcohol component *(41)*. This reaction, commonly used in the syntheses of nonpeptidic lactones, was shown to be effective in solution phase preparation of depsipeptides as well.

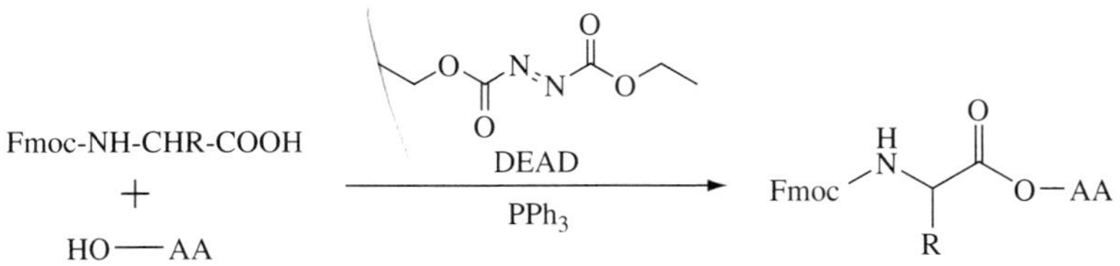

Fig. 5. Mitsunobu esterification. AA, amino acid.

This protocol was adopted from **refs. *42*** and ***43***.

1. Dissolve *N*-protected amino acid (1 eq), corresponding alcohol (1 eq) and triphenylphosphine (2 eq) in dry THF (8 mL/mmol of AA).
2. Cool the reaction mixture to 0°C under an atmosphere of dry N_2.
3. Add dropwise DEAD (2 eq) (*see* **Note 3**).
4. Stir the mixture at room temperature for 4 h.
5. Remove the solvent under reduced pressure.
6. Dissolve the residue in EtOAc and wash three times with a saturated solution of $NaHSO_4$.
7. Dry the organic layer over Na_2SO_4, concentrate the filtrate and purify by column chromatography.

3.4. Yamaguchi Esterification

The esterification reaction that requires use of 2,4,6-trichlorobenzoyl chloride for the preparation of a mixed anhydride was first reported by Yamaguchi and co-workers *(44)*. In general, the Yamaguchi esterification involves the reaction of an aliphatic acid with 2,4,6-trichlorobenzoyl chloride to form the mixed aliphatic-2,4,6-trichlorobenzoyl anhydride. The isolated mixed anhydride, upon reaction with an alcohol, in the presence of DMAP, produces the aliphatic ester with high regioselectivity (**Fig. 6**). Although commonly used in the macro-lactonization reactions, there are only few literature examples of Yamaguchi esterification in peptide chemistry *(45,46)*. Nevertheless, the Yamaguchi esterification reaction may present an interesting alternative to the Mitsunobu esterification for solution phase depsipeptide preparation.

This protocol was adopted from **refs. *45*** and ***46***.

1. Dissolve *N*-protected amino acid (1 eq), diisopropylethylamine (1.5 eq), and 2,4,6-trichlorobenzoyl chloride (1.2 eq) in dry THF (8 mL/eq of AA).
2. Stir the mixture at room temperature for 3 h.
3. Filter the reaction mixture through a pad of silica gel.

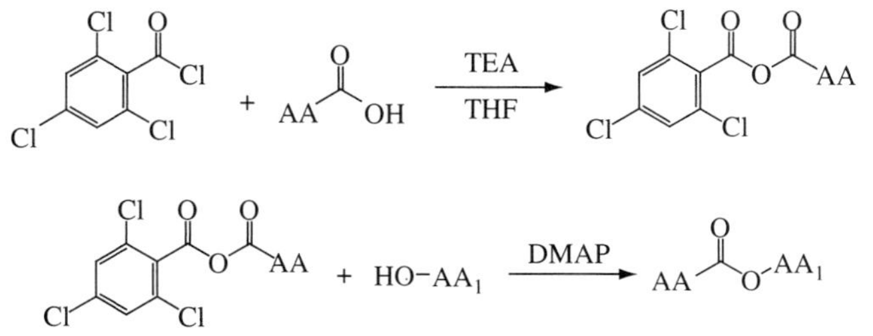

Fig. 6. Yamaguchi esterification. AA, amino acid.

4. Concentrate the filtrate under reduced pressure.
5. Dissolve the residue in toluene add corresponding alcohol (0.4 eq) and DMAP (0.8 eq).
6. Stir the mixture at room temperature for 3 h (*see* **Note 4**).
7. Dilute the reaction mixture with ethyl acetate and wash with saturated aqueous NaHCO$_3$, and brine.
8. Dry the organic layer over Na$_2$SO$_4$, concentrate the filtrate and purify by column chromatography.

3.5. Protection of Hydroxyl Groups During Solid-Phase Depsipeptide Synthesis

During synthesis of depsipeptides problem of protection/deprotection of hydroxyl groups may appear. Although it is possible to use unprotected hydroxy-amino acids such as Thr or Ser during solid phase peptide synthesis, generally, it is highly recommended that they be protected. J. S. Davis et al. *(47)* described use of a combination of *tert*-butyldimethylsilyl (TBDMS) and 9-fluorenylmethyloxycarbonyl (Fmoc) protection for the hydroxy and amino groups during solid phase synthesis of pentadepsipeptide (**Fig. 7**). However, R. Riguera et al. reported that the TBDMS group is not sufficiently stable under the esterification conditions used for coupling *(33,34)*. Instead, these authors proposed protection of hydroxyl groups as tetrahydropyranyl (THP) ethers *(33,34)* (**Fig. 8**). This method gives high yields and allows the preparation of relatively large depsipeptides. The versatility of the method has been demonstrated by the preparation of different hydroxy and amino acid-containing substrates. HFA is another protecting group to be fully compatible with depsipeptide solid phase synthesis as demonstrated by F. Albericio et al. *(48,49)*. Hexafluoroacetone is a well known protecting and activating reagent for α-hydroxy, α-amino and α-mercapto functionalized carboxylic acids *(48,49)*. The protection of hydroxy group requires one reaction step in which the α-functional

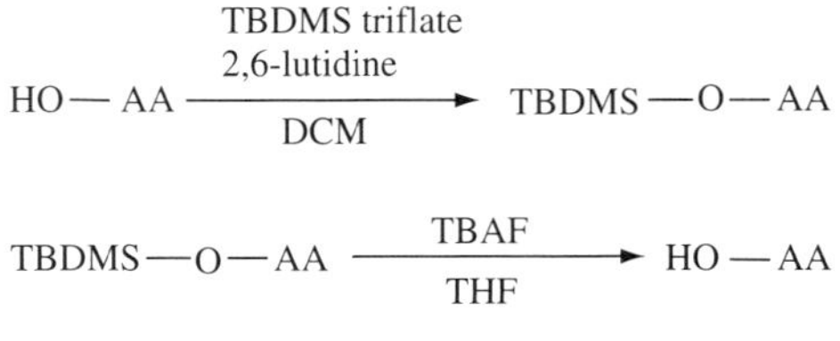

$$HO-AA \xrightarrow[\text{DCM}]{\substack{\text{TBDMS triflate} \\ \text{2,6-lutidine}}} TBDMS-O-AA$$

$$TBDMS-O-AA \xrightarrow[\text{THF}]{\text{TBAF}} HO-AA$$

Fig. 7. *tert*-Butyldimethylsilyl protection/deprotection of free hydroxyl group. AA, amino acid.

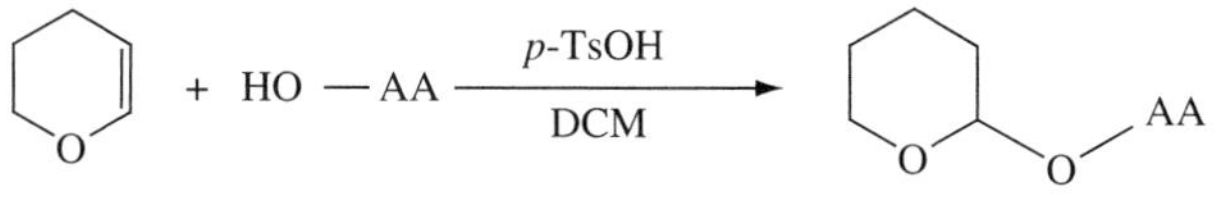

Fig. 8. Tetrahydropyranyl protection of free hydroxyl group. AA, amino acid.

group and the neighboring carboxylic group form a lactone. In this way, the carboxylic group is activated and the α-functionality is protected.

The protocols for protection of hydroxyl groups in depsipeptide solid-phase synthesis are the following:

3.5.1. TBDMS Triflate Protection Protocol

TBDMS triflate is the most efficient method for introducing the TBDMS onto free hydroxyl group (**Fig. 7**).

This protocol was adopted from **refs. *50*** and ***51***.

1. Dissolve hydroxyl amino acid (1 eq) in dry DCM (10 mL/1.2 mmol of amino acid) under inert atmosphere.
2. Cool the solution to 0°C.
3. Add sequentially 2,6-lutidine (1.5 eq) and TBDMS triflate (1.3 eq).
4. Allow reaction mixture to warm to room temperature.
5. Stir reaction mixture at room temperature for 2 h.
6. Concentrate reaction mixture under reduced pressure and purify by column chromatography.

3.5.1.1. TBDMS REMOVAL FOR SOLID-PHASE METHODOLOGY

This protocol was adopted from **ref. *45***.

1. Place the resin in dry reaction vessel.
2. Wash the resin (3 × 10 mL) with THF and remove excess of solvent.
3. Dissolve (3–4 eq, relative to the resin loading) TBAF in THF (30 mL).
4. Add above solution to the resin (20 mL/g) and allow agitating for 1 h.
5. Repeat once more **steps 2–4**.
6. Wash the resin three times sequentially with THF and DMF.

3.5.2. THP Protection for Solid-Phase Methodology

This protocol was adopted from **refs. *33*** and ***34***.

1. Dissolve hydroxy amino acid (1 eq) and *p*-TsOH (0.05 eq) in DCM (10 mL/g).
2. Add dropwise dihydropyran (1.5 eq) (*see* **Note 5**).
3. Allow reaction mixture to stir at room temperature for 1.5 h.
4. Extract reaction mixture with 0.2 *M* KOH (2 × 50 mL).

5. Combine KOH layers and acidify it with 6 N HCl to pH 3.0–4.0 and extract three times with DCM.
6. Combine DCM extracts, and wash the extracts with water.
7. Dry the organic layer over Na$_2$SO$_4$, concentrate the filtrate and purify by column chromatography.

3.5.2.1. THP Deprotection for Solid-Phase Methodology

This protocol was adopted from **refs. 33** and **34**.

1. Prepare deprotection solution of *p*-TsOH (5 mg/mL) in DCM/MeOH (97:3).
2. Wash the resin with DCM and remove excess of the solvent.
3. Add deprotection solution to the resin (15 mL/g) and allow agitating for 1 h.
4. Repeat **steps 2–3** one more time.
5. Wash the resin with DCM (3 × 10 mL), acetone (3 × 10 mL) and THF (3 × 10 mL).

3.5.3. HFA Protection for Solid Phase Methodology **(Fig. 9)**

This protocol was adopted from **refs. 52** and **53**.

1. Dissolve α-hydroxy amino acid (1 eq) in minimal amount of dimethylsulfoxide (DMSO).
2. Bubble 2 eq of HFA through the above solution (*see* **Note 6**).
3. Stir reaction mixture for 2 h at room temperature.
4. After completion of reaction pour the reaction solution into a 1:1 mixture of water/DCM.
5. Separate the organic layer and extract aqueous layer several times with DCM.
6. Combine organic layers, and wash it with water.
7. Dry the organic layer over Na$_2$SO$_4$, remove the solvent under reduced pressure and purify by crystallization from chloroform/hexanes.

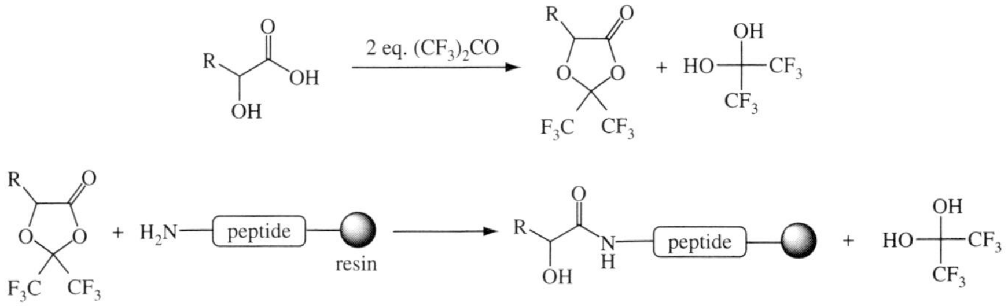

Fig. 9. Hexafluoroacetone protection and deprotection of free hydroxyl group.

3.6. Solid-Phase Reaction Monitoring

Assessment of the extent of reaction completion is crucial in the case of repetitive solid phase synthesis such as solid phase peptide synthesis. The ninhydrin or Kaiser *(54)* test is the method of choice for qualitative colorimetric monitoring of the presence or absence of free amino groups. On the other hand, for the solid phase synthesis of depsipeptides, in which amide bonds are replaced by ester bonds, it is necessary to monitor the extent of completion of ester bonds formation as well. Pomonis *(55,56)* and TCT/AliR *(57)* tests are described in the literature for detection the presence of free hydroxyl group on the solid support or in the solid support bound growing depsipeptide chain.

3.6.1. Pomonis Test

The test is based on the transformation of the hydroxyl group into its tosylate, displacement of corresponding tosylate by 4-(*p*-nitrobenzyl)pyridine (PNBP), and finally conversion of the solid supported pyridinium salt to a strongly colored internal salt by treatment with base **(Fig. 10)**.

All operations are carried out directly on the resin. When free hydroxyl groups are present, the blue to purple color is observed.
This protocol was adopted from **refs. 55** and **56**.

1. Take one drop of a suspension of resin beads in DCM (approx. 1 mg dry resin) from the reaction vessel with a Pasteur pipette and place it on a silica gel TLC plate.
2. Make one reference sample using the same procedure.

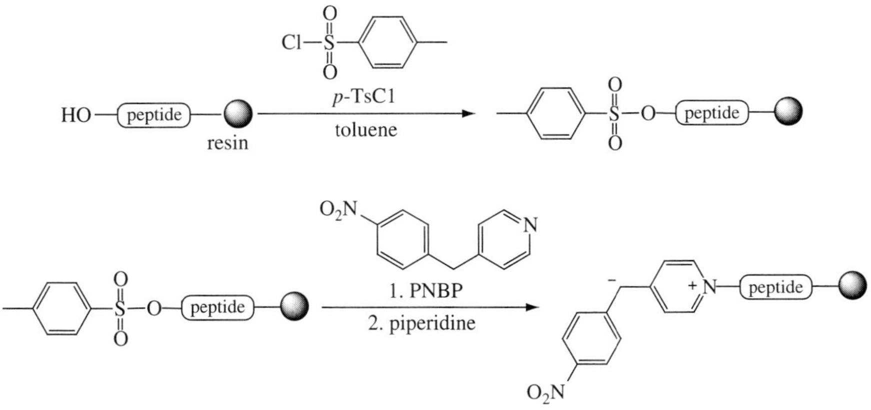

Fig. 10. Pomonis test for the presence of free hydroxyl groups on resin.

3. Spread out the sample to a thin circular film of about 0.5 cm diameter by carefully dropping DCM from a Pasteur pipet.

3. Add two drops of a toluene solution containing 0.03 M *p*-TsCl (*see* **Note 7**).

4. Add two drops of a solution of 0.075 M PNBP in toluene and heat the plate from underneath with a heat gun until the orange color that initially develops had disappeared completely (about 10–12 s) (*see* **Note 7**).

5. Add two drops of a 10% piperidine solution in chloroform to each spot followed by gentle drying of the plate with a heat gun.

6. Carefully wash samples with several drops of DCM and allow drying.

7. If no free OH are present, the test sample should appear colorless as does the negative blank. If color appears, unreacted OH groups are present in the resin.

8. Because the white silica "background" often remains colored to a certain extent in spite of the washings, an alternative is to scrape the dry resin beads from the TLC plate and to deposit them onto a white well plate. This enhances the contrast and allows easier distinction of positive and negative test results.

3.6.2. TCT/AliR Test

The TCT/AliR test was developed by M. Taddei et al. *(57)* and used for detection of free hydroxyl group on the solid support. This test is based on the activation of hydroxyl groups with TCT, followed by coupling of a carboxylic dye such as commercially available Alizarin R (AliR) or fluorescein **(Fig. 11)**. The presence of solid support bound free hydroxyl group is indicated by deep yellow-red or yellow-green stained beads.

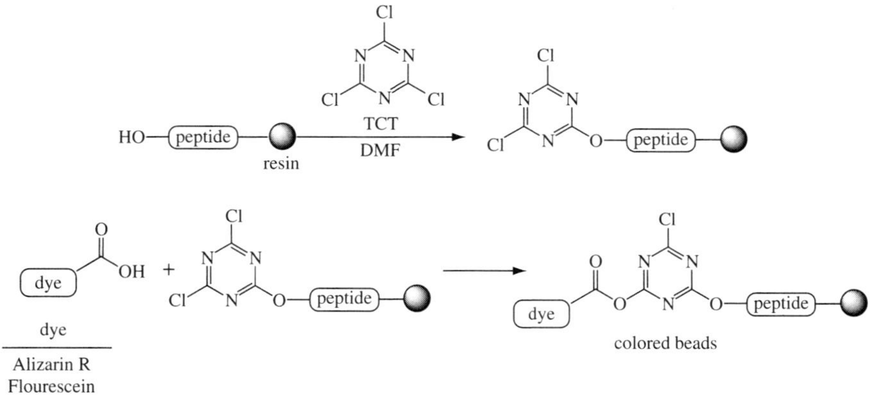

Fig. 11. 2,4,6-Trichloro-[1,3,5]-triazine /carboxylic acid dye test for the presence of free hydroxyl groups on resin.

This protocol was adopted from **ref. 57**.

1. Take some beads of the resin (swollen) and transfer them into a test tube. Wash the beads several times with DMF.
2. Add 3 mL of DMF followed by 1 mL of NMM and 5 mg of solid TCT.
3. Heat the test tube at 70°C for 20 min.
4. Remove the solution and rinse the beads several times with DMF.
5. Add 3 mL of DMF followed by 5 mg of AliR and 1 mL of NMM (*see* **Note 8**).
6. After 5 min, discard the solution and wash the beads with DMF until the solvent is clear. Wash finally with THF or DCM. Observe the color of the beads directly in the test tube or with a microscope.

4. Notes

1. DCM can also be used as a solvent instead of THF. Use of DMF may result in low reaction yields *(35)*.
2. For elongation of depsipeptide chains *via* amide bond formation using this method, *see* **ref. 39**.
3. Alternatively, diisopropylazodicarboxylate (DIAD), di-tert-butyl azodicarboxylate (DBAD), and dibenzyl azodicarboxylate (DBAD) can be used instead of DEAD.
4. After addition of an alcohol, the reaction time can be extended to 24 h.
5. Introduction of THP protecting group onto a chiral molecule results in the formation of diastereoisomers, because of additional stereogenic center present in the tetrahydropyran ring.
6. HFA is commercially available as a gas in lecture bottles or as a liquid trihydrate. The gaseous HFA is obtained upon dropwise addition of the trihydrate to concentrated sulfuric acid at 80–100°C with stirring. For safety reasons all experiments with HFA should be carried out in a fume hood! HFA protection applies to α- and β-functionalized carboxylic acids.
7. The solutions of PNBP and p-TsCl can be stored at 4°C for a few weeks only; after prolonged storage period they lose their efficacy.
8. AliR is used as the sodium salt. The method uses commercially available reagents and may also be used with different carboxylic acid dyes. In the case of fluorescein or fuchsin dyes, their 0.025% solution (3 mL) in NMP needs to be used in **Subheading 3.6.2., step 5**. These solutions must be prepared just before the use. The TCT/AliR test is not compatible with free carboxylic group due to possible lactonization of hydroxyl acids.

References

1. Grabley, S. and Thiericke, R., (1999) *Drug Discovery from Nature.* Springer-Verlag, Heidelberg.
2. Newman, D. J., Cragg, G. M., and Snader K. M. (2003) Natural products as sources of new drugs over the period 1981–2002. *J. Nat. Prod.* **66**, 1022–1037.

3. Cragg, G. M., Newman, D. J., and Snader, K. M. (1997) Natural products in drug discovery development. *J. Nat. Prod.* **60**, 52–60.
4. Bozdogan, B., Esel, D., Whitener, C., Browne, F. A., and Appelbaum P. C. (2003) Antibacterial susceptibility of a vancomycin-resistant *Staphylococcus aureus* strain isolated at the Hershey Medical Center. *J. Antimicrob. Chemother.* **52**, 864–868.
5. Loffet A. (2002) Peptides as drugs: is there a market? *J. Pept. Sci.* **8**, 1–7.
6. Loffet A. (2001) Peptides as drugs: is there a market? *Peptides: The Wave of the Future* (Lebl, M. and Hougten, R. A., eds.). American Peptide Society: pp. 214–216.
7. Andersson, L., Blomberg, L., Flegl, M., Lepsa, L., Nilsson, B., and Verlander M. (2000) Large-scale synthesis of peptides. *Biopolymers* **55**, 227–250.
8. Verlander M. (2000) Large-scale manufacturing methods for peptides—a status report. *Chim. Oggi,* **20**, 62–66.
9. Adessi, C. and Soto, C. (2002) Converting a peptide into drug: strategies to improve stability and bioavailability. *Curr. Med. Chem.* **9**, 963–978.
10. Davies, J. S. (2003) The cyclization of peptides and depsipeptides. *J. Pept. Sci.* **9**, 471–501.
11. Lambert, J. N., Mitchell, J. P., and Roberts, K. D. (2001) The synthesis of cyclic peptides. *J. Chem. Soc, Perkin Trans.* **1**, 471–484.
12. Li, P. and Roller, P. P. (2002) Cyclization strategies in peptide derived drug design. *Curr. Top. Med. Chem.* **2**, 325–341.
13. Blackburn, C. and Kates, S. A. (1997) Solid-phase synthesis of cyclic homodetic peptides. *Methods Enzymol.* **289**, 175–198.
14. Hruby, V. J. and Bonner, G. G. (1994) Design of novel synthetic peptides including cyclic conformationally and topographically constrained analogs. *Methods. Mol. Biol.* **35**, 201–240.
15. Kates, S. A., Sole, N. A., Albericio, F., and Barany, G. (1994) Solid-phase synthesis of cyclic peptides, in *Peptides: Design, Synthesis, and Biological Activity.* Brikhauser Boston: pp. 39–59.
16. Shemyakin, M. M., Shchukina, L. A., Vinogradova, E. I., Ravidel, G. A., and Ovchinnikov, Y. A. (1966) Mutual replaceability of amide and ester groups in biologically active peptide and depsipeptides. *Experimentia* **22**, 535–536.
17. Bramson, H. N., Thomas, N. E., and Kaiser, E. T. (1985) The use of *N*-methylated peptides and depsipeptides to probe the binding of heptapeptide substrates to cAMP-dependent protein kinase. *J. Biol. Chem.* **260**, 15,452–15,457.
18. Arad, O. and Goodman, M., (1990) Depsipeptide analogues of elastin repeating sequences: synthesis. *Biopolymers,* **29**, 1633–1649.
19. Coombs, G. S., Rao, M. S., Olson, A. J., Dawson, P. E., and Madison, E. L. (1999) Revisiting catalysis by chymotrypsin family serine proteases using peptide substrates and inhibitors with unnatural main chains. *J. Biol. Chem.* **274**, 24,074–24,079.

20. Davidson, B. S. (1993) Ascidians: producers of amino acid-derived metabolites. *Chem. Rev.* **93**, 1771–1791.
21. Fusetani, N. and Matsunaga, S. (1993) Bioactive sponge peptides. *Chem. Rev.* **93**, 1793–1806.
22. Simmons, T. L., Andrianasolo, E., McPhail, K., Flatt, P., and Gerwick, H. W. (2005) Marine natural products as anticancer drugs. *Mol. Chem. Ther.* **4**, 333–342.
23. Woodford, N. (2003) Novel agents for the treatment of resistant Gram-positive infections. *Expert. Opin. Investig. Drugs.* **12**, 117–137.
24. McCafferty, D. G., Cudic, P., Frankel, B. A., Barkallah, S., Kruger, R. G., and Li, W. (2002) Chemistry and biology of the ramoplanin family of peptide antibiotics. *Biopolymers* **66**, 261–284.
25. Humphrey, J. M. and Chamberlin, A. R. (1997) Chemical synthesis of natural product peptides: coupling methods for the incorporation of noncoded amino acids into peptides. *Chem. Rev.* **97**, 2243–2266.
26. Anteunis, M. O. J. and Sharma, N. K. (1988) *N,N'*-Bis(2-oxo-3-oxazolidinyl)phosphinic chloride (BOP-Cl) mediated cyclization of a linear precursor of virginiamycin S. Contra indication for using hydroxybenzotriazole as racemization suppressor. *Bull. Soc. Chim. Belg.* **97**, 281–292.
27. Kopple, K. D. (1972) Synthesis of cyclic peptides. *J. Pharm. Sci.* **61**, 1345–1356.
28. Brady, S. F., Varga, S. L., Freidinger, R. M., et al. (1979) Practical synthesis of cyclic peptides, with an example of dependence of cyclization yield upon linear sequence. *J. Org. Chem.* **44**, 3101–3105.
29. Chu, K. S., Negrete, G. R., and Konopelski, J. P. (1991) Asymmetric total synthesis of (+) jasplakinolide. *J. Org. Chem.* **56**, 5196–5202.
30. White, J. D. and Amedio, J. C. (1989) Total synthesis of geodiamolide A—a novel cyclodepsipeptide of marine origin. *J. Org Chem.* **54**, 736–738.
31. Marder, O. and Albericio, F. (2003) Industrial application of coupling reagents in peptides. *Chim. Oggi* **6**, 35–40.
32. Berry J. D., Digiovanna V. C., Metrick S. S., and Murugan R. (2001) Catalysis by 4-Dialkylaminopyridines. *Arkivoc* **i**, 201–226
33. Kuisle, O., Lolo, M., Quinoa, E., and Riguera, R., (1999) Solid Phase Synthesis of Depsides and Depsipeptides. *Tetrahedron* **55**, 14,807–14,812.
34. Kuisle, O., Quinoa, E., and Riguera, R., (1999) A general methodology for automated solid-phase synthesis of depsides and depsipeptides. Preparation of a valinomycin analogue. *J. Org Chem.* **64**, 8063–8075.
35. Stawikowski, M. and Cudic, P. (2006) A novel strategy for the solid-phase synthesis of cyclic lipodepsipeptides. *Tetrahedron Lett.* **47**, 8587–8590.
36. Murakami, N., Wang, W., Tamura, S., and Kobayashi, M. (2000) Synthesis and biological property of carba and 20-deoxo analogues of arenastatin A. *Bioorg Med. Chem. Lett.* **10**, 1823–18236.
37. Joullie, M. M., Portonovo, P., Liang, B., and Richard, D. J. (2000) Total synthesis of (−)-tamandarin B. *Tetrahedron Lett.* **41**, 9373–9376.

38. Dutton, F. E., Byung, H. L., Johnson, S. S. Coscarelli, E.M., and Lee P. H. (2003) Restricted conformation analogues of anthelmintic cyclopeptide. *J. Med. Chem.*, **46**, 2057–2073.

39. Katakai, R., Kobayashi, K., Yamada, K., Oku, H., and Emori, N. (2004) Synthesis of sequential polydepsipeptides utilizing a new approach for the synthesis of depsipeptides. *Biopolymers* **73**, 641–644.

40. Mitsunobu, O. and Yamada, M. (1967) Preparation of esters of carboxylic and phosphoric acid via quaternary phosphonium salts. *Bull.Chem. Soc. Jpn.* **40**, 2380–2382

41. Mitsunobu, O. (1981) The use of diethyl azodicarboxylate and triphenylphosphine in synthesis and transformation of natural products. *Synthesis* **1**, 1–28.

42. Boger, D. L., Keim, H., Oberhauser, B., Schreiner, E. P., and Foster, C. A. (1999) Total synthesis of HUN-7293. *J. Am. Chem. Soc.* **121**, 6197–6205

43. Grab, T. and Brase S. (2005) Efficient synthesis of lactate-containing depsipeptides by the Mitsunobu reaction of lactates. *Adv. Synth. Catal.* **347**, 1765–1768.

44. Inanaga, J., Hirata, K., Saeki, H., Katsuki, T., and Yamaguchi, M. (1979) Rapid esterification by means of mixed anhydride and its application to large-ring lactonization. *Bull. Chem. Soc. Jpn.* **52**, 7, 1989–1993.

45. Chen, J. and Forsyth, J. C. (2004) Natural product synthesis special feature: total synthesis of the marine cyanobacterial cyclodepsipeptide apratoxin A. *Proc. Natl. Acad. Sci. USA* **101**, 12,067–12,072.

46. Zou, B., Long, K, and Ma, D. (2005) Total synthesis and cytotoxicity studies of a cyclic depsipeptide with proposed structure of palau'amide. *Org. Lett.* **7**, 4237–4240.

47. Davies, J. S., Howe, J., Jayatilake J., and Riley T. (1997) Synthesis and applications of cyclopeptides and depsipeptides. *Lett. Pept. Sci.* **4**, 441–445.

48. Albericio, F., Burger, K, Ruiz-Rodrigez, J., and Spengler, J. (2005) A new strategy for solid-phase depsipeptide synthesis using recoverable building blocks. *Org. Lett.* **7**, 597–600.

49. Albericio, F., Burger, K., Cupido, T. K, Ruiz, J., and Spengler, J. (2005) Application of hexafluoroacetone as protecting and activating reagent in solid phase peptide and depsipeptide synthesis. *Arkivoc* **vi,** 191–199.

50. Corey, E. J., Cho, H., Rucker, C., and Hua, D., H. (1981) Studies with trialkylsilyltriflates: new syntheses and applications. *Tetrahedron Lett.* **22**, 3455–3458.

51. Yuan, W., Jia, Y., Tian, J., et al. (2001) Class I and III polyhydroxyalkanoate synthases from *Ralstonia eutopha* and *Allochromatium vinosum*: characterization and substrate specificity studies. *Arch. Biochem. Biophys.* **394**, 87–98.

52. Burger, K., Windeisen, E., and Pires, R., (1995) New efficient strategy for the incorporation of (*S*)-isoserine into peptides. *J. Org. Chem.* **60**, 7641–7645.

53. Radics, G., Pires, R., Koksch, B., El-Kousy, S. M., and Burger, K. (2003) New building blocks for peptide and depsipeptide synthesis: hexafluoroacetone protected L-homoserine and D,L-homocysteine derivatives. *Tetrahedron Lett.* **44**, 1059–1062.
54. Kaiser, E., Colescott, R. L., Bossinger, C. D., and Cook, P. I. (1970) Color test for detection of free terminal amino groups in the solid-phase synthesis of peptides. *Anal. Biochem.* **34**, 595–598.
55. Pomonis, J. G., Severson, R. F., and Freeman, P. J. (1969) Spot test diagnostic of hydroxyl groups. *J. Chromatog.* **40**, 78–84.
56. Kuisle, O., Lolo, M., Quinoa, E., and Riguera, R., (1999) Monitoring the solid-phase synthesis of depsides and depsipeptides. A color test for hydroxyl groups linked to a resin. *Tetrahedron* **55**, 14,807–14,812.
57. Attardi, M. E., Falchi, A., and Taddei, M. (2000) A sensitive visual test for detection of OH groups on resin. *Tetrahedron Lett.* **41**, 7395–7399.

Index